Sustainable Agriculture Reviews

Volume 40

Series Editor
Eric Lichtfouse
Aix-Marseille Université, CNRS, IRD, INRA
Coll France, CEREGE
Aix-en-Provence, France

Eric Lichtfouse

Editor

Sustainable Agriculture Reviews 40

Springer

Editor
Eric Lichtfouse
Aix-Marseille Université
CNRS, IRD, INRA, Coll France, CEREGE
Aix-en-Provence, France

ISSN 2210-4410   ISSN 2210-4429   (electronic)
Sustainable Agriculture Reviews
ISBN 978-3-030-33283-9   ISBN 978-3-030-33281-5   (eBook)
https://doi.org/10.1007/978-3-030-33281-5

This Springer imprint is published by the registered company Springer Nature Switzerland AG.
The registered company address is: Gewerbestrasse 11, 6330 Cham, Switzerland

Contents

Contents

About the Editor

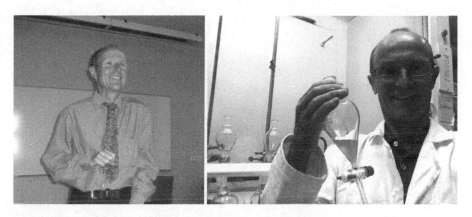

Left: Giving lectures and workshops on scientific writing. Right: Extracting soil organic compounds using CHCl₃-Methanol 3/1 v/v

Eric Lichtfouse, PhD, born in 1960, is an environmental chemist working at the University of Aix-Marseille, France. He has invented carbon-13 dating, a method allowing to measure the relative age and turnover of molecular organic compounds occurring in different temporal pools of any complex media. He is teaching scientific writing and communication, and has published the book Scientific Writing for Impact Factors, which includes a new tool – the Micro-Article – to identify the novelty of research results. He is founder and Chief Editor of scientific journals and series in environmental chemistry and agriculture. He founded the European Association of Chemistry and the Environment. He received the Analytical Chemistry Prize by the French Chemical Society, the Grand Prize of the Universities of Nancy and Metz, and a Journal Citation Award by the Essential Indicators.

Contributors

Badreddine Boudjemaa Department of Biology, Faculty of science, University of Amar Tlidji, Laghouat, Algeria

Aminata Ould El Hadj-Khelil Laboratory of Ecosystems Protection in Arid and Semi-Arid Area, University of Kasdi Merbah, Ouargla, Algeria

Hiba Gacem Epidemiology Service and Preventive Medicine, Faculty of Medicine, University of Djillali Liabes, Sidi-Bel-Abbes, Algeria

Mohamed Amine Gacem Laboratory of Ecosystems Protection in Arid and Semi-Arid Area, University of Kasdi Merbah, Ouargla, Algeria
Department of Biology, Faculty of science, University of Amar Tlidji, Laghouat, Algeria

Akankhya Guru Department of Plant Physiology, Institute of Agricultural Sciences, Banaras Hindu University, Varanasi, Uttar Pradesh, India

Akash Hidangmayum Department of Plant Physiology, Institute of Agricultural Sciences, Banaras Hindu University, Varanasi, Uttar Pradesh, India

Fayyaz Salih Hussain National Centre of Excellence in Analytical Chemistry, University of Sindh, Jamshoro, Sindh, Pakistan

Tapas Laha Metallurgical and Materials Engineering Department, Indian Institute of Technology Kharagpur, Kharagpur, West Bengal, India

Damodhara Rao Mailapalli Agricultural and Food Engineering Department, Indian Institute of Technology Kharagpur, Kharagpur, West Bengal, India

Najma Memon National Centre of Excellence in Analytical Chemistry, University of Sindh, Jamshoro, Sindh, Pakistan

Biswajit Paul Department of Environmental Science and Engineering, Indian Institute of Technology (Indian School of Mines), Dhanbad, Jharkhand, India

Chwadaka Pohshna Agricultural and Food Engineering Department, Indian Institute of Technology Kharagpur, Kharagpur, West Bengal, India

Saheli Pradhan Agricultural and Food Engineering Department, Indian Institute of Technology Kharagpur, Kharagpur, West Bengal, India

Priti Saha Department of Environmental Science and Engineering, Indian Institute of Technology (Indian School of Mines), Dhanbad, Jharkhand, India

Ankita Singh Department of Plant Physiology, Institute of Agricultural Sciences, Banaras Hindu University, Varanasi, Uttar Pradesh, India

Bansh Narayan Singh Institute of Environmental and Sustainable Development, Banaras Hindu University, Varanasi, Uttar Pradesh, India
Department of Plant Physiology, Institute of Agricultural Sciences, Banaras Hindu University, Varanasi, Uttar Pradesh, India

Gopal Shankar Singh Institute of Environmental and Sustainable Development, Banaras Hindu University, Varanasi, Uttar Pradesh, India

Bhudeo Rana Yashu Department of Plant Physiology, Institute of Agricultural Sciences, Banaras Hindu University, Varanasi, Uttar Pradesh, India

Chapter 1
Mycotoxins Occurrence, Toxicity and Detection Methods

Mohamed Amine Gacem, Aminata Ould El Hadj-Khelil, Badreddine Boudjemaa, and Hiba Gacem

Abstract Mycotoxins and their derivatives constitute a toxic group of bioproducts for human and animals' health, they induce economic losses in cereals and stored food products. The knowledge of the biosynthetic mechanisms of mycotoxins are important to improve food quality. Polyketides are the first precursors, they are synthesized by a variety of multifunctional enzymes named polyketide synthases. This review discusses the occurrence of mycotoxin in food and human biological fluids, the toxicities caused *in vitro* and *in vivo* in human and animals' organs, and finally, some conventional and recent detection methods for their detection and quantification.

Keywords Mycotoxins · Occurrence of mycotoxin · Toxicities · Detection methods

M. A. Gacem (✉)
Laboratory of Ecosystems Protection in Arid and Semi-Arid Area,
University of Kasdi Merbah, Ouargla, Algeria

Department of Biology, Faculty of science, University of Amar Tlidji, Laghouat, Algeria

A. Ould El Hadj-Khelil
Laboratory of Ecosystems Protection in Arid and Semi-Arid Area,
University of Kasdi Merbah, Ouargla, Algeria

B. Boudjemaa
Department of Biology, Faculty of science, University of Amar Tlidji, Laghouat, Algeria

H. Gacem
Epidemiology Service and Preventive Medicine, Faculty of Medicine,
University of Djillali Liabes, Sidi-Bel-Abbes, Algeria

1.1 Introduction

Mycotoxins are fungal secondary metabolites, some of which have pharmacological activities and they are used in antibiotics production, growth promoters, or in other classes of drugs (Sava et al. 2006). The majority of mycotoxins are polyketide origin (Huffman et al. 2010), polyketides (PK) are natural metabolites found in several compounds and sometimes have important biological activities (Gomes et al. 2013), their enzymatic mechanisms of biosynthesis is similar to fatty acids synthesis (Huffman et al. 2010).

Mycotoxins are produced by several fungal genera (Greco et al. 2012) and are present in a wide variety of foods, either before harvest or after industrial processing (Almeida-Ferreira et al. 2013; Schaafsma and Hooker 2007; Perši et al. 2014; Pizzolato Montanha et al. 2018). Their detections are also possible in water, drinks and wine (De Jesus et al. 2018; Mata et al. 2015). They are a very diverse chemicals group with a high toxicity for humans, animals and even plants (Peraica et al. 1999). Their toxicities may be manifested in all vital human or animal organs inducing nephrotoxicity (Palabiyik et al. 2013; Limonciel and Jennings 2014), genotoxicity (Adgigitov et al. 1984; Bárta et al. 1984; Theumer et al. 2018), teratogenicity (Celik et al. 2000; Sur and Celik 2003), neurotoxicity (Ikegwuonu 1983; Mehrzad et al. 2017), hepatotoxicity (Gagliano et al. 2006; Gayathri et al. 2015), immunotoxicity (Al-Anati and Petzinger 2006; Thuvander et al. 1995), membrane damage (Ciacci-Zanella et al. 1998), gastrointestinal toxicity (Bouhet et al. 2004; Loiseau et al. 2007), cardiotoxicity (Constable et al. 2000), pulmonary toxicity (Smith et al. 1999), and sometimes leading to cancer development (Barrett 2005; Wu and Santella 2012; Magnussen and Parsi 2013).

The destruction of these mycotoxins is very difficult because of their crystallization state or their high resistance under gamma radiation and high temperatures (Calado et al. 2018), whereas, chemical products detoxifying cause other health problems. The researchers have left the classic platforms of struggle, towards a new strategy using lactic acid bacteria (Khanafari et al. 2007), yeasts (Armando et al. 2012), actinobacteria (El Khoury et al. 2017, 2018), medicinal plants (Brinda et al. 2013). New research in progress aims to develop a new nanoparticles and nanomaterials to fight against these mycotoxins and their production sources (Hernández-Meléndez et al. 2018).

This review provides an overview of mycotoxins occurrence in food and biological fluids, the toxicities of some mycotoxins studied *in vivo* and *in vitro*, and the last section deals with the conventional and nanotechnological methods applied in the detection.

1.2 Occurrence of Mycotoxin in Food

Ochratoxin, aflatoxin, fumonisin, zearalenone and other mycotoxins are present in wheat and maize (Taheri et al. 2012; Almeida-Ferreira et al. 2013; Schaafsma and Hooker 2007; Shephard et al. 2005), they are also present in fruit, oil seeds and animal feeds that subsequently render them contaminated (Minervini and Dell'Aquila 2008), the biotransformation of mycotoxins in the animal body makes them present in cow's milk (Dehghan et al. 2014; Flores-Flores et al. 2015), meat (Perši et al. 2014; Pizzolato Montanha et al. 2018) and even eggs (Giancarlo et al. 2011).

Many drinks like wine and beer (De Jesus et al. 2018; Mateo et al. 2007), coffee and tea (García-Moraleja et al. 2015; Malir et al. 2014; Haas et al. 2013) are contaminated with mycotoxins. They are also present in bread produced from corn and wheat (Juan et al. 2008), and fermented food products (Kaymak et al. 2018; Adekoya et al. 2017). A recent study done on 84 samples identified the presence of mycotoxins in 99% of medicinal and aromatic plants, the percentage of damage depends on the toxin itself and the plant species (Santos et al. 2009). Mycotoxins are also present in herbal medicaments (Zhang et al. 2018). In bottle water, aflatoxin B2 was the most frequently detected mycotoxin with a concentration followed by aflatoxin B1, aflatoxin G1 and ochratoxin A (Mata et al. 2015). Table 1.1 shows some variety of food contaminated by mycotoxins in different countries.

1.3 Factors Controlling Mycotoxins Genesis

Mycotoxins require factors and precursors stimulating their biosynthesis, aflatoxin biosynthesis factors include nutritional factors such as nitrogen and carbon sources, and environmental factors such as pH, temperature, water activity, osmotic pressure … etc. the knowledge of these factors is very important in order to combat these poisons and their production sources. In culture medium, carbon and nitrogen are the most prominent nutritional factors fot the synthesis of aflatoxin or one of their metabolites, carbon is abundant in sugars such as glucose, sucrose, maltose, fructose, amino acids and proteins (Bennett et al. 1979). Plants, cereals and meats are also rich in these compounds in a complex form and favouring the biosynthesis of aflatoxins (Woloshuk et al. 1997). In *A. parasiticus,* researchers identified a group of genes near to aflatoxin synthesis genes, and these genes able to degrade and use sugars (Yu et al. 2000). Another gene named lipA produces substrates for aflatoxin biosynthesis, to determine it activity, this gene is cloned in *A. parasiticus* and *A. flavus*, the expression of this gene is induced by lipid substrates (Yu et al. 2003).

Table 1.1 Examples of food contamination by mycotoxins in the world

Food category	Feed type	Mycotoxins	Country	Limits of the legislation ($\mu g.kg^{-1}$)	Range or mean	References
Milk and dairy products	Cattle milk	AFM$_1$	Japan		0.005 to 0.011 $ng.g^{-1}$	Sugiyama et al. (2008)
	Cow milk		Iran	500 $ng.l^{-1}$	60.1 ± 57.4 $ng.l^{-1}$	Rahimi et al. (2010)
	Camel milk				19.0 ± 7.5 $ng.l^{-1}$	
	Sheep milk				28.1 ± 13.7 $ng.l^{-1}$	
	Goat milk				30.1 ± 18.3 $ng.l^{-1}$	
			Cameroon	0,05 $\mu g.kg^{-1}$	0.006 to 0.527 $\mu g.l^{-1}$	Tchana et al. (2010)
	Milk from traditional dairies		Morocco	0,05 $\mu g.kg^{-1}$	10 and 100 $ng.l^{-1}$	El Marnissi et al. (2012)
	Milk		Algeria	0,05 $\mu g.kg^{-1}$	9 à 103 $ng.l^{-1}$	Redouane-Salah et al. (2015)
	Milk		Brazil		50–240 $ng.l^{-1}$	Garrido et al. (2003)
	Milk		Jordan		9,71 à 288,68 $ng.kg^{-1}$	Omar (2016)
	Fresh cheese		Costa Rica		31 to 276 $ng.l^{-1}$	Chavarría et al. (2015)
	Raw bulk milk		France	50 $ng.l^{-1}$	26 $ng.l^{-1}$ or less	Boudra et al. (2007)
	Milk		Taiwan		1.17 to 54.7 $ng.l^{-1}$	Peng and Chen (2009)
	Milk		Pakistan	50 $ng.l^{-1}$	20 to 3090 $ng.l^{-1}$	Asghar et al. (2018)
Alcoholic beverages and juice	Apple juice	PAT	South Africa	50 $ng.ml^{-1}$	<10 to 75.2 $ng.ml^{-1}$	Katerere et al. (2007)
	Apple juice		Turkey		7 to 375 $\mu g.l^{-1}$	Gökmen and Acar (1998)
	Apple juice		Iran	50 $\mu g.ml^{-1}$	>3 $\mu g.l^{-1}$	Poostforoushfard et al. (2017)
	Maize based opaque beers	FB$_1$	Malawi		1522 $\mu g.kg^{-1}$	Matumba et al. (2014)
		FB$_2$			251 $\mu g.kg^{-1}$	
		FB$_3$			229 $\mu g.kg^{-1}$	

		Country	Mycotoxin	Concentration	References
			AFB$_1$	36 µg/kg^{-1}	
			AFB$_2$	4 µg/kg^{-1}	
			AFG$_1$	55 µg/kg^{-1}	
			AFG$_2$	8 µg/kg^{-1}	
		South African	Decxynivalenol (mean)	47 µg.l^{-1}	Adekoya et al. (2018)
			Neosolaniol (mean)	21	
			Fusarenon-X (mean)	167	
			AFB1 (mean)	6	
			AFG2 (mean)	5	
			Alternariol (mean)	47	
			Alternariol methyl ether (mean)	41	
			FB1 (mean)	151	
			FB2 (mean)	96	
			FB3 (mean)	36	
			Sterigmatocystin (mean)	18	
			Roquefortine C (mean)	3	
			Enniatin B (mean)	17	
Meat, egg and fish	Traditional beer	Botswana	Zearalenone	20–201 µg.l^{-1}	Nkwe et al. (2005)
	Egg	Cameroon	Aflatoxine B$_1$, B$_2$, B$_{2a}$, G$_1$ and M$_1$	0.002 to 7.604 ppb	Tchana et al. (2010)
	Chicken livers	Mozambique	AFB$_1$	1.73 µg.kg^{-1}	Sineque et al. (2017)
	Gizzards			1.07 µg.kg^{-1}	

(continued)

Table 1.1 (continued)

Food category	Feed type	Country	Mycotoxins	Limits of the legislation (μg.kg⁻¹)	Range or mean	References
	Fermented sausage 'slavonski kulen'	Croatia	AFB_1		Nd to 14.46 μg.kg⁻¹	Pleadin et al. (2017)
			OTA		Nd to 19.84 μg.kg⁻¹	
	Artisanal and industrial dry sausages	Italy	OTA		3 to 18 μg.kg⁻¹	Iacumin et al. (2009)
	Dried insects	Zambia	AFS	10 μg.kg⁻¹	2.9 to 36.8 μg.kg⁻¹	Kachapulula et al. (2018)
	Fish				0 to 20.4 μg.kg⁻¹	
	Fish	Argentina	T-2 toxin		Median 70.08 ppb	Greco et al. (2015)
			ZEA		Median 87.97 ppb	
			Aflatoxins		Median 2.82 ppb	
Fruits and vegetables fresh and processed	Date palm fruits	Egypt	AFB_1		14.4 μg.kg⁻¹	Abdallah et al. (2018)
			AFB_2		2.44 μg.kg⁻¹	
			OTA		1.48 to 6070 μg.kg⁻¹	
			OTB		0.28 to 692 μg.kg⁻¹	
			FB_1		4.99 to 16.2 μg.kg⁻¹	
	Dried vine fruits	Greek	AFB_1	2 μg.kg⁻¹	0.05 to 0.1 μg.kg⁻¹	Kollia et al. (2014)
			OTA	10 μg.kg⁻¹	5.5 to 101.5 μg.kg⁻¹	
	Dried vine fruits	China	OTA	10 μg.kg⁻¹	<0.07 to 12.83 μg.kg⁻¹	Zhang et al. (2014)
	Raisins and currants	Italy	OTA	10 μg.kg⁻¹	Nd to 15.99 μg.kg⁻¹	Fanelli et al. (2017)
	Dried fruits	United States	OTA	10 ng.g⁻¹	0.28 to 15.34 ng.g⁻¹	
	Dried fruits	Tunisia	AFB_1		45 ng.g⁻¹	Ghali et al. (2010)
			AFB_2		40 ng.g⁻¹	
			AFG_2		5 ng.g⁻¹	

Product	Type	Country	Mycotoxin	Range	Reference
Peanut butter		Zambia	AFB₃ (2012, 2013 and 2014)	>20 to 130 µg.kg⁻¹ >20 to 10,740 µg.kg⁻¹ >20 to 1000 µg.kg⁻¹	Njoroge et al. (2016)
Cereals and cereal-based products	Maize	Zimbabwe	AFS	6.1 to 247 ng.g⁻¹	Mupunga et al. (2014)
		India	AFB₁	48 to 8 µg.kg⁻¹	Mudili et al. (2014)
			FB1	76 to 123 µg.kg⁻¹	
			T-2	38 to 50 µg.kg⁻¹	
			DON	72 to 94 µg.kg⁻¹	
		Croatia	FB₁	196.8 to 1377.6 µg.kg⁻¹	Domijan et al. (2005)
			FB2	68.4 to 3084 µg.kg⁻¹	
			ZEA	0.62 to 3.22 µg.kg⁻¹	
			OTA	0.73–2.54 µg.kg⁻¹	
		Pakistan	AFE₁	0 to 30.92 µg.kg⁻¹	Shah et al. (2010)
			OTA	<0.001 to 7.32 µg.kg⁻¹	
		Brazil	AFE₁	0.2 to 129 µg.kg⁻¹	Vargas et al. (2001)
			ZEA	36.8 to 719 µg.kg⁻¹	
			FB₁	200 to 6100 µg.kg⁻¹	
	Rice	Pakistan	AFB₁	1.07 to 24.65 µg.kg⁻¹	Asghar et al. (2014)
			AFB₂	0.52 to 2.62 µg.kg⁻¹	
		Nigeria	AF	28–372 µg.kg⁻¹	Makun et al. (2011)
			OTA	134–341 µg.kg⁻¹	
		Côte d'Ivoire	AFB₁	<1.5 to 10 µg.kg⁻¹	Sangare-Tigori et al. (2006)
			OTA	0.16 to 0.92 µg.kg⁻¹	
			ZEA	20 to 200 µg.kg⁻¹	
		Sweden	AFB₁	Nd to 46.2 µg.kg⁻¹	Fredlund et al. (2009)
			AFB₂	Nd to 4.5 µg.kg⁻¹	

(continued)

Table 1.1 (continued)

Food category	Feed type	Country	Mycotoxins	Limits of the legislation (µg. kg⁻¹)	Range or mean	References
	Wheat and by-product wheat	Romania	OTA		3.88 to 11.3 ppb	Alexa et al. (2013)
			DON		294 to 3390 ppb	
			FB		960 to 1180 ppb	
		Morocco (Wheat semolina couscous)	DON		20.6 to 106.6 ng.g⁻¹	Zinedine et al. (2017)
			AFG₁		1.0 to 2.5 ng.g⁻¹	
			AFG₂		1.6 to 5.5 ng.g⁻¹	
			ENA		2.6 to 651.7 ng.g⁻¹	
		Thailand (Noodles and breads)	DON		0.14 to 1.13 µg.g⁻¹	Poapolathep et al. (2008)
	Barley	Swiss (2013)	DON		239.8 ± 56.2 µg.kg⁻¹	Schöneberg et al. (2016)
			NIV		12.5 ± 4.3 µg.kg⁻¹	
			ZEA		3.7 ± 1.0 µg.kg⁻¹	
	Sorghum	Ethiopia	AFB₁		<LOD to 33.10 µg kg⁻¹	Taye et al. (2016)
		Uruguay	AFB1		1 to 14 µg kg⁻¹	Del Palacio et al. (2016)
			FB		533 to 933 µg kg⁻¹	

Nd not determined

AFB₁ Aflatoxin B₁, *AFB₂* Aflatoxin B₂, *OTA* Ochratoxin A, *AFG₁* Aflatoxin G₁, *OTB* Ochratoxin B, *AFG2* Aflatoxin G2, *OTC* Ochratoxin C, *AFM* Aflatoxin M, *AFs* Aflatoxin, *FB* Fumonisin, *FB₁* Fumonisin B₁, *FB₂* Fumonisin B₂, *FB₃* Fumonisin B₃, *ZEA* Zearalenone, *FB₄* Fumonisin B₄

The synthesis of aflatoxins is related also to the presence of nitrogen, this compound is present in a wide variety of compounds such as aspartate, alanine, ammonium sulfate, glutamate, glutamine, asparagine, yeast extract, casitone, peptone, ammonium nitrate, ammonium nitrite (Reddy et al. 1971, 1979). Despite the presence of nitrogen in nitrates, their presence suppresses the biosynthesis of aflatoxins (Niehaus and Jiang 1989) by increasing the expression of a protein encoded by aflR gene which subsequently binds to DNA and inhibits the biosynthesis (Chang et al. 1995), tryptophan also inhibits aflatoxin formation in *A. flavus* (Wilkinson et al. 2007). Some nitrogen regulatory genes are known such as nitrate reductase gene niaD, niiA gene for nitrite reductase, and areA gene responsible for assimilation (Chang et al. 1996).

It is well known that aflatoxin biosynthesis occurs in high water activities, but the biosynthesis is also possible during drought or in the areas with low water activity (Cotty and Jaime-Garcia 2007). In the biosynthetic mechanism, temperature has a direct influence, optimal production is localized between 28 °C and 30 °C, above 37 °C, the biosynthesis is almost inhibited (O'Brian et al. 2007), the increase in the temperature affects negatively the production by refusing the transcription of transcriptional regulator gene (aflR) (Yu et al. 2011). Oxidative stress is another factor that induces the formation and production of aflatoxin in *A. parasitcus* (Jayashree and Subramanyam 2000). Regarding pH, the production of aflatoxin is optimal in acidic environments whereas in alkaline media, the biosynthesis is inhibited. The pacC gene is a factor responsible for pH homeostasis and transcription (Tilburn et al. 1995), in alkaline media, it is possible that this pacC gene binds to aflatoxin transcriptional genes and inhibits the synthesis (Espeso and Arst 2000).

It should be noted that ochratoxins biosynthesis is linked to biotic and abiotic stress. The production of these mycotoxins is sensitive to environmental factors, such as carbon and nitrogen levels (Medina et al. 2008; Hashem et al. 2015), in *A. nidulans*, the AreA factor is necessary for permeases expression and catabolic enzymes of nitrogen (Wong et al. 2007). Nutrients can also stimulate Ochratoxin A (OTA) production such as bee pollen (Medina et al. 2004). Temperature, water activity and pH also affect ochratoxin synthesis, for example, OTA production by *A. ochraceus* and *A. carbonarius* in culture medium, is affected by water activity and not by changes in pH (Kapetanakou et al. 2009), for *A. niger*, the aggregate of cells allows them to grow over a wide range of pH (Esteban et al. 2006), another study confirms that optimal conditions for OTA production differ between strains (Passamani et al. 2014).

Few studies described the role of regulators in the transmission of environmental signals to the genome in order to activate or inhibit the expression of mycotoxin biosynthesis genes. In some fungal strains such as *P. verrucosum, P. nordicum,* and *A. carbonarius*, difficult osmotic conditions activate the signal cascade of HOG MAP kinase, which in turn activates various osmoregulatory genes (Stoll et al. 2013). OTA biosynthesis is also affected by oxidative stress (Moye-Rowley 2003), under conditions of high oxidative stress, evoked by increasing concentrations of Cu^{2+} in the medium; *P. verrucosum* produces citrinin, the latter normalizes the oxidative status of fungal cells leading to an adaptation in these environmental conditions, the biosynthesis of citrinin is regulated by a cAMP/PKA signalling pathway (Schmidt-Heydt et al. 2015).

The suppression of hdaA, encoding a histone deacetylase in *A. nidulans* (HDAC), causes a subsequent increase in toxin production levels (Shwab et al. 2007). The LaeA, VeA and VelB complex is studied in fungal strains to clarify the relationship between light-dependent fungal morphology and sexual development for secondary metabolism production, the complex is functional in the dark, in addition, the deletion of laeA and veA in *A. carbonarius* caused a sharp reduction in conidial production and a decrease in OTA production, correlated with a downregulation of PKS-AcOTAnrps gene (Crespo-Sempere et al. 2013). In *A. niger*, OTA production was strongly inhibited by red and blue light compared with dark incubation, with an average reduction about 40-fold (Fanelli et al. 2012).

In addition to the proteins regulating genes transcription, other regulators are involved in the regulation of secondary metabolism, among these, the most studied are; pacC, the key factor for pH regulation in *A. nidulans* (Peñalva et al. 2008), deletion of the entire pacC coding region leads to poor growth and conidiation (Tilburn et al. 1995); CBC, which regulates the response to oxidation-reduction and iron stress; CreA, involved in the regulation of carbon metabolism; AreA and AreB, involved in nitrogen metabolism (Gallo et al. 2017).

Carbohydrates are an important source of carbon required for fumonisins biosynthesis In *F. verticillioides*, the ZFR1 gene controls the biosynthesis of FB_1 by regulating the genes involved in the perception or absorption of carbohydrates (Bluhm et al. 2008), nitrogen is also necessary for the growth of *F. verticillioides* to synthesize proteins and other compounds, the SGE1 gene increases and decreases the expression of certain genes including many effector genes encoding cell surface and transcription activator of fumonisin genes group (Brown et al. 2014).

Fumonisin production by *F. verticillioides* is dependent on water activity in the substrate (Medina et al. 2013). The optimum temperature for fumonisin production is 30 °C (Samapundo et al. 2005), but this temperature depends essentially on the species and can change (Mogensen et al. 2009) Oxidative stress also plays an important role in the modulation of fumonisin biosynthesis (Ferrigo et al. 2015). The pH significantly influences the production of FB_1, at a pH above 5, *F. proliferatum* develops normally but FB_1 is little produced. At a pH below 5, there is less growth but much more FB_1 produced. Below pH 2.5, growth and metabolism are slower with very little FB_1 produced (Keller and Hohn 1997).

Some genes are involved in the regulation of fumonisin biosynthesis and they are affected by environmental conditions. In *F. verticillioides*, the FST1 gene seems to be linked to the expression of several gene networks, particularly those involved in secondary metabolism, cell wall structure, production of conidia, virulence and ROS resistance (Niu et al. 2015), the FUG1 gene represents a new class of fungal transcription factors or genes that are otherwise involved in signal transduction, and play a role in the pathogenicity and fumonisins biosynthesis (Ridenour and Bluhm 2017).

Carbon and nitrogen are essential for the biosynthesis of zearalenone, in *G. zeae*, GzSNF1 gene is responsible for the use of carbon, it grants other biological functions in the sexual and asexual development of the species (Lee et al. 2009), while areA gene is a factor of nitrogen use in the same species (Min et al. 2012). Some mutant strain do not have Nit gene to assimilate nitrate but they are able to use nitrite (Robert et al. 1992). The temperature and water activity affect directly the synthesis of zearalenone, a large production by *F. graminearum* is observed at 25 °C with a water activity equal to 0.97 (Montani et al. 1988), above 37 °C the production of zearalenone is weakened (Jiménez et al. 1996). Under laboratory conditions, the desirable yield of mycotoxins from *F. graminearum* is obtained at an average pH of 6.86 and an incubation temperature of 17.76 °C during 28 days (Wu et al. 2017). In *F. graminearum*, the pac1 gene is responsible for the regulation of metabolite synthesis in the case of pH variations (Merhej et al. 2011), FgLaeA gene is a factor controlling metabolism, virulence and reproduction (Kim et al. 2013), FgVELB gene is a factor developing resistance under osmotic stress conditions (Jiang et al. 2012a).

1.4 Occurrence of Mycotoxins in Humans and Associated Diseases

Mycotoxin-producing fungi are almost ubiquitous in agricultural products before, during and/or after harvest, and may even be present during food processing processes. Various diets taken by humans can be contaminated by a wide range of mycotoxins, and following multiple studies in different countries of the world, these mycotoxins are the cause of several diseases, other studies have even demonstrated, *in vitro* and *in vivo*, the cytotoxicity mechanisms of these poisons.

The frequency and levels exposure of human to mycotoxins may be estimated by measuring the levels of biomarker in urine (Solfrizzo et al. 2011). During the realization of the Swedish National Nutrition Survey, the analysis done in 252 adult participants, simultaneously exposed to mycotoxins, using LC-MS/MS-based multibiomarker approaches, demonstrated that 69% of studied population was exposed to more than one toxin (Wallin et al. 2015). The detection of mycotoxins in biological fluids and human tissues is possible, a study done in patients exposed to mycotoxins with chronic fatigue syndrome (CFS) demonstrated that from 104 urine samples taken from patients, 93% were positive for at least one mycotoxin. The most frequently detected mycotoxin was OTA (83%), followed by macrocyclic trichothecenes. The results of this study demonstrated that 90% of patients were exposed to water-damaged buildings (WDB) (Brewer et al. 2013). Another study done in patients exposed to mycotoxin-producing fungi in their environment demonstrated that the levels of trichothecene in urine, sputum, biological tissues of the lung, liver and brain, after biopsy, varies from undetectable to 18 ppb. In the same tissue, aflatoxin and ochratoxin ranged from 1 to 5 ppb and from 2 to >10 ppb respectively (Hooper et al. 2009).

Application of multi-mycotoxin LC-MS/MS method for human urine analysis in the study conducted in Transkei region in South Africa, known by the high rate of esophageal cancer, demonstrated that morning urine of farmers participating in this survey contained fumonisin B_1, deoxynivalenol-15-glucuronide, zearalenone, α-zearalenol, deoxynivaleno, β-zearalenol and OTA (Shephard et al. 2013), another study with the same objective, carried out in a German population, was able to determine six mycotoxins in their urine, namely; deoxynivalenol, DON-3-glucuronide, zearalenone-14-O-glucuronide, toxin T-2, enniatin B and dihydrocitrinone, with an average daily intake of 0.52 µg.kg^{-1} (bw) for DON, which is greater than the tolerable daily intake (Gerding et al. 2014). The study conducted by Gerding and his team in 2015 approved the previous result, this latest study also demonstrated that AFM_1 is detected in urine samples in a population of Bangladesh and Haiti with a higher rate of OTA and dihydrocitrinone in Bangladesh samples (Gerding et al. 2015). In Belgium, biomarker analysis showed a clear exposure of a large segment of the Belgian population to DON, OTA and CIT (Heyndrickx et al. 2015). A more serious situation is observed, the detection of multiple mycotoxins is recorded among HIV-positive people in Cameroon, demonstrating that this situation is very serious for cases of very low immunity (Abia et al. 2013). The detection of mycotoxins or their metabolites in the blood or blood serum is also possible by the fast-multi-mycotoxin approach, the study of Osteresch and its collaborators detected enniatin B, OTA and 2'R-OTA in dried blood and serum spots (Osteresch et al. 2017).

The detection of OTA and 2'R-OTA in the blood of drinkers and non-coffee drinkers demonstrated the presence of OTA in both groups with an average concentration of 0.21 µg.l^{-1}, whereas, 2'R-OTA was present only in coffee drinkers with a maximum concentration of 0.414 µg.l^{-1}, this result is explained by the formation of 2'R-OTA from OTA during roasting process (Cramer et al. 2015), in human blood, 2'R-OTA has a biological half-life seven times higher in comparison with OTA (Sueck et al. 2018a), this long period of life in human blood is explained by the affinity of both toxins to human serum albumin (HSA), which attenuates or even suppresses the acute cellular toxicity of mycotoxins (Faisal et al. 2018a), however, another study demonstrated that OTA has more binding affinity to ASH than 2'R-OTA with binding percentages of 99.6% and 97.2% respectively, suggesting that another human blood protein with a very high affinity for 2'R-OTA is responsible for its accumulation in the blood, or the high bioavailability of 2'R-OTA compared to OTA, or differences in the metabolism of this metabolite and its transport in tissues (Sueck et al. 2018b).

Citrinin and its metabolite dihydrocitrinone are also detected in human urine, a study done in a German population showed that the levels of these two metabolites varied between 0.02 to 0.08 ng.ml^{-1} and 0.05 to 0.51 ng.ml^{-1} respectively (Ali et al. 2015a), the both mycotoxins are also detected in two populations in Bangladesh (Ali et al. 2015b). Citrinin can also bind to human serum albumin, it main binding site is in sub-domain IIA (Sudlow's Site I), this binding forming the CIT-albumin complex, it has beneficial effects by decreasing the cellular absorption of CIT and consequently, the decrease of its toxicity (Poór et al. 2015).

Zearalenone can also form with human serum albumin a complex of great biological interest (Poór et al. 2017a), the both metabolites α-zearalenol and β-zearalenol can also bind to ASH but with a lower affinity in comparison to zearalenone, this low affinity may be due to the difference in position or binding site (Faisal et al. 2018b). Citreoviridine can also bind with ASH through hydrophobic bonds (Hou et al. 2015), AFB_1 also forms stable bonds with ASH, the high affinity binding site is Sudlow's Site I of subdomain IIA, furthermore, the reactive epoxide metabolite of AFB_1 forms covalent adducts with ASH (Poór et al. 2017b), a recent study done in calf using multispectroscopic techniques, DNA fusion, viscosity measurements, and molecular Docking techniques demonstrated that AFB_1 and AFG_1 can bind with thymic DNA (Ma et al. 2017). A study done in Malawi in 230 persons from a rural population living in three districts (Kasungu, Mchinji, and Nkhotakota), receiving a diet rich in groundnuts and maize, detected AFB_1-lys adducts in 67% of blood, with an average concentration of 20.5 ± 23.4 pg.mg^{-1} of albumin (Seetha et al. 2018).

This bioaccumulation in liquids, organs and tissues is reported worldwide, and most mycotoxins have toxic effects at low concentrations (Escrivá et al. 2017), mycotoxicosis is among the adverse effects of mycotoxins, the exposure to these fungal metabolites is mainly by ingestion, or also dermal and inhalation pathway (Peraica et al. 1999), the diseases caused by mycotoxins are varied and affect a wide range of animal species and human (Richard 2007), among proven diseases *in vitro*, caused by exposure to mycotoxins; cutaneous toxicity and skin tumorigenesis in several rodent models induced by oxidative stress (Doi and Uetsuka 2014). Fumonisin also causes several diseases with variable toxicities by interference with the sphingolipids biosynthesis in several organs (Bucci et al. 1998). Aflatoxin induces hepatotoxicity, in addition, acute exposure to AFB_1 increases plasma and hepatic cholesterol, triglyceride and phospholipid levels (Rotimi et al. 2017). The Table 1.2 below shows the different toxicities caused by mycotoxins, its target organs and the mode of action.

1.5   Conventional Methods for Mycotoxins Detection

Mycotoxin detection techniques classified as conventional techniques are very expensive analytical methods and require qualified personnel for each method to achieve the analyses, analytical instruments and products used in these techniques are also expensive and very complicated without neglected their detection limits, the samples analysed must be prepared beforehand in order to extract their contents in mycotoxins. Food safety is a challenge which is not yet achieved in the world, and following this crucial point, several official methods used in laboratories are validated by the Association of Official Agricultural Chemists (AOAC) to improve detection techniques. Protect the consumer and to ensure a good quality of food.

Table 1.2 Examples of mycotoxin toxicity

Toxicity	Mycotoxin	Target organ or cells lines	Mode of action	References
Hepato-gastrointestinal toxicity	ZEA	Intestines of pregnant sows	It caused changes in the structure of jejunum with oxidative stress and an inflammatory response.	Liu et al. (2018)
	ZEA	Intestine of pregnant Sprague-Dawley.	Alterations of the bacterial numbers in cecal digesta are noted.	Liu et al. (2014b)
			It induced oxidative stress, affected the villous structure and reduced the expression of junction proteins claudin-4, occludin and connexin43 (Cx43).	
	ZEA	Rats liver	It induced an increase on the transcription of cytochrome P450 2E1 (CYP2E1).	Zhou et al. (2015) and Jiang et al. (2012b)
			It induced an increase of liver enzymes.	
	T-2 toxin	Hela, Bel-7402, and Chang liver cells	It induced apoptosis.	Zhuang et al. (2013)
		Porcine stomach and duodenum	It caused an increase of VIP-LI nerve cells and nerve fibers in both fragments of gastrointestinal tract.	Makowska et al. (2018)
		Female genital organs (animals)	Decreased in follicle integrity.	Schoevers et al. (2012)
			Reduced in the quantity of healthy follicles, which may lead to premature oocyte depletion in adulthood.	
	FA	Oesophageal SNO cells	It induced an increase in cells apoptosis.	Devnarain et al. (2017)
	HT-2 toxin	Liver	It induced oxidative stress and cell damage.	Yang et al. (2016, 2017, 2019)
			It induced changes in mRNA, protein expression and apoptosis.	
	FB	Intestine	It caused disturbance of the sphingolipid biosynthesis pathway and conducted to an alteration in intestinal epithelial cell viability and proliferation.	Bouhet and Oswald (2007)
	ENN	Intestinal cell line IPEC1 and jejunal explants	It caused a decrease in cell proliferation.	Kolf-Clauw et al. (2013)
		IPEC-J2 cells	It caused a complete disruption with declined in cells viability.	Fraeyman et al. (2018)

CB	Rat liver	It caused a decrease in albumin synthesis, glucosamine, and minimally affected glycoprotein secretion	Kaufman et al. (1988)
MPA	LS180 cell	It affected intracellular nucleotide levels, expression of structural proteins, fatty acid and lipid metabolism.	Heischmann et al. (2017)
RB	Mouse hepatic microsomes	It inhibited an enzyme activity located in the endoplasmic reticulum.	Watson and Hayes (1981)
	Dog	It is suggested as a primary etiological factor for hepatitis X.	Hayes and Williams (1978)
VIM	Mice liver	It caused necrosis, periductal fibrosis, and hypertrophy and hyperplasia of biliary epithelium.	Carlton et al. (1976)
CPA	Rat liver	It caused cells lysis and affected the endoplasmic reticulum.	Hinton et al. (1985)
	Monkey liver	It caused mild pathological changes in hepatocyte rough endoplasmic reticulum, small vessels and myocardium	Jaskiewicz et al. (1988)
AF	Rat liver	It caused an increase levels of plasma and liver cholesterol, triglycerides and phospholipids.	Rotimi et al. (2017)
	Male broiler chicks	It caused an alteration in gene expression.	Yarru et al. (2009)
	Hy-Line W36 hens	It affected the functionality of the gastrointestinal tract.	Applegate et al. (2009)
Reproductive and developmental toxicity	Female genital organs (animals).	20 mg.kg^{-1} ZEA diets increased follicle-stimulating hormone concentrations and decreased oestradiol.	Gao et al. (2017)
	Male genital organs (animals).	20 mg.kg^{-1} ZEA caused a follicular atresia and a thinning uterine layer.	
		It inhibited mRNA and protein levels of oestrogen receptor-alpha (Esr1) and 3β-hydroxysteroid dehydrogenase (HSD).	
		Neonatal exposure to ZEA could negatively influence male fertility and his spermatogenesis by the exposure of testis to ABC transporter substrates.	Koraïchi et al. (2013)

(continued)

Table 1.2 (continued)

Toxicity	Mycotoxin	Target organ or cells lines	Mode of action	References
	ZEA	Mouse TM4 Sertoli cells	It caused damage to the cytoskeletal by autophagy stimulated through PI3K-AKT-mTOR and MAPK signaling pathways via elevated oxidative stress.	Zheng et al. (2018a)
	ZEA	Murine ES cells cardiac differentiation	It caused an inhibition of cardiac differentiation with ROS accumulated in murine ES cells.	Fang et al. (2016)
	T-2 toxin	Embryonic stem cells (ES cells D3) and fibroblast 3 T3 cells,	It induced cell cycle arrest and apoptosis.	Fang et al. (2012)
			It caused the release of cytochrome C from mitochondria with the upregulation of p53, caspase-9, caspase-3 expression.	
			It induced oxidative damage and apoptosis in the cells.	
		Zebrafish embryos.	It increased the mortality and malformation rate in early developmental stages of zebrafish.	Yuan et al. (2014)
		Sertoli cells	It inhibited the cell proliferation and through ATP/AMPK pathway induced by ROS generation, the toxin induced cell apoptosis.	Zheng et al. (2018b)
	DAS	Mouse	It induced a reduction in fetal body weight and a variety of fetal malformations.	Mayura et al. (1987)
		Pig	It induced a development of multifocal, proliferative, gingival, buccal and lingual lesions.	Weaver et al. (1981)
	FB	Chicken embryos	It induced embryonic mortality.	Henry and Wyatt (2001)
	VIM	Chicken embryos	It was mildly toxic and teratogenic	Anthony et al. (1982)
	CTN		It is embryocidal and fetotoxic	Flajs and Peraica (2009)
	PAT	Growing male rats	It caused an increase in testosterone levels and a decrease in T4 levels.	Selmanoglu and Koçkaya (2004)
			It caused an increase of TLR4 protein expression and inflammatory cytokines.	
	AF	Rabbits	AFB_1 was found teratogenic for fetal development in rabbits.	Wangikar et al. (2005)

			It caused a developmental and reproductive toxicity	Malir et al. (2014)
Renal toxicity	OTA			
	ZEA	Kidney and other organs (Spleen (rate), thymus, bursa of Fabricius)	It caused a histopathological damage in the spleen and a decrease viability of splenocytes and T-cell proliferation.	Yin et al. (2014)
			It reduced the level of anti-Newcastle disease virus antibody.	Xue et al. (2010)
			It reduced the concentrations of interleukin-2 and interferon-gamma in serum.	
			It caused a histological abnormality.	
	AF	HEK 293 cells model and CD-1 mice model	It caused kidney toxicity through oxidative stress and induced downstream apoptosis.	Li et al. (2018)
	OTA	Pig kidney	It affected genes expression in kidney.	Marin et al. (2017)
	MPA	Renal transplant	Toxicity is caused by an increase of free MPA concentration.	Mudge et al. (2004)
Immunomodulatory and immunopathological effects	OTA+ T-2 toxin	Spleen, thymus, and bursa of Fabricius (yellow-feathered broiler chickens)	It caused a decrease of spleen, thymus, and bursa of Fabricius weight.	Wang et al. (2009)
			It caused a diminution of the mitogenic responses of peripheral blood lymphocytes.	
		Lymphoid organs of mice	It caused a decrease in number of lymphocytes in the thymic cortex and splenic follicles.	Shinozuka et al. (1997)
			It caused a pyknosis or karyorrhexis in lymphocytes nuclei.	
	BT-2 toxin	Bone marrow and splenic red pulp of mice	It caused a hypocellularity with a decrease in the number of myelocytes due to the loss of immature granulocytes, erythroblasts, and lymphocytes.	Shinozuka et al. (1998)
		TM3 cells	It induced an apoptosis in TM3 cells by inhibiting mTORC2/AKT and promoting Ca^{2+} production.	Wang et al. (2018)

(continued)

Table 1.2 (continued)

Toxicity	Mycotoxin	Target organ or cells lines	Mode of action	References
	GT	Immune cells	It removed immune cells responsible for tissue rejection.	Waring and Beaver (1996)
			It induced degranulation, leukotriene C4 secretion, and TNF-alpha and IL-13 production.	Niide et al. (2006)
			It induced intracellular production of superoxide.	
		Immune organ	It induced apoptosis in cells.	Sutton et al. (1994)
	ENN	Jurkat T-cell	It caused a mitochondrial alteration.	Manyes et al. (2018)
		RAW 267.4 murine macrophages	It caused an accumulation of cells in the G0/G1-phase.	Gammelsrud et al. (2012)
			It caused lysosomal damage.	
			It caused an apoptotic and necrotic cell.	
	MPA	T-lymphoblast cell line	It can modulate Rho GDI 2 levels in T lymphocytes and disrupt cell signalling pathways.	Heller et al. (2009)
	PNT	Human neutrophil granulocytes	It caused an increase in ROS production by the activation of several MAPK-signalling pathways.	Berntsen et al. (2017)
	CPA	CD34+, monocytes, THP-1 and Caco-2)	It disturbed human monocytes differentiation into macrophages.	Hymery et al. (2014)
	PA	Rat alveolar macrophages	It caused an inhibition of phagocytosis	Sorenson and Simpson (1986)
	PAT	Thymus of growing male rats	The thymus showed haemorrhage, and alteration of tissue.	Arzu Koçkaya et al. (2009)
		Dendritic cells (IDCs) of rat thymus	It caused cell apoptosis.	Ozsoy et al. (2008)

Category	Toxin	Cell type	Effect	Reference
Cytotoxicity	T-2 toxin	HepG2 and HEK293T cells	It induced an increase in the levels of mitochondrial biogenesis and ROS, by upregulation of SIRT1, which is controlled by miR-449a, whose expression was inhibited by toxin.	Ma et al. (2018)
		HeLa cells	It caused generation of ROS and increased lipid peroxidation. An apoptotic morphology with condensed chromatin and nuclear fragmentation are also noted.	Chaudhari et al. (2009)
		Human chondrocytes	It caused a decrease in chondrocytes viabilities.	Liu et al. (2014a)
	FA	Hepatocellular carcinoma (HepG2) cells	It induced apoptosis and necrosis by increasing activity of caspases.	Sheik Abdul et al. (2016)
			It induced mitochondrial stress.	
			It caused DNA damage and post-translational modifications of p53.	Ghazi et al. (2017)
			It induced NRF2 as a cytoprotective response to prevent NLRP3 (activator protein inflammasome).	Sheik Abdul et al. (2019)
	FB	U-118MG glioblastoma cells	It induced an oxidative stress, apoptosis and internucleosomal DNA fragmentation.	Stockmann-Juvala et al. (2004a)
	BEA	Chinese hamster ovary (CHO-K1) cells	It induced a disruption in mitochondrial enzymatic activity and cell proliferation and lead to the cell death.	Mallebrera et al. (2016)
		Caco-2 cells	It caused cytotoxicity and apoptosis via ROS production and mitochondrial damage	Prosperini et al. (2013)
	ENN	Balb 3 T3 and HepG2 cells	It altered cellular energy metabolism and reduced cell proliferation.	Jonsson et al. (2016)
			It caused necrotic cell death.	
		Caco-2 cells	It caused a lysosomal membrane permeabilization	Ivanova et al. (2012)
			It caused cell cycle arrest in the G2/M phase leading to the apoptosis or necrosis.	
		HepG2	It caused an apoptosis and necrosis.	Juan-García et al. (2013, 2015)

(continued)

Table 1.2 (continued)

Toxicity	Mycotoxin	Target organ or cells lines	Mode of action	References
	NIV	Human erythroleukemia cell line	It caused DNA damage and apoptosis.	Minervini et al. (2004)
		HepaRG human hepatocyte cell line	It caused hepatotoxicity.	Smith et al. (2017)
	CB	Cells lines	It induced cellular DNA fragmentation.	Kolber et al. (1990)
	MPA	Human umbilical vein endothelial cells	It inhibited the lymphocyte proliferation and decreased the expression of adhesion molecules on endothelial cells.	Raab et al. (2001)
		Human peripheral blood mononuclear cells	It induced a decrease in GTP levels and an increase of UTP.	Daxecker et al. (2001)
	CTN	Porcine kidney PK15 cells	It stimulated Hsps expressions.	Segvić Klarić et al. (2014)
			It caused alterations in cellular redox status.	
		HepG2 cell	It caused a production of ROS, DNA damage and mitochondria-mediated intrinsic apoptosis.	Gayathri et al. (2015)
			It caused a disturbance of Ca^{2+} homeostasis, cell genotoxicity and death.	Klarić et al. (2012)
	OTA	HepG2 and KK-1 cells	It caused a production of ROS.	Li et al. (2014)
		HT-29-D4 and Caco-2-14 cells	It caused an alteration of the barrier and absorption functions of the intestinal epithelium	Maresca et al. (2001)
	ZEA	Destruction of cytoskeletal structure in kidney	It caused kidney dysfunction and damage.	Jia et al. (2014)
		Cell cycle arrest and apoptosis in kidney and spleen (rate)	It induced a degenerative change in the kidney and an increase of the biochemical parameters and inflammatory cytokines.	Jia et al. (2015)
Neurotoxicity	FA	Swine	It caused vomiting and neurochemical changes.	Smith and MacDonald (1991)
		Brain and pineal gland of rats.	It caused an increase in brain serotonin, 5-hydroxyindoleacetic acid, tyrosine and dopamine, and a decrease in norepinephrine.	Porter et al. (1995)
		Pineal cell cultures	It caused an increase in melatonin levels.	Rimando and Porter (1997)

FB	Human SH-SY5Y neuroblastoma, rat C6 glioblastoma and mouse GT1-7 hypothalamic cells	It induced an oxidative stress characterised by the production of reactive oxygen species, lipid peroxidation and necrotic cell death.	Stockmann-Juvala et al. (2004b)
RB	Mouse brain	It induced a toxicity in specific brain regions by altering SOD and OGG1 activities.	Sava et al. (2018)
PNT	Rat cerebellar granule neurons	ROS production and the GABAA receptor are the causes of neuronal death.	Berntsen et al. (2013)
CPA	Rabbit	10 mg CPA.kg^{-1} (bw) produced tachycardia, tachypnoea and sedation.	Nishie et al. (1985)
AF	Zebrafish embryos and larvae	It affected the behaviors and neurodevelopment	Wu et al. (2019)
OTA	Mouse brain	It altered the proliferation and differentiation in subventricular zone.	Paradells et al. (2015)
CIV	Rat brain	It inhibited brain synaptosomal and altered the enzymatic activities.	Datta and Ghosh (1981)
Pulmonary toxicity			
FB	Swine	It caused pulmonary edema and death.	Osweiler et al. (1992)
RQF	Dog	It caused a development of aspiration pneumonia	Young et al. (2003)
OTA	RAT	It caused a damage in tissue like alveolar congestion, alveolar cell hyperplasia, and respiratory epithelial proliferation.	Okutan et al. (2004)
AF	Mousse lung	It caused a development of tumorigenesis.	Massey et al. (2000)
CIV	Rabbit	It caused respiratory arrest and death.	Nishie et al. (1988)

Some toxicity induced (in vivo and in vitro) by the different mycotoxins

AF aflatoxin, *BEA* beauvericin, *CIV* citreoviridin, *CTN* citrinin, *CPA* cyclopiazonic acid, *CB* cytochalasin, *DAS* diacetoxyscirpenol, *ENN* enniatin, *FRE* frequentin, *FB* fumonisins, *FA* fusaric acid, *GT* gliotoxin, *HT-2 toxin*, *T-2 Toxin*, *MPA* mycophenolic acid, *NIV* nivalenol, *OTA* ochratoxin A, *PAT* patulin, *PA* penicillic acid, *PNT* penitrem A, *RQF* roquefortine, *RB* rubratoxin B, *VIM* viomellein, *ZEA* zearalenone *bw* body weigh

1.5.1 Chromatographic-Based Techniques

1.5.1.1 Thin-Layer Chromatography

Thin layer chromatography can detect a wide range of mycotoxins after extraction with different solvents and at different pH, it is applicable in different types of food, after mycotoxin extraction (Gimeno 1979). In order to limit interference with lipids, pigments and other food compound, this chromatographic screening technique has undergone improvements, by using membrane cleaning (Roberts and Patterson 1975). Toluene, ethyl acetate and formic acid are the three solvents used for the separation. The revelation of the chromatogram is carried out by exposing the plate to pyridine or acetic anhydride vapors, or a mixture of both, the spots of the mycotoxins appear fluorescent under UV light at 365 nm (Goliński and Grabarkiewicz-Szczesna 1984). AFB$_1$ is detected at concentrations ranging from 0.1 to 0.3 µg.kg^{-1} (Patterson and Roberts 1979), while for others, such as diacetoxyscirpenol, ochratoxin A, patulin, penitrem A, sterigmatocystin, toxin T-2 and zearalenone are detected in a less reliable manner (Roberts and Patterson 1975).

1.5.1.2 High-Performance Thin-Layer Chromatography

After determining the role and toxicity of mycotoxins in human and animal organism, the chromatography technique has developed with more performance, high-performance thin-layer chromatography (HPLC) and Reversed-phase liquid chromatography (RPLC) coupled with a fluorescence detector are the two most used instruments for mycotoxins detection (Orti et al. 1986), other machines are currently running with wavelength variation, a fluorescence detector, and require an extraction and immunoaffinity column (IAC) for clean-up (Pena et al. 2006; Chen et al. 2017). Liquid chromatographic-tandem mass spectrometric method (LC-MS/MS) is another instrument developed for the determination of mycotoxins biomarkers in human or animal urine, the sample preparation requires several steps of clean-up (Solfrizzo et al. 2011), the latter technique is recommended for mycotoxin biomarkers analysis in biological fluids (Gerding et al. 2014). Other technique are developed such as liquid chromatography combined with heated electrospray ionization triple quadrupole tandem mass spectrometry (LC-h-ESI-MS/MS) which has been developed for the multiple detection of mycotoxin in plasma with LOD ranging from <0.01 to 0.5 ng.ml^{-1} (Devreese et al. 2012) and et Ultrasound-assisted solid-liquid extraction and immunoaffinity column clean-up coupled with high performance liquid chromatography and on-line post-column photochemical derivatization-fluorescence detection (USLE-IAC-HPLC-PCD-FLD) (Kong et al. 2013).

1.5.2 Microarrays for Mycotoxin Detection

Bio-analytical technology based on DNA microarrays are used for the detection and selection of fungal mycotoxin-producing strains, this transcriptomic technique based on hybridization is very effective for the identification of mycotoxin expression genes in complete sequences of fungal DNA (Emri et al. 2017), Several microarray oligonucleotides are developed to determine the pathogens in a single reaction, the fungal DNA must be amplified by universal primers and the PCR product after clean-up will undergo hybridization with the microarray oligonucleotide (Huang et al. 2006), there are microarray containing oligonucleotides of the biosynthetic pathways of fumonisin, aflatoxin, ochratoxin and others, even recently identified biosynthetic genes can be added to these chips (Schmidt-Heydt and Geisen 2007). These DNA microarrays reveal high specificity and stability for pathogen detection (Bouchara et al. 2009).

1.5.3 Enzyme-Linked Immunosorbent

This technique is based on the antigen-antibody immune reaction, the ELISA technique allows a highly sensitive and selective quantitative and/or qualitative analysis. For the assay, an antigen or antibody is labelled with an enzyme, hence his name comes from. The antigen is immobilized in a solid phase such as a microplate. In the immunological reaction, the antigen reacted with the antibody which will subsequently be detected by a secondary antibody labelled with an enzyme developing a color (chromogenic), the reaction time is estimated between 30 min and 1 h, at the end of the reaction, the colored products are detected using a microplate reader (Sakamoto et al. 2017). This assay technique has proved its effectiveness in detecting and measuring mycotoxins even in foods (Liu et al. 1985; Gao et al. 2012).

1.5.4 Radioimmunoassay

This technique is applicable for the detection of mycotoxins such as toxin T-2 in the liver, spleen and kidney (Hewetson et al. 1987), it can also detect T-2 toxin in biological fluids such as serum and urine with elimination of the extraction step that preceded the analysis, the sensitivity of the technique was 1 ng per analysis or 10 ng.ml^{-1} with a significant reaction (10.3%) of the toxin (Fontelo et al. 1983). RIA can also be applied for the detection of T-2 toxin in milk with a sensitivity of 2.5 ppb (Lee and Chu 1981). This radioimmunoassay (RIA), with the tracer [3H] AFB$_1$ has proven effective for detecting AFB$_1$, AFM$_1$ and a major urinary metabolite of AFB$_1$ in several species (Sizaret et al. 1982).

1.6　Nanotechnology-Based Techniques for Mycotoxin Detection

The manufacture of nanomaterials based on carbon or gold or other materials has recognized in recent years an exponential progress, by the technological development known in these nanomaterials, and their applications in different fields. Developed nanoparticles can bind to different antibodies types, which are considered as molecular recognition receptors and can trap mycotoxins due to their specificity and sensitivity. Aptamers based on synthetic oligonucleotides also have a specific affinity for mycotoxins detection. Some new methods are demonstrated in the following paragraphs.

1.6.1　Quartz Crystal Microbalance Immunosensor (QCMI)

Quartz Crystal Microbalance Immunosensor (QCMI) are widely used in the biological field, the principle of the method is based on the piezoelectric properties of quartz crystals that are used in their manufacture. These AT-cut crystals used in the QCM transducers resonate at a fixed frequency when applying an electric current, this frequency is relative to changes in the contact conditions with the surface of the crystals due to an increase in mass during the immunoreaction with anti-mycotoxin antibodies. This technique is able to detect mycotoxins by their direct immobilizations on sensor surfaces containing, for example -ethyl-3-(3-dimethylaminopropyl) carbodiimide (EDC)/N-Hydroxysuccinimide (NHS), this surface is able of be regenerated 13 times after each analysis (Pirinçci et al. 2018), quartz crystals can also be covered by gold nanoparticles (Vidal et al. 2009).

1.6.2　Surface Plasmon Resonance (SPR)

Surface plasmon resonance is one of the methods that provide rapid quantitative and qualitative detection of mycotoxins with a detection limit of about 200 Da, this method became popular in the 1990s following the commercialization of biosensors. The principle of the method is based on the phenomenon of surface plasmon resonance, the latter occurs between a very thin metal film (formed in gold) deposited between two transparent support (prism) having a different refractive index, a polarized incident light undergoes internal refraction in the prism which induces the components of the electromagnetic field of light to penetrate the gold film, the interaction of this field with free oscillating electrons on the gold film will cause the excitation of the plasmons and will cause the decrease of the intensity of the light received by an optical detector, this system detects the changes in the refraction of the surface layer in contact with the solution which is related to the molecules binding to the film (Hodnik and Anderluh 2009). The thin layer may contain gold nanoparticles, as demonstrated in the study of Hossain and his group, in the latter, the team was able to determine the T-2 toxin and T-2 toxin-3-glucoside in wheat after their extractions, on a device containing antibodies conjugated with gold nanoparticles (Hossain et al. 2018). SPR containing nanostructured films are called imaging nanoplasmonics, they can even detect several mycotoxins in same times, with regeneration surfaces after analysis (Joshi et al. 2016; Hossain and Maragos 2018). The technique undergone modifications and perfections such as indirect competitive enzyme-linked immunosorbent assay (ic-ELISA) which demonstrated its reliability (Dong et al. 2018).

1.6.3 Electrochemical Immuno-Sensor Based on Nanoparticle

Different Electrochemical Immunosensors are developed to determine a type of mycotoxin, such as electrochemical Immunosensor based on polythionine/gold nanoparticles for the determination of AFB_1 (Owino et al. 2008), but following the multi-presence of mycotoxins in foods, a design of an electrochemical immunosensor is carried out for the simultaneous detection, during a single test, of two mycotoxins namely; fumonisin B_1 (FB_1) and deoxynivalenol (DON), two electrodes were functionalized with gold nanoparticles and anti-FB_1 and anti-DON antibodies. After incubation of the electrodes in the sample solution, the immunocomplex formation generates an electrochemical signal which is then compared with the control signal to quantify and calculate the concentration of mycotoxins, the LODs were 97 pg. ml^{-1} and 35 pg.ml^{-1}, respectively for FB_1 and DON. The stability of the electrodes is marked good (Lu and Gunasekaran 2019). Another indirect competition enzyme immunoassay technique is developed to determine the presence of AFB_1 in Palm kernel cake, modified electrodes prepared from multi-walled carbon nanotubes and chitosan are prepared, the attachment of AFB_1-BSA antigens on the composites surface is activated by N-ethyl-N'-(3-dimethylaminopropyl)-carbodiimide (EDC) and N-hydroxysuccinimide (NHS). A competitive reaction takes place between AFB_1 and the fixation site with producing a signal, the detection range of the electrochemical immunosensor was between 0.0001 and 10 ng.ml^{-1} with an LOD of 0.1 pg.ml^{-1} (Azri et al. 2017). The application of these ultrasensitive methods requires improvement and optimization of several parameters such as antigens concentration, blocking agents, incubation time, temperature and reagents pH, this optimization is performed on the microplates by ELISA test (Azri et al. 2018). This detection technique is a promising approach for mycotoxin screening because it is simple, fast, extremely sensitive, specific and does not require pre-concentration of sample, it can be applied for the detection of mycotoxins in a wide variety of foods (Tan et al. 2009; Wang et al. 2017).

1.6.4 Optical Waveguide Lightmode Spectroscopy

The OWLS technique is based on the precise calculation of the resonance angle of a polarized laser light beam (632.8 nm), diffracted by a grating and coupled to a thin waveguide. Determination of the thickness of the adsorbed or bonded material layer is performed with great sensitivity from the effective refractive index determined from the coupling angle of the resonance (Székács et al. 2009). This technique is also applied for mycotoxins detection in a competitive and direct immunoassay format. The improvement of the detection signal is relative to the surface of the immune sensor formed by immobilization of gold nanoparticles (AuNPs) containing the conjugated antibodies or antigens, after fixation of the mycotoxins, an immunodetection method has been developed for the analysis (Adányi et al. 2018). Comparison of the detection results of this method with the ELISA and HPLC test demonstrated an excellent correlation indicating that the OWLS immunosensor has a good potential for the rapid determination of mycotoxins (Majer-Baranyi et al. 2016).

References

Abdallah MF, Krska R, Sulyok M (2018) Occurrence of ochratoxins, fumonisin B2, aflatoxins (B1 and B2), and other secondary fungal metabolites in dried date palm fruits from Egypt: a mini-survey. J Food Sci 83:559–564. https://doi.org/10.1111/1750-3841.14046

Abia WA, Warth B, Sulyok M et al (2013) Bio-monitoring of mycotoxin exposure in Cameroon using a urinary multi-biomarker approach. Food Chem Toxicol 62:927–934. https://doi.org/10.1016/j.fct.2013.10.003

Adányi N, Nagy ÁG, Takács B et al (2018) Sensitivity enhancement for mycotoxin determination by optical waveguide lightmode spectroscopy using gold nanoparticles of different size and origin. Food Chem 267:10–14. https://doi.org/10.1016/j.foodchem.2018.04.089

Adekoya I, Njobeh P, Obadina A, Chilaka C, Okoth S, De Boevre M, De Saeger S (2017) Awareness and prevalence of mycotoxin contamination in selected nigerian fermented foods. Toxins 9:363. https://doi.org/10.3390/toxins9110363

Adekoya I, Obadina A, Adaku CC, De Boevre M, Okoth S, De Saeger S, Njobeh P (2018) Mycobiota and co-occurrence of mycotoxins in South African maize-based opaque beer. Int J Food Microbiol 270:22–30. https://doi.org/10.1016/j.ijfoodmicro.2018.02.001

Adgigitov F, Kosichenko LP, Popandopulo PG et al (1984) Frequency of chromosome aberrations in bone marrow of monkeys and their F1 after aflatoxin B1 injection. Exp Pathol 26:163–169

Al-Anati L, Petzinger E (2006) Immunotoxic activity of ochratoxin A. J Vet Pharmacol Ther 29:79–90. https://doi.org/10.1111/j.1365-2885.2006.00718.x

Alexa E, Dehelean CA, Poiana MA et al (2013) The occurrence of mycotoxins in wheat from western Romania and histopathological impact as effect of feed intake. Chem Cent J 7:99. https://doi.org/10.1186/1752-153X-7-99

Ali N, Blaszkewicz M, Degen GH (2015a) Occurrence of the mycotoxin citrinin and its metabolite dihydrocitrinone in urines of German adults. Arch Toxicol 89:573–578. https://doi.org/10.1007/s00204-014-1363-y

Ali N, Blaszkewicz M, Mohanto NC et al (2015b) First results on citrinin biomarkers in urines from rural and urban cohorts in Bangladesh. Mycotoxin Res 31:9–16. https://doi.org/10.1007/s12550-014-0217-z

Almeida-Ferreira GC, Barbosa-Tessmann IP, Sega R et al (2013) Occurrence of zearalenone in wheat- and corn-based products commercialized in the State of Paraná, Brazil. Braz J Microbiol 44:371–375. https://doi.org/10.1590/S1517-83822013005000037

Anthony J, Delucca JJ, Dunn LS et al (1982) Toxicity, mutagenicity and teratogenicity of brevianamide, viomellein and xanthomegnin; secondary metabolites of penicillium viridicatum. J Food Saf 4:165–168. https://doi.org/10.1111/j.1745-4565.1982.tb00440.x

Applegate TJ, Schatzmayr G, Prickel K et al (2009) Effect of aflatoxin culture on intestinal function and nutrient loss in laying hens. Poult Sci 88:1235–1241. https://doi.org/10.3382/ps.2008-00494

Armando MR, Pizzolitto RP, Dogi CA et al (2012) Adsorption of ochratoxin A and zearalenone by potential probiotic Saccharomyces cerevisiae strains and its relation with cell wall thickness. J Appl Microbiol 113:256–264. https://doi.org/10.1111/j.1365-2672.2012.05331.x

Arzu Koçkaya E, Selmanoğlu G, Ozsoy N et al (2009) Evaluation of patulin toxicity in the thymus of growing male rats. Arh Hig Rada Toksikol 60:411–418. https://doi.org/10.2478/10004-1254-60-2009-1973

Asghar MA, Iqbal J, Ahmed A et al (2014) Occurrence of aflatoxins contamination in brown rice from Pakistan. Iran J Public Health 43:291–299

Asghar MA, Ahmed A, Asghar MA (2018) Aflatoxin M1 in fresh milk collected from local markets of Karachi, Pakistan. Food Addit Contam Part B Surveill 11:167–174. https://doi.org/10.1080/19393210.2018.1446459

Azri FA, Selamat J, Sukor R (2017) Electrochemical immunosensor for the detection of aflatoxin B₁ in palm kernel cake and feed samples. Sensors 17:2776. https://doi.org/10.3390/s17122776

Azri FA, Sukor R, Selamat J et al (2018) Electrochemical immunosensor for detection of aflatoxin B₁ based on indirect competitive ELISA. Toxins 10:196. https://doi.org/10.3390/toxins10050196

Barrett JR (2005) Liver cancer and aflatoxin: New information from the kenyan outbreak. Environ Health Perspect 113: A837–A838. https://doi.org/10.1289/ehp.113-a837

Bárta I, Adámková M, Markarjan D et al (1984) The mutagenic activity of aflatoxin B1 in the Cricetulus griseus hamster and Macaca mulatta monkey. J Hyg Epidemiol Microbiol Immunol 28:149–159

Bennett JW, Rubin PL, Lee LS et al (1979) Influence of trace elements and nitrogen sources on versicolorin production by a mutant strain of Aspergillus parasiticus. Mycopathologia 69:161–166. https://doi.org/10.1007/BF00452829

Berntsen HF, Wigestrand MB, Bogen IL et al (2013) Mechanisms of penitrem-induced cerebellar granule neuron death in vitro: possible involvement of GABAA receptors and oxidative processes. Neurotoxicology 35:129–136. https://doi.org/10.1016/j.neuro.2013.01.004

Berntsen HF, Bogen IL, Wigestrand MB et al (2017) The fungal neurotoxin penitrem A induces the production of reactive oxygen species in human neutrophils at submicromolar concentrations. Toxicology 392:64–70. https://doi.org/10.1016/j.tox.2017.10.008

Bluhm BH, Kim H, Butchko RA et al (2008) Involvement of ZFR1 of Fusarium verticillioides in kernel colonization and the regulation of FST1, a putative sugar transporter gene required for fumonisin biosynthesis on maize kernels. Mol Plant Pathol 9:203–211. https://doi.org/10.1111/j.1364-3703.2007.00458.x

Bouchara JP, Hsieh HY, Croquefer S et al (2009) Development of an oligonucleotide array for direct detection of fungi in sputum samples from patients with cystic fibrosis. J Clin Microbiol 47:142–152. https://doi.org/10.1128/JCeM.01668-08

Boudra H, Barnouin J, Dragacci S et al (2007) Aflatoxin M1 and ochratoxin A in raw bulk milk from French dairy herds. J Dairy Sci 90:3197–3201. https://doi.org/10.3168/jds.2006-565

Bouhet S, Oswald IP (2007) The intestine as a possible target for fumonisin toxicity. Mol Nutr Food Res 51:925–931. https://doi.org/10.1002/mnfr.200600266

Bouhet S, Hourcade E, Loiseau N et al (2004) The mycotoxin fumonisin B1 alters the proliferation and the barrier function of porcine intestinal epithelial cells. Toxicol Sci 77:165–171. https://doi.org/10.1093/toxsci/kfh006

Brewer JH, Thrasher JD, Straus DC et al (2013) Detection of mycotoxins in patients with chronic fatigue syndrome. Toxins 5:605–617. https://doi.org/10.3390/toxins5040605

Brinda R, Vijayanandraj S, Uma D et al (2013) Role of *Adhatoda vasica* (L.) *Nees* leaf extract in the prevention of aflatoxin-induced toxicity in Wistar rats. J Sci Food Agric 30:2743–2748. https://doi.org/10.1002/jsfa.6093

Brown DW, Busman M, Proctor RH (2014) *Fusarium verticillioides* SGE1 is required for full virulence and regulates expression of protein effector and secondary metabolite biosynthetic genes. Mol Plant-Microbe Interact 27:809–823. https://doi.org/10.1094/MPMI-09-13-0281-R

Bucci TJ, Howard PC, Tolleson WH et al (1998) Renal effects of fumonisin mycotoxins in animals. Toxicol Pathol 26:160–164. https://doi.org/10.1177/019262339802600119

Calado T, Fernández-Cruz ML, Cabo Verde S et al (2018) Gamma irradiation effects on ochratoxin A: degradation, cytotoxicity and application in food. Food Chem 240:463–471. https://doi.org/10.1016/j.foodchem.2017.07.136

Carlton WW, Stack ME, Eppley RM (1976) Hepatic alterations produced in mice by xanthomegnin and viomellein, metabolites of Penicillium viridicatum. Toxicol Appl Pharmacol 38:455–459. https://doi.org/10.1016/0041-008X(76)90151-4

Celik I, Oğuz H, Demet O et al (2000) Embryotoxicity assay of aflatoxin produced by *Aspergillus parasiticus* NRRL 2999. Br Poult Sci 41:401–409. https://doi.org/10.1080/713654961

Chang PK, Ehrlich KC, Yu J et al (1995) Increased expression of *Aspergillus parasiticus* aflR, encoding a sequence-specific DNA-binding protein, relieves nitrate inhibition of aflatoxin biosynthesis. Appl Environ Microbiol 61:2372–2377

Chang PK, Ehrlich KC, Linz JE et al (1996) Characterization of the *Aspergillus parasiticus* niaD and niiA gene cluster. Curr Genet 30:68–75

Chaudhari M, Jayaraj R, Bhaskar AS et al (2009) Oxidative stress induction by T-2 toxin causes DNA damage and triggers apoptosis via caspase pathway in human cervical cancer cells. Toxicology 262:153–161. https://doi.org/10.1016/j.tox.2009.06.002

Chavarría G, Granados-Chinchilla F, Alfaro-Cascante M et al (2015) Detection of aflatoxin M1 in milk, cheese and sour cream samples from Costa Rica using enzyme-assisted extraction and HPLC. Food Addit Contam Part B Surveill 8:128–135. https://doi.org/10.1080/19393210.2015.1015176

Chen F, Luan C, Wang L et al (2017) Simultaneous determination of six mycotoxins in peanut by high-performance liquid chromatography with a fluorescence detector. J Sci Food Agric 97:1805–1810. https://doi.org/10.1002/jsfa.7978

Ciacci-Zanella JR, Merrill AH Jr, Wang E et al (1998) Characterization of cell-cycle arrest by fumonisin B1 in CV-1 cells. Food Chem Toxicol 36:791–804. https://doi.org/10.1016/S0278-6915(98)00034-9

Constable PD, Smith GW, Rottinghaus GE et al (2000) Ingestion of fumonisin B1-containing culture material decreases cardiac contractility and mechanical efficiency in swine. Toxicol Appl Pharmacol 162:151–160. https://doi.org/10.1006/taap.1999.8831

Cotty PJ, Jaime-Garcia R (2007) Influences of climate on aflatoxin producing fungi and aflatoxin contamination. Int J Food Microbiol 119:109–115. https://doi.org/10.1016/j.ijfoodmicro.2007.07.060

Cramer B, Osteresch B, Muñoz KA et al (2015) Biomonitoring using dried blood spots: detection of ochratoxin A and its degradation product 2'R-ochratoxin A in blood from coffee drinkers. Mol Nutr Food Res 59:1837–1843. https://doi.org/10.1002/mnfr.201500220

Crespo-Sempere A, Marín S, Sanchis V, Ramos AJ (2013) VeA and LaeA transcriptional factors regulate ochratoxin aA biosynthesis in *Aspergillus carbonarius*. Int J Food Microbiol 166:479–486. https://doi.org/10.1016/j.ijfoodmicro.2013.07.027

Datta SC, Ghosh JJ (1981) Effect of citreoviridin, a mycotoxin from *Penicillium citreoviride*, on kinetic constants of acetylcholinesterase and ATPase in synaptosomes and microsomes from rat brain. Toxicon 19:555–562. https://doi.org/10.1016/0041-0101(81)90014-3

Daxecker H, Raab M, Cichna M et al (2001) Determination of the effects of mycophenolic acid on the nucleotide pool of human peripheral blood mononuclear cells in vitro by high-

performance liquid chromatography. Clin Chim Acta 310:81–87. https://doi.org/10.1016/
S0009-8981(01)00526-5

De Jesus CL, Bartley A, Welch AZ et al (2018) High incidence and levels of ochratoxin A in wines
sourced from the United States. Toxins 10:1. https://doi.org/10.3390/toxins10010001

Dehghan P, Pakshir K, Rafiei H et al (2014) Prevalence of ochratoxin a in human milk in the
Khorrambid Town, Fars Province, South of Iran. Jundishapur J Microbiol 7:e11220. https://
doi.org/10.5812/jjm.11220

Del Palacio A, Mionetto A, Bettucci L et al (2016) Evolution of fungal population and mycotoxins
in sorghum silage. Food Addit Contam Part A Chem Anal Control Expo Risk Assess 33:1864–
1872. https://doi.org/10.1080/19440049.2016.1244732

Devnarain N, Tiloke C, Nagiah S et al (2017) Fusaric acid induces oxidative stress and apoptosis in human cancerous oesophageal SNO cells. Toxicon 126:4–11. https://doi.org/10.1016/j.
toxicon.2016.12.006

Devreese M, De Baere S, De Backer P et al (2012) Quantitative determination of several toxicological important mycotoxins in pig plasma using multi-mycotoxin and analyte-specific high
performance liquid chromatography-tandem mass spectrometric methods. J Chromatogr A
1257:74–80. https://doi.org/10.1016/j.chroma.2012.08.008

Doi K, Uetsuka K (2014) Mechanisms of mycotoxin-induced dermal toxicity and tumorigenesis
through oxidative stress-related pathways. J Toxicol Pathol 27:1–10. https://doi.org/10.1293/
tox.2013-0062

Domijan AM, Peraica M, Cvjetković B et al (2005) Mould contamination and co-occurrence of
mycotoxins in maize grain in Croatia. Acta Pharma 55:349–356

Dong G, Pan Y, Wang Y et al (2018) Preparation of a broad-spectrum anti-zearalenone and its primary analogues antibody and its application in an indirect competitive enzyme-linked immunosorbent assay. Food Chem 247:8–15. https://doi.org/10.1016/j.foodchem.2017.12.016

El Khoury R, Mathieu F, Atoui A et al (2017) Ability of soil isolated actinobacterial strains to
prevent, bind and biodegrade ochratoxin A. Toxins 9:pii: E222. https://doi.org/10.3390/
toxins9070222

El Khoury R, Choque E, El Khoury A et al (2018) OTA prevention and detoxification by actinobacterial strains and activated carbon fibers: preliminary results. Toxins 24:pii: E137. https://
doi.org/10.3390/toxins10040137

El Marnissi B, Belkhou R, Morgavi DP et al (2012) Occurrence of aflatoxin M1 in raw milk
collected from traditional dairies in Morocco. Food Chem Toxicol 50:2819–2821. https://doi.
org/10.1016/j.fct.2012.05.031

Emri T, Zalka A, Pócsi I (2017) Detection of transcriptionally active mycotoxin gene clusters: DNA
microarray. Methods Mol Biol 1542:345–365. https://doi.org/10.1007/978-1-4939-6707-0_23

Escrivá L, Font G, Manyes L et al (2017) Studies on the presence of mycotoxins in biological
samples: an overview. Toxins 9:251. https://doi.org/10.3390/toxins9080251

Espeso EA, Arst HN Jr (2000) On the mechanism by which alkaline pH prevents expression of an
acid-expressed gene. Mol Cell Biol 20:3355–3363

Esteban A, Abarca ML, Bragulat MR et al (2006) Effect of pH on ochratoxin A production by Aspergillus niger aggregate species. Food Addit Contam 23:616–622. https://doi.
org/10.1080/02652030600599124

Faisal Z, Derdák D, Lemli B et al (2018a) Interaction of 2'R-ochratoxin A with serum albumins:
binding site, effects of site markers, thermodynamics, species differences of albumin-binding,
and influence of albumin on its toxicity in MDCK cells. Toxins 10:pii: E353. https://doi.
org/10.3390/toxins10090353

Faisal Z, Lemli B, Szerencsés D et al (2018b) Interactions of zearalenone and its reduced metabolites α-zearalenol and β-zearalenol with serum albumins: species differences, binding sites,
and thermodynamics. Mycotoxin Res 34:269–278. https://doi.org/10.1007/s12550-018-0321-6

Fanelli F, Schmidt-Heydt M, Haidukowski M et al (2012) Influence of light on growth, conidiation and the mutual regulation of fumonisin B-2 and ochratoxin A biosynthesis by Aspergillus
niger. World Mycotoxin J 5:169–176. https://doi.org/10.3920/WMJ2011.1364

Fanelli F, Cozzi G, Raiola A et al (2017) Raisins and currants as conventional nutraceuticals in Italian market: natural occurrence of ochratoxin A. J Food Sci 82:2306–2312. https://doi.org/10.1111/1750-3841.13854

Fang H, Wu Y, Guo J et al (2012) T-2 toxin induces apoptosis in differentiated murine embryonic stem cells through reactive oxygen species-mediated mitochondrial pathway. Apoptosis 17:895–907. https://doi.org/10.1007/s10495-012-0724-3

Fang H, Cong L, Zhi Y et al (2016) T-2 toxin inhibits murine ES cells cardiac differentiation and mitochondrial biogenesis by ROS and p-38 MAPK-mediated pathway. Toxicol Lett 258:259–266. https://doi.org/10.1016/j.toxlet.2016.06.2103

Ferrigo D, Raiola A, Bogialli S et al (2015) In vitro production of fumonisins by Fusarium verticillioides under oxidative stress induced by H_2O_2. J Agric Food Chem 63:4879–4885. https://doi.org/10.1021/acs.jafc.5b00113

Flajs D, Peraica M (2009) Toxicological properties of citrinin. Arh Hig Rada Toksikol 60:457–464. https://doi.org/10.2478/10004-1254-60-2009-1992

Flores-Flores ME, Lizarraga E, López de Cerain A et al (2015) Presence of mycotoxins in animal milk: a review. Food Control 53:163–176. https://doi.org/10.1016/j.foodcont.2015.01.020

Fontelo PA, Beheler J, Bunner DL et al (1983) Detection of T-2 toxin by an improved radioimmunoassay. Appl Environ Microbiol 45:640–643

Fraeyman S, Meyer E, Devreese M et al (2018) Comparative in vitro cytotoxicity of the emerging Fusarium mycotoxins beauvericin and enniatins to porcine intestinal epithelial cells. Food Chem Toxicol 121:566–572. https://doi.org/10.1016/j.fct.2018.09.053

Fredlund E, Thim AM, Gidlund A et al (2009) Moulds and mycotoxins in rice from the Swedish retail market. Food Addit Contam Part A Chem Anal Control Expo Risk Assess 26:527–533. https://doi.org/10.1080/02652030802562912

Gagliano N, Donne ID, Torri C et al (2006) Early cytotoxic effects of ochratoxin A in rat liver: a morphological, biochemical and molecular study. Toxicology 15:214–224. https://doi.org/10.1016/j.tox.2006.06.004

Gallo A, Ferrara M, Perrone G (2017) Recent advances on the molecular aspects of ochratoxin A biosynthesis. Curr Opin Food Sci 17:49–56. https://doi.org/10.1016/j.cofs.2017.09.011

Gammelsrud A, Solhaug A, Dendelé B et al (2012) Enniatin B-induced cell death and inflammatory responses in RAW 267.4 murine macrophages. Toxicol Appl Pharmacol 261:74–87. https://doi.org/10.1016/j.taap.2012.03.014

Gao Y, Yang M, Peng C et al (2012) Preparation of highly specific anti-zearalenone antibodies by using the cationic protein conjugate and development of an indirect competitive enzyme-linked immunosorbent assay. Analyst 137:229–236. https://doi.org/10.1039/c1an15487g

Gao X, Sun L, Zhang N et al (2017) Gestational zearalenone exposure causes reproductive and developmental toxicity in pregnant rats and female offspring. Toxins 9:21. https://doi.org/10.3390/toxins9010021

García-Moraleja A, Font G, Mañes J et al (2015) Analysis of mycotoxins in coffee and risk assessment in Spanish adolescents and adults. Food Chem Toxicol 86:225–233. https://doi.org/10.1016/j.fct.2015.10.014

Garrido NS, Iha MH, Santos Ortolani MR et al (2003) Occurrence of aflatoxins M(1) and M(2) in milk commercialized in Ribeirão Preto-SP, Brazil. Food Addit Contam 20:70–73. https://doi.org/10.1080/0265203021000035371

Gayathri L, Dhivya R, Dhanasekaran D et al (2015) Hepatotoxic effect of ochratoxin A and citrinin, alone and in combination, and protective effect of vitamin E: in vitro study in HepG2 cell. Food Chem Toxicol 83:151–163. https://doi.org/10.1016/j.fct.2015.06.009

Gerding J, Cramer B, Humpf HU (2014) Determination of mycotoxin exposure in Germany using an LC-MS/MS multibiomarker approach. Mol Nutr Food Res 58:2358–2368. https://doi.org/10.1002/mnfr.201400406

Gerding J, Ali N, Schwartzbord J et al (2015) A comparative study of the human urinary mycotoxin excretion patterns in Bangladesh, Germany, and Haiti using a rapid and sensitive LC-MS/MS approach. Mycotoxin Res 31:127–136. https://doi.org/10.1007/s12550-015-0223-9

Ghali R, Khlifa KH, Ghorbel H et al (2010) Aflatoxin determination in commonly consumed foods in Tunisia. J Sci Food Agric 90:2347–2351. https://doi.org/10.1002/jsfa.4069

Ghazi T, Nagiah S, Tiloke C et al (2017) Fusaric acid induces DNA damage and post-translational modifications of p53 in human hepatocellular carcinoma (HepG2) cells. J Cell Biochem 118:3866–3874. https://doi.org/10.1002/jcb.26037

Giancarlo B, Elisabetta B, Edmondo C et al (2011) Determination of ochratoxin A in eggs and target tissues of experimentally drugged hens using HPLC–FLD. Food Chem 126:1278–1282. https://doi.org/10.1016/j.foodchem.2010.11.070

Gimeno A (1979) Thin layer chromatographic determination of aflatoxins, ochratoxins, sterigmatocystin, zearalenone, citrinin, T-2 toxin, diacetoxyscirpenol, penicillic acid, patulin, and penitrem A. J Assoc Off Anal Chem 62:579–585

Gökmen V, Acar J (1998) Incidence of patulin in apple juice concentrates produced in Turkey. J Chromatogr A 815:99–102. https://doi.org/10.1016/S0021-9673(97)01280-6

Goliński P, Grabarkiewicz-Szczesna J (1984) Chemical confirmatory tests for ochratoxin A, citrinin, penicillic acid, sterigmatocystin, and zearalenone performed directly on thin layer chromatographic plates. J Assoc Off Anal Chem 67:1108–1110

Gomes ES, Schuch V, de Macedo Lemos EG (2013) Biotechnology of polyketides: new breath of life for the novel antibiotic genetic pathways discovery through metagenomics. Braz J Microbiol 44:1007–1034

Greco MV, Pardo AG, Ludemann V et al (2012) Mycoflora and natural incidence of selected mycotoxins in rabbit and chinchilla feeds. Sci World J 2012:956056. https://doi.org/10.1100/2012/956056

Greco M, Pardo A, Pose G (2015) Mycotoxigenic fungi and natural co-occurrence of mycotoxins in rainbow trout (Oncorhynchus mykiss) feeds. Toxins 7:4595–4609. https://doi.org/10.3390/toxins7114595

Haas D, Pfeifer B, Reiterich C et al (2013) Identification and quantification of fungi and mycotoxins from Pu-erh tea. Int J Food Microbiol 166:316–322. https://doi.org/10.1016/j.ijfoodmicro.2013.07.024

Hashem A, Fathi Abd-Allah E, Sultan Al-Obeed R et al (2015) Effect of carbon, nitrogen sources and water activity on growth and ochratoxin production of *Aspergillus carbonarius* (Bainier) Thom. Jundishapur J Microbiol 23:e17569. https://doi.org/10.5812/jjm.17569

Hayes AW, Williams WL (1978) Acute toxicity of aflatoxin B1 and rubratoxin B in dogs. J Environ Pathol Toxicol 1:59–70

Heischmann S, Dzieciatkowska M, Hansen K et al (2017) The immunosuppressant mycophenolic acid alters nucleotide and lipid metabolism in an intestinal cell model. Sci Rep 7:45088. https://doi.org/10.1038/srep45088

Heller T, Asif AR, Petrova DT et al (2009) Differential proteomic analysis of lymphocytes treated with mycophenolic acid reveals caspase 3-induced cleavage of rho GDP dissociation inhibitor 2. Ther Drug Monit 31:211–217. https://doi.org/10.1097/FTD.0b013e318196fb73

Henry MH, Wyatt R (2001) The toxicity of fumonisin B1, B2, and B3, individually and in combination, in chicken embryos. Poult Sci 80:401–407. https://doi.org/10.1093/ps/80.4.401

Hernández-Meléndez D, Salas-Téllez E, Zavala-Franco A et al (2018) Inhibition effect of flower-shaped zinc oxide nanostructures on growth and aflatoxin production of a highly toxigenic strain of *Aspergillus flavus* link. Materials 11:1–13. https://doi.org/10.3390/ma11081265

Hewetson JF, Pace JG, Beheler JE (1987) Detection and quantitation of T-2 mycotoxin in rat organs by radioimmunoassay. J Assoc Off Anal Chem 70:654–657

Heyndrickx E, Sioen I, Huybrechts B et al (2015) Human biomonitoring of multiple mycotoxins in the Belgian population: results of the BIOMYCO study. Environ Int 84:82–89. https://doi.org/10.1016/j.envint.2015.06.011

Hinton DM, Morrissey RE, Norred WP et al (1985) Effects of cyclopiazonic acid on the ultrastructure of rat liver. Toxicol Lett 25:211–218. https://doi.org/10.1016/0378-4274(85)90084-0

Hodnik V, Anderluh G (2009) Toxin detection by surface plasmon resonance. Sensors 9:1339–1354. https://doi.org/10.3390/s9031339

Hooper DG, Bolton VE, Guilford FT et al (2009) Mycotoxin detection in human samples from patients exposed to environmental molds. Int J Mol Sci 10:1465–1475. https://doi.org/10.3390/ijms10041465

Hossain MZ, Maragos CM (2018) Gold nanoparticle-enhanced multiplexed imaging surface plasmon resonance (iSPR) detection of Fusarium mycotoxins in wheat. Biosens Bioelectron 101:245–252. https://doi.org/10.1016/j.bios.2017.10.033

Hossain MZ, McCormick SP, Maragos CM (2018) An imaging surface plasmon resonance biosensor assay for the detection of T-2 toxin and masked T-2 toxin-3-glucoside in wheat. Toxins 10:119. https://doi.org/10.3390/toxins10030119

Hou H, Qu X, Li Y et al (2015) Binding of citreoviridin to human serum albumin: multispectroscopic and molecular docking. Biomed Res Int 2015:162391. https://doi.org/10.1155/2015/162391

Huang A, Li JW, Shen ZQ et al (2006) High-throughput identification of clinical pathogenic fungi by hybridization to an oligonucleotide microarray. J Clin Microbiol 44:3299–3305. https://doi.org/10.1128/JCM.00417-06

Huffman J, Gerber R, Du L (2010) Recent advancements in the biosynthetic mechanisms for polyketide-derived mycotoxins. Biopolymers 93:764–776. https://doi.org/10.1002/bip.21483

Hymery N, Masson F, Barbier G et al (2014) Cytotoxicity and immunotoxicity of cyclopiazonic acid on human cells. Toxicol In Vitro 28:940–947. https://doi.org/10.1016/j.tiv.2014.04.003

Iacumin L, Chiesa L, Boscolo D et al (2009) Moulds and ochratoxin A on surfaces of artisanal and industrial dry sausages. Food Microbiol 26:65–70. https://doi.org/10.1016/j.fm.2008.07.006

Ikegwuonu FI (1983) The neurotoxicity of aflatoxin B1 in the rat. Toxicology 28:247–259. https://doi.org/10.1016/0300-483X(83)90121-X

Ivanova L, Egge-Jacobsen WM, Solhaug A et al (2012) Lysosomes as a possible target of enniatin B-induced toxicity in Caco-2 cells. Chem Res Toxicol 25:1662–1674. https://doi.org/10.1021/tx300114x

Jaskiewicz K, Close PM, Thiel PG et al (1988) Preliminary studies on toxic effects of cyclopiazonic acid alone and in combination with aflatoxin B1 in non-human primates. Toxicology 52:297–307. https://doi.org/10.1016/0300-483X(88)90134-5

Jayashree T, Subramanyam C (2000) Oxidative stress as a prerequisite for aflatoxin production by Aspergillus parasiticus. Free Radic Biol Med 29:981–985. https://doi.org/10.1016/S0891-5849(00)00398-1

Jia Z, Liu M, Qu Z et al (2014) Toxic effects of zearalenone on oxidative stress, inflammatory cytokines, biochemical and pathological changes induced by this toxin in the kidney of pregnant rats. Environ Toxicol Pharmacol 37:580–591. https://doi.org/10.1016/j.etap.2014.01.010

Jia Z, Yin S, Liu M et al (2015) Modified halloysite nanotubes and the alleviation of kidney damage induced by dietary zearalenone in swine. Food Addit Contam Part A Chem Anal Control Expo Risk Assess 32:1312–1321. https://doi.org/10.1080/19440049.2015.1048748

Jiang J, Yun Y, Liu Y et al (2012a) FgVELB is associated with vegetative differentiation, secondary metabolism and virulence in Fusarium graminearum. Fungal Genet Biol 49:653–662. https://doi.org/10.1016/j.fgb.2012.06.005

Jiang SZ, Yang ZB, Yang WR et al (2012b) Effect on hepatonephric organs, serum metabolites and oxidative stress in post-weaning piglets fed purified zearalenone-contaminated diets with or without Calibrin-Z. J Anim Physiol Anim Nutr 96:1147–1156. https://doi.org/10.1111/j.1439-0396.2011.01233.x

Jiménez M, Máñez M, Hernández E (1996) Influence of water activity and temperature on the production of zearalenone in corn by three Fusarium species. Int J Food Microbiol 29:417–421. https://doi.org/10.1016/0168-1605(95)00073-9

Jonsson M, Jestoi M, Anthoni M et al (2016) Fusarium mycotoxin enniatin B: cytotoxic effects and changes in gene expression profile. Toxicol In Vitro 34:309–320. https://doi.org/10.1016/j.tiv.2016.04.017

Joshi S, Segarra-Fas A, Peters J et al (2016) Multiplex surface plasmon resonance biosensing and its transferability towards imaging nanoplasmonics for detection of mycotoxins in barley. Analyst 141:1307–1318. https://doi.org/10.1039/C5AN02512E

Juan C, Pena A, Lino C et al (2008) Levels of ochratoxin A in wheat and maize bread from the central zone of Portugal. Int J Food Microbiol 127:284–289. https://doi.org/10.1016/j.ijfoodmicro.2008.07.018

Juan-García A, Manyes L, Ruiz MJ et al (2013) Involvement of enniatins-induced cytotoxicity in human HepG2 cells. Toxicol Lett 218:166–173. https://doi.org/10.1016/j.toxlet.2013.01.014

Juan-García A, Ruiz MJ, Font G et al (2015) Enniatin A1, enniatin B1 and beauvericin on HepG2: evaluation of toxic effects. Food Chem Toxicol 84:188–196. https://doi.org/10.1016/j.fct.2015.08.030

Kachapulula PW, Akello J, Bandyopadhyay R et al (2018) Aflatoxin contamination of dried insects and fish in Zambia. J Food Prot 81:1508–1518. https://doi.org/10.4315/0362-028X.JFP-17-527

Kapetanakou AE, Panagou EZ, Gialitaki M et al (2009) Evaluating the combined effect of water activity, pH and temperature on ochratoxin A production by *Aspergillus ochraceus* and *Aspergillus carbonarius* on culture medium and Corinth raisins. Food Control 20:725–732. https://doi.org/10.1016/j.foodcont.2008.09.008

Katerere DR, Stockenström S, Thembo KM et al (2007) Investigation of patulin contamination in apple juice sold in retail outlets in Italy and South Africa. Food Addit Contam 24:630–634. https://doi.org/10.1080/02652030601137668

Kaufman SS, Tuma DJ, Park JH et al (1988) Effects of cytochalasin B on the synthesis and secretion of plasma proteins by developing rat liver. J Pediatr Gastroenterol Nutr 7:107–114

Kaymak T, Koca E, Atak M et al (2018) Determination of aflatoxins and ochratoxin A in traditional turkish cereal-based fermented food by multi-affinity column cleanup and LC fluorescence detection: single-laboratory validation. J AOAC Int 102(1):156–163. https://doi.org/10.5740/jaoacint.17-0490

Keller NP, Hohn TM (1997) Metabolic pathway gene clusters in filamentous fungi. Fungal Genet Biol 21:17–29. https://doi.org/10.1006/fgbi.1997.0970

Khanafari A, Soudi H, Miraboulfathi M et al (2007) An *in vitro* investigation of aflatoxin B1 biological control by *Lactobacillus plantarum*. Pak J Biol Sci 10:2553–2556. https://doi.org/10.3923/pjbs.2007.2553.255

Kim HK, Lee S, Jo SM et al (2013) Functional roles of FgLacA in controlling secondary metabolism, sexual development, and virulence in *Fusarium graminearum*. PLoS One 8:e68441. https://doi.org/10.1371/journal.pone.0068441

Klarić MS, Zelježić D, Rumora L et al (2012) A potential role of calcium in apoptosis and aberrant chromatin forms in porcine kidney PK15 cells induced by individual and combined ochratoxin a and citrinin. Arch Toxicol 86:97–107. https://doi.org/10.1007/s00204-011-0735-9

Kolber MA, Broschat KO, Landa-Gonzalez B (1990) Cytochalasin B induces cellular DNA fragmentation. FASEB J 4:3021–3027. https://doi.org/10.1096/fasebj.4.12.2394319

Kolf-Clauw M, Sassahara M, Lucioli J et al (2013) The emerging mycotoxin, enniatin B1, down-modulates the gastrointestinal toxicity of T-2 toxin in vitro on intestinal epithelial cells and ex vivo on intestinal explants. Arch Toxicol 87:2233–2241. https://doi.org/10.1007/s00204-013-1067-8

Kollia E, Kanapitsas A, Markaki P (2014) Occurrence of aflatoxin B1 and ochratoxin A in dried vine fruits from Greek market. Food Addit Contam Part B Surveill 7:11–16. https://doi.org/10.1080/19393210.2013.825647

Kong WJ, Liu SY, Qiu F et al (2013) Simultaneous multi-mycotoxin determination in nutmeg by ultrasound-assisted solid-liquid extraction and immunoaffinity column clean-up coupled with liquid chromatography and on-line post-column photochemical derivatization-fluorescence detection. Analyst 138:2729–2739. https://doi.org/10.1039/C3AN00059A

Koraïchi F, Inoubli L, Lakhdari N et al (2013) Neonatal exposure to zearalenone induces long term modulation of ABC transporter expression in testis. Toxicology 310:29–38. https://doi.org/10.1016/j.tox.2013.05.002

Lee S, Chu FS (1981) Radioimmunoassay of T-2 toxin in biological fluids. J Assoc Off Anal Chem 64:684–688

Lee SH, Lee J, Lee S et al (2009) GzSNF1 is required for normal sexual and asexual development in the ascomycete *Gibberella zeae*. Eukaryot Cell 8:116–127. https://doi.org/10.1128/EC.00176-08

Li Y, Zhang B, He X et al (2014) Analysis of individual and combined effects of ochratoxin A and zearalenone on HepG2 and KK-1 cells with mathematical models. Toxins 6:1177–1192. https://doi.org/10.3390/toxins6041177

Li H, Xing L, Zhang M et al (2018) The toxic effects of aflatoxin B1 and aflatoxin M1 on kidney through regulating L-proline and downstream apoptosis. Biomed Res Int 2018:9074861. https://doi.org/10.1155/2018/9074861

Limonciel A, Jennings P (2014) A review of the evidence that ochratoxin A is an Nrf2 inhibitor: implications for nephrotoxicity and renal carcinogenicity. Toxins 6:371–379. https://doi.org/10.3390/toxins6010371

Liu MT, Ram BP, Hart LP et al (1985) Indirect enzyme-linked immunosorbent assay for the mycotoxin zearalenone. Appl Environ Microbiol 50:332–336

Liu J, Wang L, Guo X et al (2014a) The role of mitochondria in T-2 toxin-induced human chondrocytes apoptosis. PLoS One 9:e108394. https://doi.org/10.1371/journal.pone.0108394

Liu M, Gao R, Meng Q et al (2014b) Toxic effects of maternal zearalenone exposure on intestinal oxidative stress, barrier function, immunological and morphological changes in rats. PLoS One 9:e106412. https://doi.org/10.1371/journal.pone.0106412

Liu M, Zhu D, Guo T et al (2018) Toxicity of zearalenone on the intestines of pregnant sows and their offspring and alleviation with modified halloysite nanotubes. J Sci Food Agric 98:698–706. https://doi.org/10.1002/jsfa.8517

Loiseau N, Debrauwer L, Sambou T et al (2007) Fumonisin B1 exposure and its selective effect on porcine jejunal segment: sphingolipids, glycolipids and trans-epithelial passage disturbance. Biochem Pharmacol 74:144–152. https://doi.org/10.1016/j.bcp.2007.03.031

Lu L, Gunasekaran S (2019) Dual-channel ITO-microfluidic electrochemical immunosensor for simultaneous detection of two mycotoxins. Talanta 194:709–716. https://doi.org/10.1016/j.talanta.2018.10.091

Ma L, Wang J, Zhang Y (2017) Probing the characterization of the interaction of aflatoxins B1 and G1 with calf thymus DNA *in vitro*. Toxins 9:pii: E209. https://doi.org/10.3390/toxins9070209

Ma S, Zhao Y, Sun J et al (2018) miR449a/SIRT1/PGC-1α is necessary for mitochondrial biogenesis induced by T-2 toxin. Front Pharmacol 8:954. https://doi.org/10.3389/fphar.2017.00954

Magnussen A, Parsi MA (2013) Aflatoxins, hepatocellular carcinoma and public health. World J Gastroenterol 19:1508–1512. https://doi.org/10.3748/wjg.v19.i10.1508

Majer-Baranyi K, Zalán Z, Mörtl M et al (2016) Optical waveguide lightmode spectroscopy technique-based immunosensor development for aflatoxin B1 determination in spice paprika samples. Food Chem 211:972–977. https://doi.org/10.1016/j.foodchem.2016.05.089

Makowska K, Obremski K, Gonkowski S (2018) The impact of T-2 toxin on vasoactive intestinal polypeptide-like immunoreactive (VIP-LI) nerve structures in the wall of the porcine stomach and duodenum. Toxins 10:138. https://doi.org/10.3390/toxins10040138

Makun HA, Dutton MF, Njobeh PB et al (2011) Natural multi-occurrence of mycotoxins in rice from Niger State, Nigeria. Mycotoxin Res 27:97–104. https://doi.org/10.1007/s12550-010-0080-5

Malir F, Ostry V, Pfohl-Leszkowicz A et al (2014) Transfer of ochratoxin A into tea and coffee beverages. Toxins 6:3438–3453. https://doi.org/10.3390/toxins6123438

Mallebrera B, Juan-Garcia A, Font G et al (2016) Mechanisms of beauvericin toxicity and antioxidant cellular defense. Toxicol Lett 246:28–34. https://doi.org/10.1016/j.toxlet.2016.01.013

Manyes L, Escrivá L, Ruiz MJ et al (2018) Beauvericin and enniatin B effects on a human lymphoblastoid Jurkat T-cell model. Food Chem Toxicol 115:127–135. https://doi.org/10.1016/j.fct.2018.03.008

Maresca M, Mahfoud R, Pfohl-Leszkowicz A et al (2001) The mycotoxin ochratoxin A alters intestinal barrier and absorption functions but has no effect on chloride secretion. Toxicol Appl Pharmacol 176:54–63. https://doi.org/10.1006/taap.2001.9254

Marin DE, Braicu C, Gras MA et al (2017) Low level of ochratoxin A affects genome-wide expression in kidney of pig. Toxicon 136:67–77. https://doi.org/10.1016/j.toxicon.2017.07.004

Massey TE, Smith GB, Tam AS (2000) Mechanisms of aflatoxin B1 lung tumorigenesis. Exp Lung Res 26:673–683. https://doi.org/10.1080/01902140150216756

Mata AT, Ferreira JP, Oliveira BR et al (2015) Bottled water: analysis of mycotoxins by LC-MS/MS. Food Chem 176:455–464. https://doi.org/10.1016/j.foodchem.2014.12.088

Mateo R, Medina A, Mateo EM et al (2007) An overview of ochratoxin A in beer and wine. Int J Food Microbiol 20:79–83. https://doi.org/10.1016/j.ijfoodmicro.2007.07.029

Matumba LU, Van Poucke CU, Biswick T et al (2014) A limited survey of mycotoxins in traditional maize based opaque beers in Malawi. Food Control 36:253–256. https://doi.org/10.1016/j.foodcont.2013.08.032

Mayura K, Smith EE, Clement BA et al (1987) Developmental toxicity of diacetoxyscirpenol in the mouse. Toxicology 45:245–255. https://doi.org/10.1016/0300-483X(87)90016-3

Medina A, González G, Sáez JM et al (2004) Bee pollen, a substrate that stimulates ochratoxin A production by *Aspergillus ochraceus* Wilh. Syst Appl Microbiol 27:261–267. https://doi.org/10.1078/072320204322881880

Medina A, Mateo EM, Valle-Algarra FM et al (2008) Influence of nitrogen and carbon sources on the production of ochratoxin A by ochratoxigenic strains of *Aspergillus spp.* isolated from grapes. Int J Food Microbiol 122:93–99. https://doi.org/10.1016/j.ijfoodmicro.2007.11.055

Medina A, Schmidt-Heydt M, Cárdenas-Chávez DL et al (2013) Integrating toxin gene expression, growth and fumonisin B1 and B2 production by a strain of *Fusarium verticillioides* under different environmental factors. J R Soc Interface 10:20130320. https://doi.org/10.1098/rsif.2013.0320

Mehrzad J, Malvandi AM, Alipour M et al (2017) Environmentally relevant level of aflatoxin B1 elicits toxic pro-inflammatory response in murine CNS-derived cells. Toxicol Lett 279:96–106. https://doi.org/10.1016/j.toxlet.2017.07.902

Merhej J, Richard-Forget F, Barreau C (2011) The pH regulatory factor Pac1 regulates Tri gene expression and trichothecene production in *Fusarium graminearum*. Fungal Genet Biol 48:275–284. https://doi.org/10.1016/j.fgb.2010.11.008

Min K, Shin Y, Son H et al (2012) Functional analyses of the nitrogen regulatory gene areA in *Gibberella zeae*. FEMS Microbiol Lett 334:66–73. https://doi.org/10.1111/j.1574-6968.2012.02620.x

Minervini F, Dell'Aquila ME (2008) Zearalenone and reproductive function in farm animals. Int J Mol Sci 9:2570–2584. https://doi.org/10.3390/ijms9122570

Minervini F, Fornelli F, Flynn KM (2004) Toxicity and apoptosis induced by the mycotoxins nivalenol, deoxynivalenol and fumonisin B1 in a human erythroleukemia cell line. Toxicol In Vitro 18:21–28. https://doi.org/10.1016/S0887-2333(03)00130-9

Mogensen JM, Nielsen KF, Samson RA et al (2009) Effect of temperature and water activity on the production of fumonisins by *Aspergillus niger* and different *Fusarium* species. BMC Microbiol 9:281. https://doi.org/10.1186/1471-2180-9-281

Montani ML, Vaamonde G, Resnik SL et al (1988) Influence of water activity and temperature on the accumulation of zearalenone in corn. Int J Food Microbiol 6:1–8. https://doi.org/10.1016/0168-1605(88)90078-5

Moye-Rowley WS (2003) Regulation of the transcriptional response to oxidative stress in fungi: similarities and differences. Eukaryot Cell 2:381–389. https://doi.org/10.1128/EC.2.3.381-389.2003

Mudge DW, Atcheson BA, Taylor PJ et al (2004) Severe toxicity associated with a markedly elevated mycophenolic acid free fraction in a renal transplant recipient. Ther Drug Monit 26:453–455

Mudili V, Siddaih CN, Nagesh M et al (2014) Mould incidence and mycotoxin contamination in freshly harvested maize kernels originated from India. J Sci Food Agric 94:2674–2683. https://doi.org/10.1002/jsfa.6608

Mupunga I, Lebelo SL, Mngqawa P et al (2014) Natural occurrence of aflatoxins in peanuts and peanut butter from Bulawayo, Zimbabwe. J Food Prot 77:1814–1818. https://doi.
org/10.4315/0362-028X.JFP-14-129

Niehaus WG Jr, Jiang WP (1989) Nitrate induces enzymes of the mannitol cycle and suppresses
versicolorin synthesis in Aspergillus parasiticus. Mycopathologia 107:131–137. https://doi.
org/10.1007/BF00707550

Niide O, Suzuki Y, Yoshimaru T et al (2006) Fungal metabolite gliotoxin blocks mast cell activation by a calcium- and superoxide-dependent mechanism: implications for immunosuppressive
activities. Clin Immunol 118:108–116. https://doi.org/10.1016/j.clim.2005.08.012

Nishie K, Cole RJ, Dorner JW (1985) Toxicity and neuropharmacology of cyclopiazonic acid.
Food Chem Toxicol 23:831–839. https://doi.org/10.1016/0278-6915(85)90284-4

Nishie K, Cole RJ, Dorner JW (1988) Toxicity of citreoviridin. Res Commun Chem Pathol
Pharmacol 59:31–52

Niu C, Payne GA, Woloshuk CP (2015) Transcriptome changes in Fusarium verticillioides caused
by mutation in the transporter-like gene FST1. BMC Microbiol 15:90. https://doi.org/10.1186/
s12866-015-0427-3

Njoroge SM, Matumba L, Kanenga K et al (2016) A case for regular aflatoxin monitoring in peanut
butter in sub-Saharan Africa: lessons from a 3-year survey in Zambia. J Food Prot 79:795–800.
https://doi.org/10.4315/0362-028X.JFP-15-542

Nkwe DO, Taylor JE, Siame BA (2005) Fungi, aflatoxins, fumonisin Bl and zearalenone contami-
nating sorghum-based traditional malt, wort and beer in Botswana. Mycopathologia 160:177–
186. https://doi.org/10.1007/s11046-005-6867-9

O'Brian GR, Georgianna DR, Wilkinson JR et al (2007) The effect of elevated temperature on
gene transcription and aflatoxin biosynthesis. Mycologia 99:232–239

Okutan H, Aydin G, Ozcelik N (2004) Protective role of melatonin in ochratoxin a toxicity in rat
heart and lung. J Appl Toxicol 24:505–512. https://doi.org/10.1002/jat.1010

Omar SS (2016) Aflatoxin M1 levels in raw milk, pasteurised milk and infant formula. Ital J Food
Saf 5:5788. https://doi.org/10.4081/ijfs.2016.5788

Orti DL, Hill RH Jr, Liddle JA et al (1986) High performance liquid chromatography of mycotoxin
metabolites in human urine. J Anal Toxicol 100:41–45

Osteresch B, Viegas S, Cramer B et al (2017) Multi-mycotoxin analysis using dried blood
spots and dried serum spots. Anal Bioanal Chem 409:3369–3382. https://doi.org/10.1007/
s00216-017-0279-9

Osweiler GD, Ross PF, Wilson TM et al (1992) Characterization of an epizootic of pulmonary
edema in swine associated with fumonisin in corn screenings. J Vet Diagn Investig 4:53–59.
https://doi.org/10.1177/104063879200400112

Owino JH, Arotiba OA, Hendricks N et al (2008) Electrochemical immunosensor based on poly-
thionine/gold nanoparticles for the determination of aflatoxin B₁. Sensors 8:8262–8274. https://
doi.org/10.3390/s8128262

Ozsoy N, Selmanoğlu G, Koçkaya EA et al (2008) Effect of patulin on the interdigitating dendritic
cells (IDCs) of rat thymus. Cell Biochem Funct 26:192–196. https://doi.org/10.1002/cbf.1431

Palabiyik SS, Erkekoglu P, Zeybek ND et al (2013) Protective effect of lycopene against ochratoxin A induced renal oxidative stress and apoptosis in rats. Exp Toxicol Pathol 65:853–861.
https://doi.org/10.1016/j.etp.2012.12.004

Paradells S, Rocamonde B, Llinares C et al (2015) Neurotoxic effects of ochratoxin A on the
subventricular zone of adult mouse brain. J Appl Toxicol 35:737–751. https://doi.org/10.1002/
jat.3061

Passamani FR, Hernandes T, Lopes NA et al (2014) Effect of temperature, water activity, and
pH on growth and production of ochratoxin A by Aspergillus niger and Aspergillus carbonarius from Brazilian grapes. J Food Prot 77:1947–1952. https://doi.org/10.4315/0362-028X.
JFP-13-495

Patterson DS, Roberts BA (1979) Mycotoxins in animal feedstuffs: sensitive thin layer chromatographic detection of aflatoxin, ochratoxin A, sterigmatocystin, zearalenone, and T-2 toxin. J Assoc Off Anal Chem 62:1265–1267

Pena A, Seifrtová M, Lino C et al (2006) Estimation of ochratoxin A in portuguese population: new data on the occurrence in human urine by high performance liquid chromatography with fluorescence detection. Food Chem Toxicol 44:1449–1454. https://doi.org/10.1016/j.fct.2006.04.017

Peñalva MA, Tilburn J, Bignell E et al (2008) Ambient pH gene regulation in fungi: making connections. Trends Microbiol 16:291–300. https://doi.org/10.1016/j.tim.2008.03.006

Peng KY, Chen CY (2009) Prevalence of aflatoxin M1 in milk and its potential liver cancer risk in Taiwan. J Food Prot 72:1025–1029. https://doi.org/10.4315/0362-028X-72.5.1025

Peraica M, Radić B, Lucić A et al (1999) Toxic effects of mycotoxins in humans. Bull World Health Organ 77:754–766. https://www.ncbi.nlm.nih.gov/pmc/articles/PMC2557730/pdf/10534900.pdf

Perši N, Pleadin J, Kovačević D et al (2014) Ochratoxin A in raw materials and cooked meat products made from OTA-treated pigs. Meat Sci 96:203–210. https://doi.org/10.1016/j.meatsci.2013.07.005

Pirinçci ŞŞ, Ertekin Ö, Laguna DE et al (2018) Label-free QCM immunosensor for the detection of ochratoxin A. Sensors 18:1161. https://doi.org/10.3390/s18041161

Pizzolato Montanha F, Anater A, Burchard JF et al (2018) Mycotoxins in dry-cured meats: a review. Food Chem Toxicol 111:494–502. https://doi.org/10.1016/j.fct.2017.12.008

Pleadin J, Zadravec M, Brnić D et al (2017) Moulds and mycotoxins detected in the regional speciality fermented sausage 'slavonski kulen' during a 1-year production period. Food Addit Contam Part A Chem Anal Control Expo Risk Assess 34:282–290. https://doi.org/10.1080/19440049.2016.1266395

Poapolathep A, Poapolathep S, Klangkaew N et al (2008) Detection of deoxynivalenol contamination in wheat products in Thailand. J Food Prot 71:1931–1933. https://doi.org/10.4315/0362-028X-71.9.1931

Poór M, Lemli B, Bálint M et al (2015) Interaction of citrinin with human serum albumin. Toxins 7:5155–5166. https://doi.org/10.3390/toxins7124871

Poór M, Kunsági-Máté S, Bálint M et al (2017a) Interaction of mycotoxin zearalenone with human serum albumin. J Photochem Photobiol B 170:16–24. https://doi.org/10.1016/j.jphotobiol.2017.03.016

Poór M, Bálint M, Hetényi C et al (2017b) Investigation of non-covalent interactions of aflatoxins (B1, B2, G1, G2, and M1) with serum albumin. Toxins 9:339. https://doi.org/10.3390/toxins9110339

Poostforoushfard A, Pishgar AR, Berizi E et al (2017) Patulin contamination in apple products marketed in Shiraz, Southern Iran. Curr Med Mycol 3:32–35. https://doi.org/10.29252/cmm.3.4.32

Porter JK, Bacon CW, Wray EM et al (1995) Fusaric acid in *Fusarium moniliforme* cultures, corn, and feeds toxic to livestock and the neurochemical effects in the brain and pineal gland of rats. Nat Toxins 3:91–100. https://doi.org/10.1002/nt.2620030206

Prosperini A, Juan-García A, Font G et al (2013) Beauvericin-induced cytotoxicity via ROS production and mitochondrial damage in Caco-2 cells. Toxicol Lett 222:204–211. https://doi.org/10.1016/j.toxlet.2013.07.005

Raab M, Daxecker H, Karimi A et al (2001) In vitro effects of mycophenolic acid on the nucleotide pool and on the expression of adhesion molecules of human umbilical vein endothelial cells. Clin Chim Acta 310:89–98. https://doi.org/10.1016/S0009-8981(01)00527-7

Rahimi E, Bonyadian M, Rafei M et al (2010) Occurrence of aflatoxin M1 in raw milk of five dairy species in Ahvaz, Iran. Food Chem Toxicol 48:129–131. https://doi.org/10.1016/j.fct.2009.09.028

Reddy TV, Viswanathan L, Venkitasubramanian TA (1971) High aflatoxin production on a chemically defined medium. Appl Microbiol 22:393–396

Reddy TV, Viswanathan L, Venkitasubramanian TA (1979) Factors affecting aflatoxin production by *Aspergillus parasiticus* in a chemically defined medium. J Gen Microbiol 114:409–413. https://doi.org/10.1099/00221287-114-2-409

Redouane-Salah S, Morgavi DP, Arhab R et al (2015) Presence of aflatoxin M1 in raw, reconstituted, and powdered milk samples collected in Algeria. Environ Monit Assess 187:375. https://doi.org/10.1007/s10661-015-4627-y

Richard JL (2007) Some major mycotoxins and their mycotoxicoses–an overview. Int J Food Microbiol 20:3–10. https://doi.org/10.1016/j.ijfoodmicro.2007.07.019

Ridenour JB, Bluhm BH (2017) The novel fungal-specific gene FUG1 has a role in pathogenicity and fumonisin biosynthesis in *Fusarium verticillioides*. Mol Plant Pathol 18:513–528. https://doi.org/10.1111/mpp.12414

Rimando AM, Porter JK (1997) Fusaric acid increases melatonin levels in the weanling rat and in pineal cell cultures. J Toxicol Environ Health 50:275–284. https://doi.org/10.1080/009841097160483

Robert L, Bowden F, John FL (1992) Nitrate-nonutilizing mutants of *Gibberella zeae* (*Fusarium graminearum*) and their use in determining vegetative compatibility. Exp Mycol 16:308–315. https://doi.org/10.1016/0147-5975(92)90007-E

Roberts BA, Patterson DS (1975) Detection of twelve mycotoxins in mixed animal feedstuffs, using a novel membrane cleanup procedure. J Assoc Off Anal Chem 58:1178–1181

Rotimi OA, Rotimi SO, Duru CU et al (2017) Acute aflatoxin B1 – induced hepatotoxicity alters gene expression and disrupts lipid and lipoprotein metabolism in rats. Toxicol Rep 4:408–414. https://doi.org/10.1016/j.toxrep.2017.07.006

Sakamoto S, Putalun W, Vimolmangkang S et al (2017) Enzyme-linked immunosorbent assay for the quantitative/qualitative analysis of plant secondary metabolites. J Nat Med 72:32–42. https://doi.org/10.1007/s11418-017-1144-z

Samapundo S, Devliehgere F, De Meulenaer B et al (2005) Effect of water activity and temperature on growth and the relationship between fumonisin production and the radial growth of *Fusarium verticillioides* and *Fusarium proliferatum* on corn. J Food Prot 68:1054–1059. https://doi.org/10.4315/0362-028X-68.5.1054

Sangare-Tigori B, Moukha S, Kouadio HJ et al (2006) Co-occurrence of aflatoxin B1, fumonisin B1, ochratoxin A and zearalenone in cereals and peanuts from Côte d'Ivoire. Food Addit Contam 23:1000–1007

Santos L, Marín S, Sanchis V et al (2009) Screening of mycotoxin multicontamination in medicinal and aromatic herbs sampled in Spain. J Sci Food Agric 89:1802–1807. https://doi.org/10.1002/jsfa.3647

Sava V, Reunova O, Velasquez A et al (2006) Can low level exposure to ochratoxin-A cause parkinsonism? J Neurol Sci 249:68–75. https://doi.org/10.1016/j.jns.2006.06.006

Sava V, Mosquera D, Song S et al (2018) Rubratoxin B elicits antioxidative and DNA repair responses in mouse brain. Gene Expr 11:211–219. https://doi.org/10.3727/000000003783992261

Schaafsma AW, Hooker DC (2007) Climatic models to predict occurrence of *Fusarium* toxins in wheat and maize. Int J Food Microbiol 119:116–125. https://doi.org/10.1016/j.ijfoodmicro.2007.08.006

Schmidt-Heydt M, Geisen R (2007) A microarray for monitoring the production of mycotoxins in food. Int J Food Microbiol 117:131–140. https://doi.org/10.1016/j.ijfoodmicro.2007.01.014

Schmidt-Heydt M, Stoll D, Schütz P et al (2015) Oxidative stress induces the biosynthesis of citrinin by *Penicillium verrucosum* at the expense of ochratoxin. Int J Food Microbiol 192(192):1–6. https://doi.org/10.1016/j.ijfoodmicro.2014.09.008

Schoevers EJ, Santos RR, Colenbrander B et al (2012) Transgenerational toxicity of Zearalenone in pigs. Reprod Toxicol 34:110–119. https://doi.org/10.1016/j.reprotox.2012.03.004

Schöneberg T, Martin C, Wettstein FE et al (2016) Fusarium and mycotoxin spectra in Swiss barley are affected by various cropping techniques. Food Addit Contam Part A Chem Anal Control Expo Risk Assess 33:1608–1619. https://doi.org/10.1080/19440049.2016.1219071

Seetha A, Monyo ES, Tsusaka TW et al (2018) Aflatoxin-lysine adducts in blood serum of the Malawian rural population and aflatoxin contamination in foods (groundnuts, maize) in the corresponding areas. Mycotoxin Res 34:195–204. https://doi.org/10.1007/s12550-018-0314-5

Segvić Klarić M, Medić N, Hulina A et al (2014) Disturbed Hsp70 and Hsp27 expression and thiol redox status in porcine kidney PK15 cells provoked by individual and combined ochratoxin a and citrinin treatments. Food Chem Toxicol 71:97–105. https://doi.org/10.1016/j.fct.2014.06.002

Selmanoglu G, Koçkaya EA (2004) Investigation of the effects of patulin on thyroid and testis, and hormone levels in growing male rats. Food Chem Toxicol 42:721–727. https://doi.org/10.1016/j.fct.2003.12.007

Shah HU, Simpson TJ, Alam S et al (2010) Mould incidence and mycotoxin contamination in maize kernels from Swat Valley, North West Frontier Province of Pakistan. Food Chem Toxicol 48:1111–1116. https://doi.org/10.1016/j.fct.2010.02.004

Sheik Abdul N, Nagiah S, Chuturgoon AA (2016) Fusaric acid induces mitochondrial stress in human hepatocellular carcinoma (HepG2) cells. Toxicon 119:336–344. https://doi.org/10.1016/j.toxicon.2016.07.002

Sheik Abdul N, Nagiah S, Chuturgoon AA (2019) Fusaric acid induces NRF2 as a cytoprotective response to prevent NLRP3 activation in the liver derived HepG2 cell line. Toxicol In Vitro 55:151–159. https://doi.org/10.1016/j.tiv.2018.12.008

Shephard GS, van der Westhuizen L, Gatyeni PM et al (2005) Do fumonisin mycotoxins occur in wheat? J Agric Food Chem 53:9293–9296. https://doi.org/10.1021/jf052101s

Shephard GS, Burger HM, Gambacorta L et al (2013) Multiple mycotoxin exposure determined by urinary biomarkers in rural subsistence farmers in the former Transkei, South Africa. Food Chem Toxicol 62:217–225. https://doi.org/10.1016/j.fct.2013.08.040

Shinozuka J, Li G, Kiatipattanasakul W et al (1997) T-2 toxin-induced apoptosis in lymphoid organs of mice. Exp Toxicol Pathol 49:387–392. https://doi.org/10.1016/S0940-2993(97)80124-8

Shinozuka J, Suzuki M, Noguchi N et al (1998) T-2 toxin-induced apoptosis in hematopoietic tissues of mice. Toxicol Pathol 26:674–681. https://doi.org/10.1177/019262339802600512

Shwab EK, Bok JW, Tribus M et al (2007) Histone deacetylase activity regulates chemical diversity in *Aspergillus*. Eukaryot Cell 6:1656–1664. https://doi.org/10.1128/EC.00186-07

Sineque AR, Macuamule CL, Dos Anjos FR (2017) Aflatoxin B1 contamination in chicken livers and gizzards from industrial and small abattoirs, measured by ELISA technique in Maputo, Mozambique. Int J Environ Res Public Health 14:951. https://doi.org/10.3390/ijerph14090951

Sizaret P, Malaveille C, Montesano R et al (1982) Detection of aflatoxins and related metabolites by radioimmunoassay. J Natl Cancer Inst 69:1375–1381

Smith TK, MacDonald EJ (1991) Effect of fusaric acid on brain regional neurochemistry and vomiting behavior in swine. J Anim Sci 69:2044–2049. https://doi.org/10.2527/1991.6952044x

Smith GW, Constable PD, Tumbleson ME et al (1999) Sequence of cardiovascular changes leading to pulmonary edema in swine fed culture material containing fumonisin. Am J Vet Res 60:1292–1300

Smith MC, Hymery N, Troadec S et al (2017) Hepatotoxicity of fusariotoxins, alone and in combination, towards the HepaRG human hepatocyte cell line. Food Chem Toxicol 109:439–451. https://doi.org/10.1016/j.fct.2017.09.022

Solfrizzo M, Gambacorta L, Lattanzio VM et al (2011) Simultaneous LC-MS/MS determination of aflatoxin M1, ochratoxin A, deoxynivalenol, de-epoxydeoxynivalenol, α and β-zearalenols and fumonisin B1 in urine as a multi-biomarker method to assess exposure to mycotoxins. Anal Bioanal Chem 401:2831–2841. https://doi.org/10.1007/s00216-011-5354-z

Sorenson WG, Simpson J (1986) Toxicity of penicillic acid for rat alveolar macrophages in vitro. Environ Res 41:505–513. https://doi.org/10.1016/S0013-9351(86)80145-1

Stockmann-Juvala H, Mikkola J, Naarala J et al (2004a) Fumonisin B1-induced toxicity and oxidative damage in U-118MG glioblastoma cells. Toxicology 202:173–183. https://doi.org/10.1016/j.tox.2004.05.002

Stockmann-Juvala H, Mikkola J, Naarala J et al (2004b) Oxidative stress induced by fumonisin B1 in continuous human and rodent neural cell cultures. Free Radic Res 38:933–942. https://doi.org/10.1080/10715760412331273205

Stoll D, Schmidt-Heydt M, Geisen R (2013) Differences in the regulation of ochratoxin A by the HOG pathway in *Penicillium* and *Aspergillus* in response to high osmolar environments. Toxins 5:1282–1298. https://doi.org/10.3390/toxins5071282

Sueck F, Cramer B, Czeschinski P et al (2018a) Human study on the kinetics of 2'R-ochratoxin A in the blood of coffee drinkers. Mol Nutr Food Res 63(4):e1801026. https://doi.org/10.1002/mnfr.201801026

Sueck F, Poór M, Faisal Z et al (2018b) Interaction of ochratoxin A and its thermal degradation product 2'R-ochratoxin A with human serum albumin. Toxins 10:pii: E256. https://doi.org/10.3390/toxins10070256

Sugiyama K, Hiraoka H, Sugita-Konishi Y (2008) Aflatoxin M1 contamination in raw bulk milk and the presence of aflatoxin B1 in corn supplied to dairy cattle in Japan. Shokuhin Eiseigaku Zasshi 49:352–355. https://doi.org/10.3358/shokueishi.49.352

Sur E, Celik I (2003) Effects of aflatoxin B1 on the development of the bursa of *Fabricius* and blood lymphocyte acid phosphatase of the chicken. Br Poult Sci 44:558–566. https://doi.org/10.1080/00071660310001618352

Sutton P, Newcombe NR, Waring P et al (1994) *In vivo* immunosuppressive activity of gliotoxin, a metabolite produced by human pathogenic fungi. Infect Immun 62:1192–1198

Székács A, Adányi N, Székács I et al (2009) Optical waveguide light-mode spectroscopy immunosensors for environmental monitoring. Appl Opt 48:B151–B158. https://doi.org/10.1364/AO.48.00B151

Taheri N, Semnani S, Roshandel G et al (2012) Aflatoxin contamination in wheat flour samples from golestan province, northeast of iran. Iran J Public Health 41:42–47

Tan Y, Chu X, Shen GL et al (2009) A signal-amplified electrochemical immunosensor for aflatoxin B(1) determination in rice. Anal Biochem 387:82–86. https://doi.org/10.1016/j.ab.2008.12.030

Taye W, Ayalew A, Chala A et al (2016) Aflatoxin B1 and total fumonisin contamination and their producing fungi in fresh and stored sorghum grain in East Hararghe, Ethiopia. Food Addit Contam Part B Surveill 9:237–245. https://doi.org/10.1080/19393210.2016.1184190

Tchana AN, Moundipa PF, Tchouanguep FM (2010) Aflatoxin contamination in food and body fluids in relation to malnutrition and cancer status in Cameroon. Int J Environ Res Public Health 7:178–188. https://doi.org/10.3390/ijerph7010178

Theumer MG, Henneb Y, Khoury L et al (2018) Genotoxicity of aflatoxins and their precursors in human cells. Toxicol Lett 287:100–107. https://doi.org/10.1016/j.toxlet.2018.02.007

Thuvander A, Breitholtz-Emanuelsson A, Olsen M (1995) Effects of ochratoxin A on the mouse immune system after subchronic exposure. Food Chem Toxicol 33:1005–1011. https://doi.org/10.1016/0278-6915(95)00075-5

Tilburn J, Sarkar S, Widdick DA et al (1995) The *Aspergillus* PacC zinc finger transcription factor mediates regulation of both acid-and alkaline-expressed genes by ambient pH. EMBO J 14:779–790. https://doi.org/10.1002/j.1460-2075.1995.tb07056.x

Vargas EA, Preis RA, Castro L et al (2001) Co-occurrence of aflatoxins B1, B2, G1, G2, zearalenone and fumonisin B1 in Brazilian corn. Food Addit Contam 18:981–986. https://doi.org/10.1080/02652030110046190

Vidal JC, Duato P, Bonel L et al (2009) Use of polyclonal antibodies to ochratoxin A with a quartz-crystal microbalance for developing real-time mycotoxin piezoelectric immunosensors. Anal Bioanal Chem 394:575–582. https://doi.org/10.1007/s00216-009-2736-6

Wallin S, Gambacorta L, Kotova N et al (2015) Biomonitoring of concurrent mycotoxin exposure among adults in Sweden through urinary multi-biomarker analysis. Food Chem Toxicol 83:133–139. https://doi.org/10.1016/j.fct.2015.05.023

Wang GH, Xue CY, Chen F et al (2009) Effects of combinations of ochratoxin A and T-2 toxin on immune function of yellow-feathered broiler chickens. Poult Sci 88:504–510. https://doi. org/10.3382/ps.2008-00329

Wang C, Qian J, An K et al (2017) Magneto-controlled aptasensor for simultaneous electrochemical detection of dual mycotoxins in maize using metal sulfide quantum dots coated silica as labels. Biosens Bioelectron 89:802–809. https://doi.org/10.1016/j.bios.2016.10.010

Wang J, Yang C, Yuan Z et al (2018) Toxin exposure induces apoptosis in TM3 cells by inhibiting mammalian target of rapamycin/serine/threonine protein kinase(mTORC2/AKT) to promote Ca2+production. Int J Mol Sci 19:3360. https://doi.org/10.3390/ijms19113360

Wangikar PB, Dwivedi P, Sinha N et al (2005) Effects of aflatoxin B1 on embryo fetal development in rabbits. Food Chem Toxicol 43:607–615. https://doi.org/10.1016/j.fct.2005.01.004

Waring P, Beaver J (1996) Gliotoxin and related epipolythiodioxopiperazines. Gen Pharmacol 27:1311–1316. https://doi.org/10.1016/S0306-3623(96)00083-3

Watson SA, Hayes AW (1981) Binding of rubratoxin B to mouse hepatic microsomes and in vitro effects of the mycotoxin on polysome binding to microsomal membranes as measured by the activity of an enzyme catalyzing disulphide interchange. Toxicon 19:509–516. https://doi. org/10.1016/0041-0101(81)90009-X

Weaver GA, Kurtz HJ, Bates FY et al (1981) Diacetoxyscirpenol toxicity in pigs. Res Vet Sci 31:131–135. https://doi.org/10.1016/S0034-5288(18)32480-9

Wilkinson JR, Yu J, Bland JM et al (2007) Amino acid supplementation reveals differential regulation of aflatoxin biosynthesis in *Aspergillus flavus* NRRL 3357 and *Aspergillus parasiticus* SRRC 143. Appl Microbiol Biotechnol 74:1308–1319. https://doi.org/10.1007/ s00253-006-0768-9

Woloshuk CP, Cavaletto JR, Cleveland TE (1997) Inducers of aflatoxin biosynthesis from colonized maize kernels are generated by an amylase activity from *Aspergillus flavus*. Phytopathology 87:164–169. https://doi.org/10.1094/PHYTO.1997.87.2.164

Wong KH, Hynes MJ, Todd RB et al (2007) Transcriptional control of nmrA by the bZIP transcription factor MeaB reveals a new level of nitrogen regulation in *Aspergillus nidulans*. Mol Microbiol 66:534–551. https://doi.org/10.1111/j.1365-2958.2007.05940.x

Wu HC, Santella R (2012) The role of aflatoxins in hepatocellular carcinoma. Hepat Mon 12:e7238. https://doi.org/10.5812/hepatmon.7238

Wu L, Qiu L, Zhang H et al (2017) Optimization for the production of deoxynivalenoland zearalenone by *Fusarium graminearum* using response surface methodology. Toxins 9:pii: E57. https://doi.org/10.3390/toxins9020057

Wu TS, Cheng YC, Chen PJ et al (2019) Exposure to aflatoxin B1 interferes with locomotion and neural development in zebrafish embryos and larvae. Chemosphere 217:905–913. https://doi. org/10.1016/j.chemosphere.2018.11.058

Xue CY, Wang GH, Chen F et al (2010) Immunopathological effects of ochratoxin A and T-2 toxin combination on broilers. Poult Sci 89:1162–1166. https://doi.org/10.3382/ps.2009-00609

Yang L, Yu Z, Hou J et al (2016) Toxicity and oxidative stress induced by T-2 toxin and HT-2 toxin in broilers and broiler hepatocytes. Food Chem Toxicol 87:128–137. https://doi.org/10.1016/j. fct.2015.12.003

Yang L, Tu D, Zhao Z et al (2017) Cytotoxicity and apoptosis induced by mixed mycotoxins (T-2 and HT-2 toxin) on primary hepatocytes of broilers in vitro. Toxicon 129:1–10. https://doi. org/10.1016/j.toxicon.2017.01.001

Yang L, Tu D, Wang N et al (2019) The protective effects of DL-Selenomethionine against T-2/ HT-2 toxins-induced cytotoxicity and oxidative stress in broiler hepatocytes. Toxicol In Vitro 54:137–146. https://doi.org/10.1016/j.tiv.2018.09.016

Yarru LP, Settivari RS, Antoniou E et al (2009) Toxicological and gene expression analysis of the impact of aflatoxin B1 on hepatic function of male broiler chicks. Poult Sci 88:360–371. https://doi.org/10.3382/ps.2008-00258

Yin S, Zhang Y, Gao R et al (2014) The immunomodulatory effects induced by dietary zearalenone in pregnant rats. Immunopharmacol Immunotoxicol 36:187–194. https://doi.org/10.3109/089 23973.2014.909847

Young KL, Villar D, Carson TL et al (2003) Tremorgenic mycotoxin intoxication with penitrem a and roquefortine in two dogs. J Am Vet Med Assoc 222:52–3, 35. https://doi.org/10.2460/ javma.2003.222.52

Yu J, Chang P, Bhatnagar D et al (2000) Cloning of a sugar utilization gene cluster in *Aspergillus parasiticus*. Biochim Biophys Acta 1493:211–214. https://doi.org/10.1016/ S0167-4781(00)00148-2

Yu J, Mohawed SM, Bhatnagar D et al (2003) Substrate-induced lipase gene expression and aflatoxin production in *Aspergillus parasiticus* and *Aspergillus flavus*. J Appl Microbiol 95:1334–1342. https://doi.org/10.1046/j.1365-2672.2003.02096.x

Yu J, Fedorova ND, Montalbano BG et al (2011) Tight control of mycotoxin biosynthesis gene expression in *Aspergillus flavus* by temperature as revealed by RNA-Seq. FEMS Microbiol Lett 322:145–149. https://doi.org/10.1111/j.1574-6968.2011.02345.x

Yuan G, Wang Y, Yuan X et al (2014) T-2 toxin induces developmental toxicity and apoptosis in zebrafish embryos. J Environ Sci 26:917–925. https://doi.org/10.1016/S1001-0742(13)60510-0

Zhang X, Li J, Zong N et al (2014) Ochratoxin A in dried vine fruits from Chinese markets. Food Addit Contam Part B Surveill 7:157–161. https://doi.org/10.1080/19393210.2013.867365

Zhang L, Dou X-W, Zhang C et al (2018) A review of current methods for analysis of mycotoxins in herbal medicines. Toxins 10:65. https://doi.org/10.3390/toxins10020065

Zheng W, Wang B, Si M et al (2018a) Zearalenone altered the cytoskeletal structure via ER stress-autophagy- oxidative stress pathway in mouse TM4 Sertoli cells. Sci Rep 8:3320. https://doi. org/10.1038/s41598-018-21567-8

Zheng WL, Wang BJ, Wang L et al (2018b) ROS-mediated cell cycle arrest and apoptosis induced by zearalenone in mouse sertoli cells via ER stress and the ATP/AMPK pathway. Toxins 10:24. https://doi.org/10.3390/toxins10010024

Zhou C, Zhang Y, Yin S et al (2015) Biochemical changes and oxidative stress induced by zearalenone in the liver of pregnant rats. Hum Exp Toxicol 34:65–73. https://doi. org/10.1177/0960327113504972

Zhuang Z, Yang D, Huang Y et al (2013) Study on the apoptosis mechanism induced by T-2 toxin. PLoS One 8:e83105. https://doi.org/10.1371/journal.pone.0083105

Zinedine A, Fernández-Franzón M, Mañes J et al (2017) Multi-mycotoxin contamination of couscous semolina commercialized in Morocco. Food Chem 214:440–446. https://doi. org/10.1016/j.foodchem.2016.07.098

Chapter 2
Nanopesticides for Pest Control

Saheli Pradhan and Damodhara Rao Mailapalli

Abstract The excessive use of pesticides results in poor targeted delivery and off-target waste accumulation that induces mutation in genetic make-up in the target pest, followed by pesticidal resistance. As a consequence, most of the pesticides are lost because of pesticide drift, leading to environment hazards. The active components released from the pesticide formulations are either biodegraded or hydrolysed in the environment. They can even contaminate the surroundings by leaching. Therefore, an urgent attention is required to protect resources. Nanopesticides appear as an alternative because they can be used as 'smart delivery systems' for the release of the pesticides in timely but controlled manner, for a desired time-span. This would reduce the risk of environmental pollution and its associated hazards. Physicochemical properties of nanopesticides along with their efficacy against target and non-target organisms need to be studied. Since nanomaterials might exhibit non-specific toxicity to both targeted and non-targeted organisms, toxicological analyses should be performed.

Keywords Nanopesticide · Insect resistance · Mortality · Pest control · Nanoencapsulation · Nanocarriers · Targeted delivery · Toxicity · Nanomaterials · Crop protection

2.1 Introduction

In this current era of precision farming and sustainable agricultural practices, augmentation of food production for the ever-increasing global population's demand is one of the prime concerns. It is really a hard task to increase the crop yield by monitoring interrelated environmental variables with minimum, controlled, and balanced

S. Pradhan (✉) · D. R. Mailapalli
Agricultural and Food Engineering Department, Indian Institute of Technology Kharagpur,
Kharagpur, West Bengal, India

© Springer Nature Switzerland AG 2020

E. Lichtfouse (ed.), *Sustainable Agriculture Reviews 40*, Sustainable Agriculture
Reviews 40, https://doi.org/10.1007/978-3-030-33281-5_2

target actions in accordance with every situation. Henceforth, when the population and its demand for food have been growing apace under a deregulated regime, maximum crop yield with minimum waste would be the main focus of agricultural sustainability (Chhipa 2017; Ayoub et al. 2017). Due to the limitation of occurrences of natural resources (land, water, soil fertility), synthetic pesticide has been applied abruptly without following proper rules and regulations, which causes serious environmental disturbances. Since the cost of the pesticides has also increased at an alarming rate, which has become less affable to the farmers. However, it is essential to control the pest in order to maximize output by conserving the protective resources without affecting the surrounding environment under the sustainable agricultural practices (Lichtfouse et al. 2005; Rahaie and Rahaie 2015; Srivastava et al. 2018). Nevertheless, system application of chemical pesticide is important to achieve the primary objectives of food and energy. However, conventional application of synthetic pesticides leads to ground water contamination causing environmental pollution. The toxic substances released and/or degraded from the pesticides have degenerated in the environment via biodegradation, photo-degradation, and hydrolysis; even they have also contaminated the surroundings by leaching or in the run-off by rainfall leading to eutrophication as shown in Fig. 2.1. Due to all of these issues, only 1% or less than 1% of the applied pesticide reaches the targeted sites and remaining residues adversely interact with the non-target organisms (Memarizadeh et al. 2014a, b). In order to combat these problems, the use of botanical extracts, synthetic specialized pesticides and genetically modified disease-resistant crops have also been introduced in farming for the past five decades. Although the food production has much been improved, the quality of the food and soil fertility have been compromised. Pathogen and pest resistance appear to be developing, along with a sharp decline in soil biodiversity (Ghormade et al. 2011). To address the potential implications of chemical pesticides to the non-targeted organisms and environment, a careful insight should be taken into account. A balance technology has to be introduced in the farming which must have the least negative impact on the surrounding environment. Besides, it should be highly species-specific, yet cost-effective to farmers. Therefore, development of novel targeted pesticide with low toxicity to non-target species and low pesticidal residues is very much in demand for sustainable agricultural practices (Liu et al. 2014).

Engineered nanomaterials (ENMs) with their characteristic properties like small size and large surface to volume ratio, greater permeability, thermal stability, solubility, and biodegradability are becoming more widely accepted for use in the agricultural sector (Nair et al. 2008; Lin et al. 2014; Ranjan et al. 2018). Because of the unique properties of nanoparticles, these materials can be utilized to encapsulate the agrochemicals in a more stable and safer ways to improve crop yield and productivity in a sustainable manner. By using these nanoparticles efficiently, damage to the environment and human health may be minimized (Yata et al. 2018). There have been a lot of reports regarding the application of ENMs in agricultural sectors, such as the development of nanofertilizers and nanopesticides; as the effective remedial measure in the detection of pesticides (Zhu et al. 2008; Lin and Xing 2007; Noji et al. 2011; Pradhan et al. 2013a, b, 2014, 2015; Ghafariyan et al. 2013; Chandra

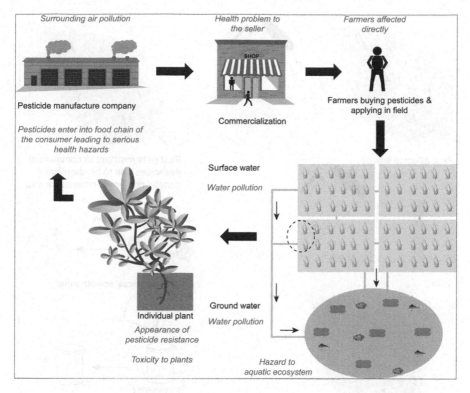

Fig. 2.1 Effect of pesticides in the environment. Conventional application of synthetic pesticides leads to plant toxicity as well as ground water contamination, causing environmental pollution. Pesticides are transformed via biodegradation, photo-degradation, and hydrolysis

et al. 2014; Kim et al. 2015; Servin et al. 2015; Dubey and Mailapalli 2016), heavy metals and chemical contaminants (Yu et al. 2007; Liu et al. 2008; Qu et al. 2009; Kang et al. 2010; Kumaravel and Chandrasekaran 2011; Zhao et al. 2011; Guo et al. 2015; Talbert et al. 2016), as a detoxifying agent of harmful pollutants (Seitz et al. 2012; Xu et al. 2014; Pang et al. 2015; Hou et al. 2016). Proper utilization of nanopesticides is an important aspect of agricultural practices, however, over application of nanopesticides might result in one of the unintentional diffuse inputs of ENMs into the environment. Thereafter, a thoughtful evaluation of risk and benefit associated to the ENMs is essential prior to commercialization. Besides, detailed documentation of synthetic procedures and testing protocols of nanopesticides is essential for future reference. Unfortunately, only a handful of reports are available in this regard. This article reviews some of the strategic ways that nanomaterials may be utilized as a suitable alternative to commercial pesticides (Fig. 2.2). In order to understand the possible application and any additions to the modes of action when using nanopesticides, a short overview of present scenario of the conventional pesticides is also reviewed. To get a comprehensive structure of present-day

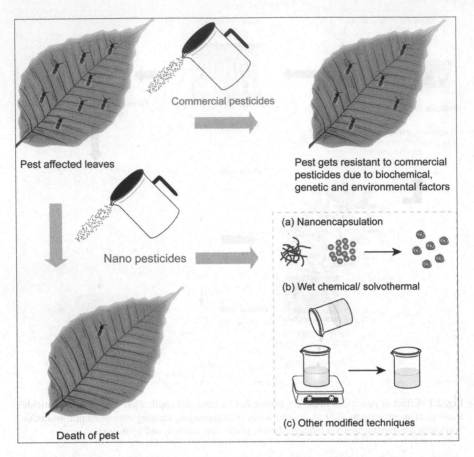

Fig. 2.2 Comparison of the effect of current commercial pesticides and alternative nano-pesticides on pests. Indiscriminate use of commercial pesticides results in unpremeditated pesticidal resistance. Nanopesticides could be effectively used on the targeted and non-targeted pests, without hampering the immediate ecosystem. Nanopesticides could be a suitable alternative for the commercial pesticides, since they show slow yet targeted delivery of the desired pesticides to the target organisms

nanopesticides, works of past 10 years are also being documented. Finally, a critical analysis of the present and future generations of pesticides is elucidated to get a better picture of sustainable agricultural practices.

2.2   Classification and Traditional Applications of Pesticides

Insects, which make up approximately the two-thirds of the known species of animals, can feed on almost all kinds of plants, like crop plants, forest trees, medicinal plants, herbs, and weeds. They not only destroy agricultural crops but also stored grains and eventually deteriorate the quality of the food. Insects that cause more than 5% of the damage to the crop and/or its food grains are called pests (Rai and Ingle 2012). Any compound that is designed to prevent, destroy, repel or mitigate any pests is called pesticide. Along with that, they can act as the plant regulator, defoliants or desiccant. Under the rules and regulations of International Pest Management (IPM), they should be environment-friendly and cost-effective to the farmers. On the basis of the chemical structure and mode of action, pesticides are classified into different subgroups (Pereira et al. 2015; Sparks and Nauen 2015). Details of the each of the groups are summarized in Table 2.1. It is very important to understand the mode of action of the chemical compounds in order to evaluate their activity against the target pests without undesirable economic damage. Pest control strategy must be carefully incorporated into the crop management system without minimum risks to the beneficial and non-target organism and environment.

Application of pesticides into sustainable long-term crop protection had been in practice and continually developing for thousands of years. It is believed that Sumerians applied sulfur compounds in order to protect crops from insect and mite attacks 4500 years ago (Unsworth 2010). Pyrethrum, a compound extracted from the dried flower of *Chrysanthemum cinerariaefolium* was used as an insecticide approximately 2000 years ago. Traditionally, salt water or sea water was used in ancient India as a g. However, during World War II, there was an urgent need to enhance food production. During 1940s, many synthetic pesticides like dichloro diphenyl trichloroethane (DDT) and 2,4-dichlorophenoxyacetic acid (2, 4-D) were developed (Gupta 2007). A new era of food revolution was developed, where there was no limited concern of chemical pesticides to the environment and human health. Starting from that time, pesticides have been applied to the field as sprays as a part of proper cropping methods like crop rotation, land, water and post-harvest management (Yu et al. 2017). They are initially sprayed to deposit onto the crop foliage and then reach the target site of the pest attack via diffusion, uptake and/or transfer processes, leading to pest poisoning or contact attack (Nuruzzaman et al. 2016). However, continuous use of chemical pesticides has bolstered the evolution of pesticide resistance in the targeted pest along with the risk of bio-magnification. The active constituent present in chemical pesticide often interfere the metabolic pathway by inhibiting the enzymatic activities (Pandey et al. 2016). The sensitivity of high yielding crops to diseases, insects and biotic factors has protracted the use of chemical pesticides.

2.3 Emergence of Biopesticides for Biological Control

Use of chemical pesticides has continued to persist in the agricultural sectors for decades and persistence, and biomagnification of some of these complex compounds in the environment remained on the back burner. Later on, scientists had tried to imbibe a new approach, called "biological control of the pest", where natural enemies of the major pests had been introduced to the farming in order to destroy

Table 2.1 Classification of pesticides according to Insect Resistance Action Committee (IRAC) on the basis of mode of action of the pesticides. This classification system includes all chemical, biological or other materials, used to control insects and acarines on crops and/or the environment. It is an useful manual for the selection of insecticides and acaricides for insect resistance management

Group	Name of the pesticides	Chemical nature	Mode of action	Target site of action
I	Organophosphate	They are ester of phosphoric acid $R_L-\overset{\overset{O}{\|}}{\underset{\underset{OR_2}{\|}}{P}}-OR_3$	Acetyl choline esterase inhibitor	Nerve and muscle
	Carbamate	They are derived from carbamic acid; functional groups present in this compound are carbamate group, carbamate esters and carbamic acids $\underset{O}{}\overset{R}{\underset{N}{\|}}\overset{}{\underset{H}{}}{}^{CH_3}$		
II	Cyclodienes	Compound derived from Hexachlorocyclopentadiene, an organochloride pesticide	GABA gated chlorine channel antagonist	
	Fiproles	Compound belongs to the phenylpyrazole chemical family		
III	Pyrethrins	Organic compounds normally derived from *Chrysanthemum cinerariifolium*	Voltage gated sodium channel modulator	
	DDT and analogs	They are the crystalline organochloride		

(continued)

Table 2.1 (continued)

Group	Name of the pesticides	Chemical nature	Mode of action	Target site of action
IV	Neonicotinoids	A class of neuro-active insecticides chemically similar to nicotine	Nicotinic acetyl choline receptor agonist	
	Butenolides	A class of lactones with a four-carbon heterocyclic ring structure		
V	Spinosyns		Nicotinic acetyl choline receptor allosteric	
VI	Avermectins	They are a 16-membered macrocyclic lactone	CC activators	
	Milbemycins	A group of macrolides chemically related to the avermectins		

(continued)

Table 2.1 (continued)

Group	Name of the pesticides	Chemical nature	Mode of action	Target site of action
VII	Pymetrozine	A chemical compound derived from the group of pyridine – azomethines	Modulation of chordotonal organs	
	Flonicamid	Derived from the group of nicotinoids		
VIII	Nereistoxin analog	Toxins derived from 4-N,N-dimethylamino-1,2-dithiolane	Nicotinic acetyl choline receptor blockers	
IX	Formamidines		Octopamine receptor	
X	Oxadiazines	An unsaturated six-membered heterocycle	Voltage gated sodium channel blocker	
XI	Diamides	Two amide groups are present	Ryanodine receptor allosteric	
XII	Juvenoids	Juvenile hormone analogue	Juvenile hormone receptor agonist	Growth and development
	Fenoxycarb	A carbamate insect growth regulator		
	Pyriproxyfen	A pyridine-based pesticide		
XIII	Oxazoles	Parent compound of heterocyclic aromatic organic compounds	MGI	

(continued)

Table 2.1 (continued)

Group	Name of the pesticides	Chemical nature	Mode of action	Target site of action
XIV	Benzoylureas	Chemical derivatives of *N*-benzoyl-*N'*-phenylurea	Chitin synthesis inhibitor	
XV	Cyromazine	A triazine insect growth regulator	Chitin synthesis inhibitor	
XVI	Diacylhydrazine		Ecdysone receptor agonist	
XVII	Tetramic acid		Acetyl CoA carboxylase respiratory organ	
XVIII	Diafenthiuron	A chemical compound from the group of thioureas	ATP synthase	
XIX	Chlorfenapyr	A pro-insecticide derived from a class of microbially produced compounds known as halogenated pyrroles	Oxidative phosphorylation uncoupler	
XX	Rotenone	Non-selective pesticide	Mitochondrial electron transport I inhibitors	
XXI	Phosphine	A group of organophosphorus compounds with the formula R_3P (R = organic derivative)	Mitochondrial electron transport IV inhibitors	

(continued)

Table 2.1 (continued)

Group	Name of the pesticides	Chemical nature	Mode of action	Target site of action
XXII	B-Ketonitrile derivatives		Mitochondrial electron transport II inhibitors	
XXII	*Bacillus thuringiensis*	A Gram-positive, soil-dwelling bacterium, commonly used as a biological pesticide	Midgut membrane	
XXIV	Miscellaneous (alkyl halides, borates)	–	–	
XXV	Unknown mode of action (Azadiractin, benzoximate, dicofol)	–	–	

the pest population causing economic injury (Seiber et al. 2014). Since these organisms are self-perpetuating and do not interfere the activity of non-target organisms, this approach of pest control immediately gained acceptability. However, the use of biological control demands skilled professionals. The biological control pests may become susceptible to insect resistance in the germplasm and introgression of unwanted harmful traits from them.

2.4 Transgenic Approach

A transgenic approach is more species-specific, where insecticidal proteins, such as *Bacillus thuringiensis* endotoxin, plant protease inhibitors, chitinase, lectin, and biotin-binding proteins have been designed to be produced in the insect-resistant transgenic plants (Thungrabeab and Tongma 2007). The rapid emergence of insecticidal resistance to such endotoxins along with the collateral interaction with the non-targeted environment and micro-environment limits the application of this approach. Sub-optimal expression of such endotoxin, mutation of the targeted gene of the insect pest, loss of the target midgut protease and change of the membrane integrity of the target organisms may be additional reasons for such an outbreak. Henceforth, there is an urgent need for exploring another possible avenue to obviate direct and/or indirect effect of the agricultural pest.

2.4.1 RNAi Technology: Sequence-Specific Silencing of Targeted Gene

RNA interference (RNAi) technology is the new-age technology for controlling pest population through post-translational silencing by dsRNA mediated down-regulation of specific genes by breaking down its subsequent messenger RNA (mRNA) (Mamta and Rajam 2017). RNAi pathway entails development of small interfering molecules like small interfering RNA (siRNA) and miRNA, with the help of dicer enzymes. These molecules are loaded onto RNA induced silencing complex (RISC) comprising Argonaute protein (AGO). RISC then directs the interfering molecules to the active site where homology-based cleavage of targeted mRNA occurs. Insects can eventually take up SiRNA directly from the environment and transfer the signal from cell to cell. Injecting or feeding the bacteria expressing dsRNA has been adopted to silence different target genes. In host-RNAi interaction, the host plant is engineered with hairpin RNAi vector to synthesize dsRNA against the targeted gene of the insect pest. Upon feeding on plant parts, dsRNA enters into the insect gut, leading to the induction of RNAi machinery and then silencing the gene of choice in the insect pest (Kitzmann et al. 2013; Gillet et al. 2017). However, the lifetime of the dsRNA entirely depends upon its own persistence in the hemolymph of the pest, where nucleic acid degrading enzymes, i.e., nuclease are found abundantly and they are prone to degrade the substrate of choice. The major drawbacks of this approach are its poor solubility and uptake into insect cell vicinity along with the sensitivity of the dsRNA to the nuclease, which limits them to be expressive and therefore exhibits functional redundancy at the time of gene silencing (Joga et al. 2016; Katoch et al. 2013). An ideal pest control strategy should be employed which would be economic, environmental and farmer friendly. It should be species-specific and negate all the limitation of previously mentioned pest control programmes.

2.5 Development of Nanopesticides

Nanopesticide is referred to those tiny molecules which are solely constituent of pest control derivatives and/or entraps the active constituent of pesticide into a protective nanocarrier (Kookana et al. 2014). It must ensure the improvement of precision farming through "smart field management". Because of its higher surface to volume ratio and quantum effects due to small size, unusual phase transformation, and stabilization (Bakshi et al. 2015; Kuswandi 2018), ENMs can minimize photodegradation and improvise physicochemical stability of the materials (de Oliveira et al. 2014). Therefore, a new diffusion-, erosion- and swelling controlled nanodevice can be customized in order to deliver active pesticidal component to the targeted agricultural pest with enhanced durability and efficiency; without any environmental contamination hazards (Choudhury et al. (2012); Chowdhury et al.

2017). Depending upon the nature of the pesticidal structure and chemical constituents, different types of nano pesticidal formulations have been reported which is documented in Table 2.2.

The primary objective of the development of nanopesticide is to maximize the entomotoxicity of the targeted drugs by increasing solubility of the insoluble and/or sparingly soluble pesticides in the vicinity of the site of action; slow yet controlled release of the active constituent of pesticides without any premature degradation (Atta et al. 2015; Villaseñor and Ríos 2018). In order to circumscribe such environmental burden, different approaches of nanopesticides have been reported in literature; although most of the works have been carried out at very preliminary levels. Nonetheless, this could be used as a baseline for the future researchers related to crop protection management.

2.6 Pesticidal Delivery by Nanoencapsulation

2.6.1 β-Cyclodextrin (β-CD) Grafted Magnetic Nanoparticles Loaded with Diuron

Another group of pesticide, diuron, has also been loaded into hydrophobic cavities of hydrophilic oligosaccharide, β-cyclodextrin grafted magnetic nanoparticles (Kah and Hofmann 2014). This newly designed compound has shown better water solubility, the feasibility of controlled release of diuron, and to have increased the bioavailability of the active compound to the site of action. This particle also has been found to be toxic to soil microbes when analyzed both micro-colorimetrically and enzymatically. Similarly, another pesticidal compound, ricinoleic acid has been encapsulated into carboxymethyl chitosan matrix to achieve spherical shaped, water-soluble nanopesticides (Feng and Zhang 2011).

2.6.2 Fluorescent-Tagged 2,4-Dichlorophenoxyacetic Acid Nanoparticles

Different pesticides, like 2,4-dichlorophenoxyacetic acid have been fluorescent-tagged and then trapped into organic nanoparticles of perylene-3-butylmethanol (Atta et al. 2015). 2,4-dichlorophenoxyacetic acid has been fluorescent-tagged in order to understand morphological changes developed inside the plant after NP application. Bioassay experiments have carried out in order to study the herbicidal activity of the nanoformulation in the model plants.

Table 2.2 Examples of nanopesticides for target and non-target pests

Nano-pesticides	Active compound	Chemical nature	Target organism	Mode of action	References
β-Cyclodextrin-magnetite nanoparticle-diuron	Diuron	Better solubility	Soil microbial activity	Controlled release	Liu et al. (2014)
Carboxy-methyl-cellulose nanoparticle-atrazine	Atrazine	Better solubility	–	Controlled release	Rahaie and Rahaie (2015)
Silica nanoparticle	Silica nanoparticle	175 nm, amorphous	Cotton leaf worm (Spodoptera littoralis)	Mortality due to dehydration	Ayoub et al. (2017)
Carboxy-methyl-chitoan-ricinoleic acid nanoparticle	Ricinoleic acid	Spherical with negative surface charge	–	–	Feng and Zhang (2011)
Chitosan-beauvericin nanoparticle	Beauvericin	160–230 nm, spherical	Spodoptera litura	Better entomotoxicity	Bharani et al. (2014)
Silver nanoparticles (ag NP)	Silver nanoparticle	14–28 nm, uniform, stable	Spodoptera frugiperda, SF-21 cell line	Drastic reduction of gut microflora, extracellular enzymes	Bharani et al. (2017)
Poly-(citric acid)-poly-(ethylene glycol)-poly(citric acid)-g-titanium dioxide nanoparticle	Titanium dioxide nanoparticle	10–12 nm	5th instar larvae of Glyphodes pyloalis	Controlled release	Memarizadeh et al. (2014a, b)
PCL-azadiractin nanoparticle	Azadiractin	Encapsulation efficiency (EE) 98%, 245 nm, spherical	Diamond black moth	Controlled release	Forim et al. (2013)
PEG-, PVP-, sodium alginate loaded rotenone nanoparticle	Rotenone	EE 98–100%, 600 nm–1.5 μm		Controlled release	Martin et al. (2013)
Carvacrol-chitosan/TPP nanoparticle	Carvacrol	EE 14–31%, spherical, 40–80 nm	–	Controlled release	Woranuch, and Yoksan (2013)

(continued)

Table 2.2 (continued)

Nano-pesticides	Active compound	Chemical nature	Target organism	Mode of action	References
2-HP-β-CD-eugenol nanoparticle	Eugenol	EE 89–100%, 170–900 nm	–	Controlled release	Choi et al. (2009)
Zein-curcumin nanoparticle	Curcumin	EE 85–90%, 170–900 nm	–	Controlled release	Gomez-Estaca et al. (2012)
PEG- essential oil of garlic nanoparticle	Essential oil	240 nm, spherical, EE 80%	Adult beetles	Controlled release	Yang et al. (2009)
Chitosan-pepper-rosmarin nanoparticle	Thymol	335–558 nm	*Aedes aegypti*	Controlled release	Abreu et al. (2012)
Graphene oxide-Cu₂₋ₓSe-chlorpyrifos nanoparticle	Chlorpyrifos	20 nm	*Pieris rapae*	80% mortality	Sharma et al. (2017)
Abamectin-PLA nanoparticle	Abamectin	Spherical, 450 nm, EE 50%	Soil bacteria	Entomotoxicity	Yu et al. (2017)
Iron oxide nanoparticle	Iron oxide nanoparticle	–	Transgenic and non-transgenic Bt cotton	Entomotoxicity	Nhan et al. (2016)
Calcium carbonate nanoparticle	Calcium carbonate nanoparticle	60 nm	California red scale and oriental fruit fly	Entomotoxic	Hua et al. (2015)
PCL-essential oil nanoparticle	Essential oil (*Zanthaxylum rhoifolium*)	500 nm, EE 96%	*Bemisia tabaci*	Entomotoxic	Christofoli et al. (2015)

Nickel nanoparticle	Nickel nanoparticle	Cubic, 47 nm	Callasobruchus macukates, Helicoverpa armigera	97.31% mortality	Elango et al. (2016)
Chitosan/TPP-Paraquat nanoparticle	Paraquat	EE 62.66%, 390–420 nm, spherical	Allium cepa	–	Grillo et al. (2014)
Gold nanoparticle- ferbam	Ferbam	30 nm	Tea leaves	Better surface adhesion	Hou et al. (2016)
Functionalized fluorescence dendrimer-thiamethoxam nanoparticle	Thiamethoxam	–	2nd instar larvae of Heliothis armigera	Entomotoxic	Kumar et al. (2015)
PEG encapsulated essential oil nanoparticle	Essential oil	235 nm, EE 75%	Tribolium castaneum, Rhizopertha Dominica	Entomotoxic	González et al. (2014)
Pyrethrin nanoparticle	Pyrethrin	–	Larval stage of Coccinella septempunctata; fourth instar stage of Macrolophus pygmaeus	Entomotoxic	Papanikolaou et al. (2017)
Oleoyl-CM-chitosan-rotenone nanoparticle	Rotenone	Spherical, EE 97%	–	–	Kamari et al. (2016)

2.6.3 Poly(Citric Acid)-Poly(Ethylene Glycol)-Poly(Citric Acid) Encapsulated Indoxacarb Nanoparticles

Photodegradable and biocompatible nano indoxacarb has synthesized by encapsulating poly (citric acid)-poly (ethylene glycol)-poly (citric acid) (PCA-PEG-PCA) dendritic copolymers with nano titanium oxide and without titanium oxide (Memarizadeh et al. 2014a, b). Particles are 10–12 nm in size as observed by transmission electron microscopic analysis (TEM). Nano indoxacarb has shown better protection and stability under normal and UV light compared to its bulk counterpart. The nanoformulation has shown better insecticidal activity against fifth instar larvae of *Glyphodes pyloalis* in leaf dip bioassay.

2.6.4 Poly-lactic Acid Encapsulated Abam Nanoparticles

Another group of pesticide, abamectin (Abam), has been functionalized with biopolymer, poly-lactic acid (PLA) in order to enhance better adhesion to the foliage of cucumber and effective utilization of the application of the rate of Abam (Yu et al. 2017). Abam is a mixture of abamectins and utilized widely in order to control a wide range of pests of agronomic, vegetable and fruit crops. However, short half-time and restricted utilization rate are the major drawbacks which facilities to encapsulate this pesticide in a biodegradable polymeric matrix. That is why Abam-poly-lactic acid nanoparticle has shown better adhesion property on the cucumber foliage since it interacts strongly with the foliage with hydrogen, electrostatic and covalent bonds.

2.6.5 Thiamethoxam Nanoparticles

Another hydrophobic drug of neonicotinoid group, thiamethoxam, also has been encapsulated into cationic dendrimer in order to resolve the problem of low solubility and efficacy against Homoptera pests, such as aphids, leafhoppers, and planthoppers (Liu et al. 2015). Nanoparticles has been added to insect diet and fed to second instar larvae of *Heliothis armigera*. The mortality rate of nanoparticle treatment has elevated much fold compared to the control, which enables this novel nanoparticle to come out as a novel pesticide.

2.6.6 Essential Oil Encapsulated Poly Ethylene Glycol Nanoparticle

The essential oil has been encapsulated into biopolymer matrix of poly ethylene glycol (PEG) to evaluate entomotoxicity against stored product pests, *Tribolium castaneum* and *Rhizopertha dominica* (González et al. 2014). Nanoparticles are approximately 235 nm in size with less than 75% of loading efficiency. The pest population has been shown to decline significantly after nanoparticle treatment, however, no chemical deformities have been observed. Increased mortality rate after nanoparticle treatment develops mostly due to contact toxicity and altered nutritional physiology of the stored product pests (González et al. 2014).

Similarly, pyrethrin formulated water-in-oil emulsion has been tested against target pest, viz., *Aphis gossypii*, and non-target predators, larvae of *Coccinella septempunctata L.*, and fourth instar nymph of *Macrolophus pygmaeus* (Papanikolaou et al. 2017). Nanoformulation of pyrethrin has shown increased toxicity to the target pest, while retaining their neutral behavioral effect on the non-target pests.

2.6.7 Chitosan-Rotenone Nanoparticle

Another important herbicide, rotenone, also has loaded into hydrophilic oleoyl-carboxymethyl chitosan (Kamari et al. 2016). NPs become water-soluble with a critical micellar concentration of 0.096 mg/mL. The micellar structures are self-aggregated, spherical in shape and 35.5–66.4 nm in size with the encapsulation efficiency of 97%. The particle has been shown to have slow release rates and in a controlled manner, releasing within 50 h of loading.

2.6.8 Atrazine Nanoparticle

Atrazine is known to be used as a control agent against post-emergence of broad leaf and grassy weeds. The residual problem associated with rotenone accurses the application of the herbicide and restricts the choice of crop rotation. Application of silver nanoparticle (Ag nanoparticle) modified magnetite nanoparticle stabilized with carboxy methyl cellulose (CMC) has shown 88% degradation of atrazine under controlled environmental condition might be an effective remedial measure of bioremediation over a short span of time (Mehrazar et al. 2015). When Atrazine-loaded polycaprolactone nanoparticles have tested on mustard plants, there has been tenfold increase in effectiveness in the control of mustard plants (Oliveira et al. 2015). However, atrazine has been found to be toxic to non-target plant, i.e., maize. Nanoatrazine formulation has been tested in the maize plants in this regard, in order to know whether it shows the same phytotoxicity to the maize plants. There has been no or little change in the maize plants compared to control plants with post-nanoparticle treatment. Henceforth, it can be concluded that atrazine loaded poly-caprolactone nanoparticle does not impair any phytotoxicity to the non-target plant as it does to the target plant and thus might come out as a better alternative of weed control agent in the near future.

2.7 Delivery of Pesticides by Nanocarriers

Pesticide loaded nanocarriers are designed to be formulated in such a way that act as a "smart device for crop control management", which is expected to contain pesticide with plant binding properties that eventually resists drift loss. Pesticide should be released to the targeted pest in a controlled, timely manner. Pesticidal toxicity is mostly dependent upon the constituent and concentration of the active component of the pesticide and also the size of the particle entrapped in it. The nanocarrier can potentially modify the biological distribution and persistence of the chemical of choice, relative to the micro-sized particle of same constituents. Several studies have been reported in accordance with this hypothesis, where the pesticide of choice has been introduced to the nanosized biocompatible carrier in order to develop novel surface-tuned engineered nanopesticides. Metsulfuron methyl loaded pectin NPs has been reported to have greater herbicidal efficacy against *Chenopodium alba* plant (Kumar et al. 2017). Particles are 50–90 nm in size with zeta potential value of −35.9 mV. This nano-herbicide has found to be safe when tested in the cell lines. It has observed that application of metsulfuron methyl loaded pectin nanoparticles might minimize the herbicidal application with improved efficacy and better environmental safety.

2.7.1 Silica-Fipronil Nanoparticle

Fipronil, a systemic insecticide, restricts its application in the crop protection management because of its low solubility, volatility, and mobility in the soil leading to groundwater contamination. On the other hand, silica is amorphous biocompatible material that could be used as the nanocarrier for the pesticide delivery in the agricultural field. Fipronil loaded silica nanoparticles (8–44 nm) showed 73% encapsulation efficiency along with prolonged sustained release *in vitro* (Wibowo et al. 2014). When tested against subterranean termite, *Coptotermes acinaciformis*, nearly 100% mortality has been observed with prolonged insecticidal effect as compared with commercial fipronil. This resulted in improved area-wide control of the large population of termite colonies.

2.7.2 Solid Lipid Nanoparticles Loaded Atrazine and Simazine Nanoparticles

Solid lipid nanoparticles (SLNPs) loaded atrazine and simazine herbicides have been synthesized and characterized along with the evaluation of in vitro release kinetics and herbicidal efficacy analyses (de Oliveira et al. 2015). Newly synthesized NPs showed better solubility with prolonged release of active constituent without any premature degradation. Particles have found to be more effective in the target organism, *Raphanus raphanistrum,* and do not show any inhibitory effects to non-target plant, *Zea mays.*

2.7.3 α-Pinene and Linalool Loaded Silica Nanoparticles

Terpene compounds like α-pinene and linalool are known to have insect antifeedant and toxic properties to the herbivorous insects. They have been incorporated into silica nanoparticles in order to increase their shelf life, suspension stability, and bioactivity. They have been tested against tobacco cutworm (*Spodoptera litura*) and castor semilooper (*Achaea janata*) under laboratory condition (Rani et al. 2014).

2.7.4 Graphene Oxide -$Cu_{2-x}Se$-Chlorpyrifos Nanoparticle

Specialized nanoparticles like copper selenide grafted in nano-graphene oxide has been synthesized by the arrested precipitation method and the composite (graphene oxide-$Cu_{2-x}Se$) has been loaded with the pesticide chlorpyrifos by a solvent evaporation method (Sharma et al. 2017). The final product, i.e., graphene oxide -$Cu_{2-x}Se$-chlorpyrifos, has been designed to overcome the problem of drift loss by runoff. When sprayed on cauliflower leaf, the composite is shown to have more adhesion to the leaf surface (57%) compared to the commercial chp (34%) as observed from the standard gas chromatography- mass spectroscopic assay. This is mainly due to the strong carbon-carbon binding of the composite to the foliar surface of the cauliflower. Along with this, the irregularly layered lamellar surface of graphene oxide and the protuberance developed on the graphene oxide layer at time of coating might support the composite to anchor for longer time on the leaf surface. In addition, the piercing effect by sharp graphene oxide surface provides an extra support to resist runoff of the chlorpyrifos from the leaf surface. The nanocomposite has shown greater mortality (80%) compared to conventional chp (50%) when tested against diuron pest, *Pieris rapae.* After treatment, the nanoparticle treated larvae have shown reduction in size and developmental disorder followed by death. The reason for such a high entomotoxicity may be due to the ability of the nanocomposite to withstand the runoff, enhanced drug uptake and controlled chlorpyrifos delivery at the site of the action.

2.7.5 Paraquat Loaded Chitosan-Sodium Tripolyphosphate Nanoparticle

Another group of the researcher has developed a nano-vehicle to deliver paraquat, a fast-acting non-selective contact herbicide, in targeted maize plant and non-targeted mustard plant (Grillo et al. 2014). Paraquat has bound with chitosan-sodium tripoly-phosphate nanocarrier by electrostatic interaction in such a way that a fraction of the paraquat remains unassociated with the chitosan-sodium tripolyphosphate nano-matrix. This free part of the component provides initial eradication of the weeds with subsequent control provided by slower release from nanoparticles. The synthe-sized nanoparticles are more or stable, spherical in shape and 390–420 nm in size. When tested in maize plants, leaf necrosis has been observed (Grillo et al. 2014). This may be due to the strong interaction of active part of the nanocomposite to the targeted site of the leaf. However, a less pronounced effect of the nanocomposite has been shown in the non-target mustard plant, which may indicate species-specific characteristics for the nanocomposite.

2.7.6 Beauvercin Loaded Chitosan Nanoparticle

An insecticidal cyclodepsipeptide, beauvercin, has been loaded to chitosan nanopar-ticle by ionic gelation method (Bharani et al. 2014) to test its efficacy against *Spodoptera litura* under the controlled environment. Particles are 160–230 nm in size with loading efficiency of 82% and entrapment efficiency of 85%. When tested against the targeted pest, nanoparticle showed the detrimental effect on the entire life cycle, though early larval instar stage has found to be the most susceptible one.

2.7.7 Botanical Insecticides

Plant secondary metabolites like alkaloids, phenolics, and terpenoids, have been used as plant defense agents and/or insect repellents for many years. Because of the complex structure, botanical extracts exhibit toxicity to the target organisms impair-ing developmental changes including sterility, reduction in growth and altered behaviour (de Oliveira et al. 2014). There are several reports showcasing the ento-motoxic nature of the botanic extracts of siam weed, tobacco, castor oil plants, *Azadirachta indica etc* (Boursier et al. 2011; Gomez-Estaca et al. 2012; Amoabeng et al. 2014; Kim and Lee 2014). These have been used from the ancient ages against various pests, like, *Plutella xylostella, Brevicoryne brassiceae, Bemisia tabaci, Spodoptera litoralis*. However, a majority of the plant metabolites are been encapsulated in micro- and/or nano domain, since these metabolites are poor in the physicochemical stability, high volatility, and thermal instability.

Keeping this in mind, several research groups have aimed to design micro and/or nanocapsules of secondary plant metabolites in order to make it environmental friendly yet target specific. Entrapment of botanical extracts into the nanocapsules potentially can improve biological distribution and persistence of the active component at the site of action (Forim et al. 2013). Since size can amplify the pharmacokinetics of the particle in terms of adsorption, uptake, and translocation, metabolism, and excretion, encapsulation of the botanical extract is believed to preclude entomotoxicity to the target pest (Da Costa et al. 2014).

2.7.7.1 Azadiractin Nanoparticle

Farim et al. (2013) has developed poly(ε-caprolactone) (PCL) encapsulated azadiractin nanoparticle and tested against *P xylostella* by spray drying process. Particles are 245 nm in size, spherical in shape with 98% of encapsulation efficiency. The synthesized product has been found to be stable under UV radiation and has shown 100% mortality against treated diamond black moth. Similarly, Da Costa et al. (2014) has also designed azadiractin encapsulated nanocapsules and tested against bean weevil, *Zabrotes subfasciatus*. Nanoparticle has shown to have only 20% degradation after 14 days and to be more effective and stable than the commercial product.

2.7.7.2 Rotenone Nanoparticle

Rotenone, a botanical insecticide extracted from the roots of rhizomes of leguminous family, has been restricted in use because of its water insolubility, instability under UV light and extreme toxicity to non-target organisms like fishes. Martin et al. (2013) has encapsulated rotenone into biodegradable polymers like sodium alginate, polyethylene glycol, and poly-vinyl pyrrolidone. Particles are more or less spherical in shape, 600–1.5 μm in size with 98–100% of encapsulation efficiency.

2.7.7.3 Carvacrol Nanoparticle

Carvacrol, a phenolic monoterpenoid extracted from the thymes and oregano, is known to have bactericidal and insecticidal activities. It has been loaded into chitosan-pentasodium tripolyphosphate matrix in order to investigate the release kinetics prior to analysis (Woranuch and Yoksan 2013). Encapsulated particles are 40–80 nm in size, spherical in shape, cationic in nature. They show Fickian diffusion mechanism, since 53%, 23%, 33% of active components are released from the nanoencapsulation complexes for acetate, phosphate and alkaline phosphate buffer respectively as demonstrated in 30 days of in vitro study.

2.7.7.4 Eugenol Nanoparticle

Eugenol is extracted from plants of Myrtaceae family like *Syzygium aromaticum*. Eugenol has encapsulated into PCL, β-CD, and 2-hydroxypropyl- β-CD (2-HP-β-CD) irrespectively (Choi et al. 2009). Particles are 320 nm in size with the encapsulation efficiency of 89.1–100%.

2.7.8   Encapsulation of Essential Oil

Lai et al. (2006) has loaded essential oil of *Artemisia arborescens* into solid lipid nanoparticles. Particles are 200 nm in size and stable for 2 months. Solid lipid nanoparticles help to reduce the evaporation of the essential oil which might increase the potential of the oil for use as a botanical pesticide in agriculture. Similarly, Yang et al. (2009) has encapsulated essential oil of garlic into polyethylene glycol via fusion-dispersion method and tested against adult beetles under controlled conditions. Nanoparticles are spherical in shape and 240 nm in size with 80% of encapsulation efficiency. Nanoparticles have shown better entomotoxicity (80% mortality) compared to essential oil (11% mortality). Likewise, Abreu et al. (2012) have developed chitosan-pepper-rosmarin encapsulation complex (335–558 nm). When tested against *Aedes aegypti*, 75% mortality has been observed after 48 h of treatment, whereas 90% mortality has documented after 72 h of treatment.

2.8   Elemental and Specialized Nanoparticles as Nanopesticides

The main characteristics of nanopesticides are the small size and high surface area by volume ratio which allow chemical reactions to be performed more efficiently and precisely in the biological domain. These promote a high diffusion rate of an active component of the nanopesticide; enhanced elasticity; surface free energy; and compressibility (even at elevated temperature) (Chen et al. 2009). Simultaneously, nanoparticle-mediated adsorption enhances the crystal growth and decreases surface free energy resulting the nanopesticides to be more stable (Zhang et al. 2009). Therefore, several specialized nanopesticides with novel properties have been synthesized within the last 5 years in order to reduce the high agricultural burden.

2.8.1 Silica Nanoparticle

Amorphous silica nanoparticle is considered as a biocompatible and biosafe material as approved by US Food and Drug Administration (Ayoub et al. 2017). Therefore it could be used as a good pest control agent against cotton leaf worm (*Spodoptera littoralis*). Particles have shown better entomotoxicity compared to commercial silica, as tested following topical application, residual or surface contact and immersion feeding bioassay protocols. A high mortality rate has been observed after nanoparticle treatment; 29.6% after 1st day treatment to 96% after the 3rd day of exposure. Nanoparticles treated pupa have shown malformed structures and even adult emergence has found to be restricted (Ayoub et al. 2017). The main reason for such a response is mainly due to impairment of the digestive tract and surface enlargement of the insect integument resulting in dehydration. Reduction in both carbohydrate and protein content of the nanoparticles treated larvae results in pupal malformation; as a consequence of this, treated insects are unlikely to be genetically selected.

2.8.2 Silver Nanoparticles and Nanoalumina

Silver nanoparticles (14–28 nm), synthesized from pomegranate peel extract, have tested against *Spodoptera litura* as well as on SF-21 cell line (Bharani and Namasivayam 2017). Effective growth reduction in dose-dependent manner has been observed in nanoparticles-treated cell line. Besides, the mortality rate of silver nanoparticles treated pests has been found to be varied between 16.4% and 86.4%, 8.41% and 78.4%, 4.3% and 61.3%, 3.7% and 58.2% for the second, third, fourth, fifth and sixth instars of larvae respectively, at 10–100 μg of silver nanoparticles concentrations. A similar type of dose dependency has been observed when a mixture of silver nanoparticles, zinc oxide (ZnO) nanoparticles and titanium oxide nanoparticles (TiO$_2$ NP) has been applied to the *S litura* (Rouhani et al. 2012). There has been a reduced larval, pupal period, adult emergence and adult longevity with respect to control. A similar effect has been seen when nano alumina (150 nm), synthesized by the glycine-nitrate combustion process, has been tested on store grain pests, viz., *Sitophilus oryzae, Rhyzopertha dominica*. A greater mortality rate has been observed in *S oryzae* than *R dominica* (Buteler et al. 2015). A similar pattern of results has been observed, when silver nanoparticle, biosynthesized from the foliar extract of *Ficus religiosa* and *F benghalensis,* has been tested on insecticidal resistant gram caterpillar, *Helicoverpa armigera* (Kantrao et al. 2017). A decrease in larval weight and longevity has been observed after nanoparticles treatment. The silver nanoparticles have shown to inhibit the gut protease activity at higher concentration. Silver nanoparticle facilitates the conformational changes in the gut protease due to exposure of more tyrosine residue into the gut digestive cocktails leading to no substrate available for the binding of the protease (Kantrao et al. 2017). This might lead to the innovation of new pest control management strategy in the course of sustainable agricultural practices.

2.8.3 Iron Oxide Nanoparticle

Iron oxide nanoparticles have tested on Bt-transgenic cotton plant along with the conventional cotton plant for 10 days to understand the physiological and insecticidal activities of the nanoparticles on the targeted model plants (Nhan et al. 2016). Nanoparticles treated transgenic plants have shown reduction in plant height and root length, whereas nanoparticle has promoted root hair growth and biomass in non-transgenic plants. Interestingly, Bt toxin content in both leaves and roots of the transgenic plants has been found to be elevated after nanoparticles treatment. Iron oxide nanoparticles are taken up by the roots and subsequently transported to the shoot in both transgenic and non-transgenic plants as observed in transmission electron microscope.

Calcium carbonate nanoparticles have come out as a potent nanopesticide, as it has shown the detrimental effect on California red scale (*Aonidiella auranti*) and oriental fruit flies (*Bactrocera dorsalis*) with respect to colloidal calcium carbonate (Hua et al. 2015). Nickel nanoparticles have isolated from the methanolic extract of *Cocos nucifera* and tested against *Callasobruchus maculate* (Elango et al. 2016). About 97.31% of mortality has been observed and the result is favourably better than commercially available standard azadiractin.

2.8.4 Gold-Ferbam Nanoparticles

Ferbam, a non-systematic pesticide, has been conjugated with gold nanoparticles and sprayed on the upper surface of the tea leaves (Hou et al. 2016).It was observed that 30 nm gold nanoparticle-ferbam penetrates more rapidly compared to the commercial ferbam to a depth of 190 μm.

Herein, we have discussed the unprecedented advantages of nanopesticides in the agricultural perspectives. However, little attention has been paid to the mode of action studies of these nanoparticles with respect to insect physiology. In the invertebrate system, pH of the gut digestive cocktail ranges from 6 to 11 (Khandelwal et al. 2015). This is the reason why the majority of the commercial pesticides fail to retain their insecticidal properties for a long time. They are prone to degrade in the milieu leading to the development of resistant to the targeted pest. 'Smart' pH-responsive targeted nanopesticide can easily cross the barrier of pH gradient inside the insect gut and resides in the oxidation-reduction state in the gut environment for a certain period of time (Khandelwal et al. 2015). Inside the gut, nanoparticle inevitably interacts with the serine protease (active at alkaline pH) or cysteine protease (active in acidic pH) and forms a "corona" like structure that protects from the undesirable degradation (Nel et al. 2009). Physicochemical properties of nanoparticles (size, shape, surface charge, hydrophobicity and surface chemistry) mostly affect the selectivity and specificity of the nanoparticles-protein interaction inside the gut environment. As a whole, this interaction of nanoparticles with the protease enzymes within the gut microenvironment determines the release, retention, and toxicity of the nanopesticides.

2.9 Actual Concerns about Nanopesticides

The main concern with these new-age pesticides arises whether engineered nanopesticides cause specific interaction within the ecosystem and therefore renders adverse effect in plants as well as on the ecosystem and/or disturbs the equilibrium of the food chain (Kah et al. 2013; Dev et al. 2018; Kaphle et al. 2018). Presence of engineered nanomaterials in the human tissues might develop adverse effects like developmental abnormalities, infertility, and disturbance in brain and muscle activity, stress and endocrine-related problems in mammalian systems (Meredith et al. 2016; Kumar et al. 2017). Proper physiochemical characterizations have to be documented for the safe manufacturing practices and product testing prior to the commercialization in order to avoid any undesirable side-effects on the environment as well as human health (Sardoiwala et al. 2018; Chowdhury et al. 2017). Therefore, it is essential to quantify as well as qualify the risks associated with the application of pesticides in the environment beforehand. Rules and regulation of pest management systems must be reformed in respect to newly developed nano-pesticidal compounds in the light of: (i) potential risk of the newly developed compounds to the mammalian system and environment, (ii) if not the existing model system is adequate to analyse all the criteria, new model system and/or new protocols and guidelines must be introduced, (iii) detailed characterization of the compound must be published by the agrochemical industries, so that farmers should be aware of the chemical and take precautions before its application. It is very important to note that engineered nanomaterials often undergo chemical modifications on the basis of dispersion and agglomeration over time. However, the concentration of the nanoparticles as well as several environmental factors like pH, ionic strength of the medium play important role in it. Thus, documentation of physicochemical characterization of nanopesticides at different levels of the environmental life cycle and detailed analysis of fate and effect studies are mandatory for the betterment of agricultural sustainable practices.

2.10 Conclusion

Introduction of engineered nanomaterials as nanopesticides into the agricultural practice provides unprecedented advantages over past crop protection tactics. Whether traditional methods of crop protection (crop rotation) or application of different types of plant protection products like chemical pesticides, botanical extracts, biological control agents and/or transgenic and genetic engineering approaches, agricultural pests inevitably quell the challenges and revert back to their original state of action. In accordance with the current scenario, it is highly recommended to design a novel class of compounds with a unique mode of action so that they can circumscribe all the hindrances reinforced by the agricultural pest. In this regard, nanopesticides have shown detrimental effects against target pathogens and protects the agricultural crops from further attack pathogen. However, it is essential to note that toxicological aspect of the newly synthesized pesticides should be carefully evaluated prior to commercialization. Integration of nanopesticides into the multiple pest suppression and crop protection techniques must be carried out for the development of sustainable long-term crop protection management.

Acknowledgement This work was generously supported by major grant from the Food security-MHRD, Government of India (Grant No: 4-25/2013-TS-1) for providing financial support.

References

Abreu FOMS, Oliveira EF, Paula HCB, de Paula RCM (2012) Chitosan/cashew gum nanogels for essential oil encapsulation. Carbohydr Polym 89:1277–1282. https://doi.org/10.1016/j. carbpol.2012.04.048

Amoabeng BW, Gurr GM, Gitau CW, Stevenson PC (2014) Cost: benefit analysis of botanical insecticide use in cabbage: implications for smallholder farmers in developing countries. Crop Prot 57:71–76. https://doi.org/10.1016/j.cropro.2013.11.019

Atta S, Bera M, Chattopadhyay T, Paul A, Ikbal M, Maiti MK, Singh NDP (2015) Nano-pesticide formulation based on fluorescent organic photoresponsive nanoparticles: for controlled release of 2,4-D and real time monitoring of morphological changes induced by 2,4-D in plant systems. RSC Adv 5:86990–86996. https://doi.org/10.1039/C5RA17121K

Ayoub HA, Khairy M, Rashwan FA, Abdel-Hafez HF (2017) Synthesis and characterization of silica nanostructures for cotton leaf worm control. J Nanostruct Chem 7:91–100. https://doi. org/10.1007/s40097-017-0229-2

Bakshi S, He Z, Harris WG (2015) Natural nanoparticles: implications for environment and human health. Crit Rev Environ Sci Technol 45:861–904. https://doi.org/10.1080/10643389.2014.92 1975

Bharani RSA, Namasivayam SKR (2017) Biogenic silver nanoparticles mediated stress on developmental period and gut physiology of major lepidopteran pest *Spodoptera litura* (Fab.) (Lepidoptera: Noctuidae)—an eco-friendly approach of insect pest control. J Environ Chem Eng 5:453–467

Bharani RSA, Namasivayam SKR, Shankar SS (2014) Biocompatible chitosan nanoparticles incorporated pesticidal protein Beauvericin (Csnp-Bv) preparation for the improved Pesticidal activity against major groundnut defoliator *Spodoptera Litura* (Fab.) (Lepidoptera; Noctuidae). Int J ChemTech Res 6:5007–5012

Boursier CM, Bosco D, Coulibaly A, Negre M (2011) Are traditional neem extract preparations as efficient as a commercial formulation of azadirachtin a? Crop Prot 30:318–322. https://doi. org/10.1016/j.cropro.2010.11.022

Buteler M, Sofie SW, Weaver DK, Driscoll D, Muretta J, Stadler T (2015) Development of nanoalumina dust as insecticide against *Sitophilus oryzae* and *Rhyzopertha dominica*. Int J Pest Manage 61:80–89. https://doi.org/10.1080/09670874.2014.1001008

Chandra S, Pradhan S, Mitra S, Patra P, Bhattacharya A, Pramanik P, Goswami A (2014) High throughput electron transfer from aminated carbon dots to the chloroplast: a rationale of enhanced photosynthesis. Nanoscale 6:3647–3655. https://doi.org/10.1039/c3nr06079a

Chen B, Zhang H, Dunphy-Guzman KA, Spagnoli D, Kruger MB, Muthu DVS, Kunz M, Fakra S, Hu JZ, Guo QZ, Banfield JF (2009) Size-dependent elasticity of nanocrystalline titania. Phys Rev B 79:125406. https://doi.org/10.1103/PhysRevB.79.125406

Chhipa H (2017) Nanofertilizers and nanopesticides for agriculture. Environ Chem Lett 15:15–22. https://doi.org/10.1007/s10311-016-0600-4

Choi MJ, Soottitantawat A, Nuchuchua O, Min SG, Ruktanonchai U (2009) Physical and light oxidative properties of eugenol encapsulated by molecular inclusion and emulsion–diffusion method. Food Res Int 42:148–156. https://doi.org/10.1016/j.foodres.2008.09.011

Choudhury SR, Pradhan S, Goswami A (2012) Preparation and characterisation of acephate nano-encapsulated complex. Nanosci Methods 1:9–15. https://doi.org/10.1080/17458080.2010.533443

Chowdhury P, Gogoi M, Borchetia S, Bandyopadhyay T (2017) Nanotechnology applications and intellectual property rights in agriculture. Environ Chem Lett 15:413–419. https://doi.org/10.1007/s10311-017-0632-4

Christofoli M, Costa ECC, Bicalho KU, Domingues VC, Peixoto MF, Alves CCF, Araújo WL, Cazal CM (2015) Insecticidal effect of nanoencapsulated essential oils from *Zanthoxylum rhoifolium* (Rutaceae) in *Bemisia tabaci* populations. Ind Crop Prod 70:301–308. https://doi.org/10.1016/j.indcrop.2015.03.025

Da Costa JT, Forim MR, Costa ES, De Souza JR, Mondego JM, Boiça Junior AL (2014) Effects of different formulations of neem oil-based products on control *Zabrotes subfasciatus* (Coleoptera: Bruchidae) on beans. J Stored Prod Res 56:49–53. https://doi.org/10.1016/j.jspr.2013.10.004

Dev A, Srivastava AK, Karmakar S (2018) Nanomaterial toxicity for plants. Environ Chem Lett 16:85–100. https://doi.org/10.1007/s10311-017-0667-6

Dubey A, Mailapalli DR (2016) Nanofertilisers, nanopesticides, nanosensors of pest and nanotoxicity in agriculture. Sustain Agric Rev. Chapter 07 19:307–330

Elango G, Roopan SM, Dhamodaran KI, Elumalai K, Al-Dhabi NA, Arasu MV (2016) Spectroscopic investigation of biosynthesized nickel nanoparticles and its larvicidal, pesticidal activities. J Photochem Photobiol B 162:162–167. https://doi.org/10.1016/j.jphotobiol.2016.06.045

Farim MR, Costa ES, da Silva MF, Das GF, Fernandes JB, Mondego JM, Boiça Junior AL (2013) Development of a new method to prepare nano-/microparticles loaded with extracts of *Azadirachta indica*, their characterization and use in controlling *Plutella xylostella*. J Agric Food Chem 61:9131–9139. https://doi.org/10.1021/jf403187y

Feng B, Zhang Z (2011) Carboxymethy chitosan grafted Ricinoleic acid group for Nanopesticide carriers. Adv Mater Res 36:1783–1788. https://doi.org/10.4028/www.scientific.net/AMR.236-238.1783

Ghafariyan MH, Malakouti MJ, Dadpour MR, Stroeve P, Mahmoudi M (2013) Effects of magnetite nanoparticles on soybean chlorophyll. Environ Sci Technol 47:10645–10652. https://doi.org/10.1021/es402249b

Ghormade V, Deshpande MV, Paknikar KM (2011) Perspectives for nano-biotechnology enabled protection and nutrition of plants. Biotechnol Adv 29:792–803. https://doi.org/10.1016/j.biotechadv.2011.06.007

Gillet FX, Garcia RA, Macedo LLPP, Albuquerque EVS, Silva MCM, Grossi-de-Sa MF (2017) Investigating engineered ribonucleoprotein particles to improve oral RNAi delivery in crop insect pests. Front Physiol 8:1–14. https://doi.org/10.3389/fphys.2017.00256

Gomez-Estaca J, Balaguer MP, Gavara R, Hernandez-Munoz P (2012) Formation of zein nanoparticles by electrohydrodynamic atomization: effect of the main processing variables and suitability for encapsulating the food coloring and active ingredient curcumin. Food Hydrocoll 28:82–91. https://doi.org/10.1016/j.foodhyd.2011.11.013

González JOW, Gutiérrez MM, Ferrero AA, Band BB (2014) Essential oils nanoformulations for stored-product pest control – characterization and biological properties. Chemosphere 100:130–138. https://doi.org/10.1016/j.chemosphere.2013.11.056

Grillo R, Pereira AES, Nishisaka CS, de Lima R, Oehlke K, Greiner R, Fraceto LF (2014) Chitosan/tripolyphosphate nanoparticles loaded with paraquat herbicide: an environmentally safer alternative for weed control. J Hazard Mater 278:163–171. https://doi.org/10.1016/j. jhazmat.2014.05.079

Guo P, Sikdar D, Huang X, Si KJ, Xiong W, Gong S, Yap LW, Premaratne M, Cheng W (2015) Plasmonic core–shell nanoparticles for SERS detection of the pesticide thiram: size- and shape-dependent Raman enhancement. Nanoscale 7:2862–2868. https://doi.org/10.1039/C4NR06429A

Gupta PK (2007) Toxicity of herbicides. In: Gupta RC (ed) Veterinary toxicology: basic and clinic principles. Elsevier, Boston, pp 567–586

Hou R, Zhang Z, Pang S, Yang T, Clark JM, He L (2016) Alteration of the nonsystemic behavior of the pesticide ferbam on tea leaves by engineered gold nanoparticles. Environ Sci Technol 50:6216–6223. https://doi.org/10.1021/acs.est.6b01336

Hua K, Wang H, Chung R, Hsu J (2015) Calcium carbonate nanoparticles can enhance plant nutrition and insect pest tolerance. J Pestic Sci 40:208–213. https://doi.org/10.1584/jpestics. D15-025

Joga MR, Zotti MJ, Smagghe G, Christiaens O (2016) RNAi efficiency, systemic properties, and novel delivery methods for pest insect control: what we know so far. Front Physiol 7:1–14. https://doi.org/10.3389/fphys.2016.00553

Kah M, Hofmann T (2014) Nanopesticide research: current trends and future priorities. Environ Int 63:224–235. https://doi.org/10.1016/j.envint.2013.11.015

Kah M, Beulke S, Tiede K, Hofmann T (2013) Nanopesticides: state of knowledge, environmental fate, and exposure modeling. Crit Rev Environ Sci Technol 43:1823–1867. https://doi.org/10. 1080/10643389.2012.671750

Kamari A, Aljafree NFA, Yusoff SNM (2016) Oleoyl-carboxymethyl chitosan as a new carrier agent for the rotenone pesticide. Environ Chem Lett 14:417–422. https://doi.org/10.1007/ s10311-016-0550-x

Kang TF, Wang F, Lu LP, Zhang Y, Liu TS (2010) Methyl parathion sensors based on gold nanoparticles and Nafion film modified glassy carbon electrodes. Sensor Actuat B-Chem 145:104–109. https://doi.org/10.1016/j.snb.2009.11.038

Kantrao S, Ravindra MA, Akbar SMD, Jayanthi PDK, Venkataraman A (2017) Effect of biosynthesized silver nanoparticles on growth and development of Helicoverpa armigera (Lepidoptera: Noctuidae): interaction with midgut protease. J Asia Pac Entomol 20:583–589. https://doi. org/10.1016/j.aspen.2017.03.018

Kaphle A, Navya PN, Umapathi A, Daima HK (2018) Nanomaterials for agriculture, food and environment: applications, toxicity and regulation. Environ Chem Lett 16:43–58. https://doi. org/10.1007/s10311-017-0662-y

Katoch R, Sethi A, Thakur N, Murdock LL (2013) RNAi for insect control: current perspective and future challenges. Appl Biochem Biotechnol 171:847–873. https://doi.org/10.1007/ s12010-013-0399-4

Khandelwal N, Doke DS, Khandare JJ, Jawale PV, Biradar AV, Giri AP (2015) Bio-physical evaluation and in vivo delivery of plant proteinase inhibitor immobilized on silica nanospheres. Colloids Surf B Biointerfaces 130:84–92. https://doi.org/10.1016/j.colsurfb.2015.03.060

Kim SI, Lee DW (2014) Toxicity of basil and orange essential oils and their components against two coleopteran stored products insect pests. J Asia Pac Entomol 17:13–17. https://doi. org/10.1016/j.aspen.2013.09.002

Kim J, Oh Y, Yoon H, Hwang I, Chang Y (2015) Iron nanoparticle-induced activation of plasma membrane H+-ATPase promotes stomatal opening in Arabidopsis thaliana. Environ Sci Technol 49:1113–1119. https://doi.org/10.1021/es504375t

Kitzmann P, Schwirz J, Schmitt-Engel C, Bucher G (2013) RNAi phenotypes are influenced by the genetic background of the injected strain. BMC Genome:14. https://doi. org/10.1186/1471-2164-14-5

Kookana RS, Boxall ABA, Reeves PT, Ashauer R, Beulke S, Chaudhry Q, Cornelis G, Fernandes TF, Gan J, Kah M, Lynch I, Ranville J, Sinclair C, Spurgeon D, Tiede K, Van den Brink PJ

(2014) Nanopesticides: guiding principles for regulatory evaluation of environmental risks. J Agric Food Chem 62:4227–4240. https://doi.org/10.1021/jf500232f

Kumar S, Chauhan N, Gopal M, Kumar R, Dilbaghi N (2015) Development and evaluation of alginate–chitosan nanocapsules for controlled release of acetamiprid. Int J Biol Macromol 81:631–637. https://doi.org/10.1016/j.ijbiomac.2015.08.062

Kumar S, Bhanjana G, Sharma A, Dilbaghi N, Sidhu MC, Kim K (2017) Development of nanoformulation approaches for the control of weeds. Sci Total Environ 586:1272–1278. https://doi.org/10.1016/j.scitotenv.2017.02.138

Kumaravel A, Chandrasekaran M (2011) A biocompatible nano TiO$_2$/nafion composite modified glassy carbon electrode for the detection of fenitrothion. J Electroanal Chem 650:163–170. https://doi.org/10.1016/j.jelechem.2010.10.013

Kuswandi B (2018) Nanobiosensor approaches for pollutant monitoring. Environ Chem Lett. https://doi.org/10.1007/s10311-018-00853-x

Lai F, Wissing SA, Müller RH, Fadda AM (2006) Artemisia arborescens L. essential oil-loaded solid lipid nanoparticles for potential agricultural application: preparation and characterization. AAPS PharmSciTech 7:E10–E18. https://doi.org/10.1208/pt070102

Lichtfouse E, Schwarzbauer J, Robert D (2005) Environmental chemistry green chemistry and pollutants in ecosystems. Springer, Berlin. https://doi.org/10.1007/b137751

Lin D, Xing B (2007) Phytotoxicity of nanoparticles: inhibition of seed germination and root growth. Environ Pollut 150:243–250. https://doi.org/10.1016/j.envpol.2007.01.016

Lin P, Lin S, Wang PC, Sridhar R (2014) Techniques for physicochemical characterization of nanomaterials. Biotechnol Adv 32:711–726. https://doi.org/10.1016/j.biotechadv.2013.11.006

Liu S, Yuan L, Yue X, Zheng Z, Tang Z (2008) Recent advances in nanosensors for organophosphate pesticide detection. Adv Powder Technol 19:419–441. https://doi.org/10.1016/S0921-8831(08)60910-3

Liu W, Yao J, Cai M, Chai H, Zhang C, Sun J, Chandankere R, Masakorala K (2014) Synthesis of a novel nanopesticide and its potential toxic effect on soil microbial activity. J Nanopart Res 16:1–13. https://doi.org/10.1007/s11051-014-2677-7

Liu X, He B, Xu Z, Yin M, Yang W, Zhang H, Cao J, Shen J (2015) A functionalized fluorescent dendrimer as a pesticide nanocarrier: application in pest control. Nanoscale 7:445–449. https://doi.org/10.1039/C4NR05733C

Mamta B, Rajam MV (2017) RNAi technology: a new platform for crop pest control. Physiol Mol Biol Plants. https://doi.org/10.1007/s12298-017-0443-x

Martin L, Liparoti S, Della Porta G, Adami R, Marqués JL, Urieta JS (2013) Rotenone coprecipitation with biodegradable polymers by supercritical assisted atomization. J Supercrit Fluids 81:48–54. https://doi.org/10.1016/j.supflu.2013.03.032

Mehrazar E, Rahaie M, Rahaie S (2015) Application of nanoparticles for pesticides, herbicides, fertilisers and animals feed management. Int J Nanoparticles 8:1–19. https://doi.org/10.1504/IJNP.2015.070339

Memarizadeh N, Ghadamyari M, Adeli M, Talebi K (2014a) Linear-dendritic copolymers/indoxacarb supramolecular systems: biodegradable and efficient nano-pesticides. Environ Sci: Processes Impacts 16:2380–2389. https://doi.org/10.1039/c4em00321g

Memarizadeh N, Ghadamyari M, Adeli M, Talebi K (2014b) Preparation, characterization and efficiency of nanoencapsulated imidacloprid under laboratory conditions. Ecotoxicol Environ Saf 107:77–83. https://doi.org/10.1016/j.ecoenv.2014.05.009

Meredith AN, Harper B, Harper SL (2016) The influence of size on the toxicity of an encapsulated pesticide: a comparison of micron- and nano-sized capsules. Environ Int 6:68–74. https://doi.org/10.1016/j.envint.2015.10.012

Nair RR, Blake P, Grigorenko AN, Novoselov KS, Booth TJ, Stauber T, Peres NMR, Geim AK (2008) Fine structure constant defines visual transparency of graphene. Science 320:1308. https://doi.org/10.1126/science.1156965

Nel AE, Mädler L, Velegol D, Xia T, Hoek EM, Somasundaran P, Klaessig F, Castranova V, Thompson M (2009) Understanding biophysicochemical interactions at the nanobio interface. Nat Mater 8:543–557. https://doi.org/10.1038/nmat2442

Nhan LV, Ma C, Rui Y, Cao W, Deng Y, Liu L, Xing B (2016) The effects of Fe_2O_3 nanoparticles on physiology and insecticide activity in non-transgenic and Bt-transgenic cotton. Front Plant Sci 6:1–12. https://doi.org/10.3389/fpls.2015

Noji T, Suzuki H, Gotoh T, Iwai M, Ikeuchi M, Tomo T, Noguchi T (2011) Photosystem II-gold nanoparticle conjugate as a nanodevice for the development of artificial light-driven water-splitting systems. J Phys Chem Lett 2:2448–2452. https://doi.org/10.1021/jz201172y

Nuruzzaman M, Rahman MM, Liu Y, Naidu R (2016) Nanoencapsulation, nano-guard for pesticides: a new window for safe application. J Agric Food Chem 64:1447–1483. https://doi.org/10.1021/acs.jafc.5b05214

de Oliveira JL, Campos EVR, Bakshi M, Abhilash PC, Fraceto LF (2014) Application of nanotechnology for the encapsulation of botanical insecticides for sustainable agriculture: prospects and promises. Biotechnol Adv 32:1550–1561. https://doi.org/10.1016/j.biotechadv.2014.10.010

de Oliveira JL, Campos EVR, da Silva CMG, Pasquoto T, Lima R, Fraceto LF (2015) Solid lipid nanoparticles co-loaded with simazine and atrazine: preparation, characterization, and evaluation of herbicidal activity. J Agric Food Chem 63:422–432. https://doi.org/10.1021/jf5059045

Oliveira HC, Stolf-Moreira R, Martinez CBR, Sousa GFM, Grillo R, deJesus MB, Fraceto LF (2015) Evaluation of the side effects of poly(epsilon-caprolactone) nanocapsules containing atrazine toward maize plants. Front Chem 3:1–9. https://doi.org/10.3389/fchem.2015.00061

Pandey S, Giri K, Kumar R, Mishra G, Rishi RR (2016) Nanopesticides: opportunities in crop protection and associated environmental risks. Proc Natl Acad Sci India Sect B Biol Sci. https://doi.org/10.1007/s40011-016-0791-2

Pang Z, Hu CJ, Fang RH, Luk BT, Gao W, Wang F, Chuluun E, Angsantikul P, Thamphiwatana S, Lu W, Jiang X, Zhang L (2015) Detoxification of organophosphate poisoning using nanoparticle bioscavengers. ACS Nano 9:6450–6458. https://doi.org/10.1021/acsnano.5b02132

Papanikolaou NE, Kalaitzaki A, Karamaouna F, Michaelakis A, Papadimitriou V, Dourtoglou V, Papachristo DP (2017) Nano-formulation enhances insecticidal activity of natural pyrethrins against Aphis gossypii (Hemiptera: Aphididae) and retains their harmless effect to non-target predators. Environ Sci Pollut Res. https://doi.org/10.1007/s11356-017-8596-2

Pereira LC, de Souza AO, Franco Bernardes MF, Pazin M, Tasso MJ, Pereira PH, Dorta DJ (2015) A perspective on the potential risks of emerging contaminants to human and environmental health. Environ Sci Pollut Res Int 22:13800–13823. https://doi.org/10.1007/s11356-015-4896-6

Pradhan S, Patra P, Das S, Chandra S, Mitra S, Dey K, Akbar S, Palit P, Goswami A (2013a) A detailed molecular biochemical and biophysical study of manganese nanoparticles, a new nano modulator of photochemistry on plant model, Vigna radiata and its biosafety assessment. Environ Sci Technol 47:13122–13131. https://doi.org/10.1021/es402659t

Pradhan S, Roy I, Lodh G, Patra P, Choudhury SR, Samanta A, Goswami A (2013b) Entomotoxicity and biosafety assessment of PEGylated acephate nanoparticles: a biologically safe alternative to neurotoxic pesticides. J Environ Sci Health B 8:559–569. https://doi.org/10.1080/0360123 4.2013.774891

Pradhan S, Patra P, Mitra S, Dey KK, Jain S, Sarkar S, Roy S, Palit P, Goswami A (2014) Manganese nanoparticle: impact on non-nodulated plant as a potent enhancer in nitrogen metabolism and toxicity study both in vivo and in vitro. J Agric Food Chem 62:8777–8785. https://doi.org/10.1021/jf502716c

Pradhan S, Patra P, Mitra S, Dey KK, Basu S, Chandra S, Palit P, Goswami A (2015) Physiological, biochemical and biophysical assessment in Vigna radiata by CuNP nanochain array: a new approach for crop improvement. J Agric Food Chem 63:2606–2617. https://doi.org/10.1021/jf504614w

Qu F, Zhou X, Xu J, Li H, Xie G (2009) Luminescence switching of CdTe quantum dots in presence of p-sulfonatocalix[4] arene to detect pesticides in aqueous solution. Talanta 78:1359–1363. https://doi.org/10.1016/j.talanta.2009.02.013

Rahaie M, Rahaie S (2015) Application of nanoparticles for pesticides, herbicides, fertilizers and animals feed management. Int J Nanoparticles 8:1–19. https://doi.org/10.1504/IJNP.2015.070339

Rai M, Ingle A (2012) Role of nanotechnology in agriculture with special reference to management of insect pests. Appl Microbiol Biotechnol 94:287–293. https://doi.org/10.1007/s00253-012-3969-4

Rani UP, Madhusudhanamurthy J, Sreedhar B (2014) Dynamic adsorption of a-pinene and linalool on silica nanoparticles for enhanced antifeedant activity against agricultural pests. J Pest Sci 87:191–200. https://doi.org/10.1007/s10340-013-0538-2

Ranjan S, Dasgupta N, Singh S, Gandhi M (2018) Toxicity and regulations of food nanomaterials. Environ Chem Lett. https://doi.org/10.1007/s10311-018-00851-z

Rouhani M, Samih MA, Kalantari S (2012) Insecticide effect of silver and zinc nanoparticles against *Aphis nerii* Boyer De Fonscolombe (Hemiptera: Aphididae). Chil J Agric Res 72:590–594

Sardoiwala MN, Kaundal B, Choudhury SR (2018) Toxic impact of nanomaterials on microbes, plants and animals. Environ Chem Lett 16:147–160. https://doi.org/10.1007/s10311-017-0672-9

Seiber JN, Coats J, Duke SO, Gross AD (2014) Biopesticides: state of the art and future opportunities. J Agric Food Chem 62:11613–11619. https://doi.org/10.1021/jf504252n

Seitz F, Bundschuh M, Dabrunz A, Bandow N, Schaumann GE, Schulz R (2012) Titanium dioxide nanoparticles detoxify pirimicarb under UV irradiation at ambient intensities. Environ Toxicol Chem 31:518–523. https://doi.org/10.1002/etc.1715

Servin A, Elmer W, Mukherjee A, De la Torre-Roche R, Hamdi H, White JC, Bindraban P, Dimkpa C (2015) A review of the use of engineered nanomaterials to suppress plant disease and enhance crop yield. J Nanopart Res 17:92. https://doi.org/10.1007/s11051-015-2907-7

Sharma S, Singh S, Ganguli AK, Shanmugam V (2017) Anti-drift nano-stickers made of graphene oxide for targeted pesticide delivery and crop pest control. Carbon 15:781–790. https://doi.org/10.1016/j.carbon.2017.01.075

Sparks TC, Nauen R (2015) IRAC: mode of action classification and insecticide resistance management. Pest Biochem Physiol 121:122–128. https://doi.org/10.1016/j.pestbp.2014.11.014

Srivastava AK, Dev A, Karmakar S (2018) Nanosensors and nanobiosensors in food and agriculture. Environ Chem Lett 16:161–182. https://doi.org/10.1007/s10311-017-0674-7

Talbert W, Jones D, Morimoto J, Levine M (2016) Turn-on detection of pesticides via reversible fluorescence enhancement of conjugated polymer nanoparticles and thin films. New J Chem 40:7273–7277. https://doi.org/10.1039/C6NJ00690F

Thungrabeab M, Tongma S (2007) Effect of entomopathogenic fungi, *Beauveria bassiana* (Balsam) and *Metarhizium anisopliae* (Metsch) on non-target insects. KMITL Sci Technol 7:8–12

Unsworth 2010 History of pesticide use. International Union of pure and applied chemistry (IUPAC). (2010) http://agrochemicals.iupac.org/index.php?option=com_sobi2& sobi2Task=sobi2Details&catid=3&sobi2Id=31

Villaseñor MJ, Ríos Á (2018) Nanomaterials for water cleaning and desalination, energy production, disinfection, agriculture and green chemistry. Environ Chem Lett 16:11–34. https://doi.org/10.1007/s10311-017-0656-9

Wibowo D, Zhao C, Peters BC, Middelberg APJ (2014) Sustained release of fipronil insecticide in vitro and in vivo from biocompatible silica nanocapsules. J Agric Food Chem 62:12504–12511. https://doi.org/10.1021/jf504455x

Woranuch S, Yoksan R (2013) Eugenol-loaded chitosan nanoparticles: I. thermal stability improvement of eugenol through encapsulation. Carbohydr Polym 96:578–585. https://doi.org/10.1016/j.carbpol.2012.08.117

Xu P, Guo S, Yu H, Li X (2014) Mesoporous silica nanoparticles (MSNs) for detoxification of hazardous organophorous chemicals. Small 10:2404–2412. https://doi.org/10.1002/smll.201303633

Yang FL, Li XG, Zhu F, Lei CL (2009) Structural characterization of nanoparticles loaded with garlic essential oil and their insecticidal activity against *Tribolium castaneum* (Herbst)

(Coleoptera: Tenebrionidae). J Agric Food Chem 57:10156–10162. https://doi.org/10.1021/jf9023118

Yata VK, Tiwari BC, Ahmad I (2018) Nanoscience in food and agriculture: research, industries and patents. Environ Chem Lett 16:79–84. https://doi.org/10.1007/s10311-017-0666-7

Yu B, Zeng J, Gong L, Zhang M, Zhang L, Xi C (2007) Investigation of the photocatalytic degradation of organochlorine pesticides on a nano-TiO_2 coated film. Talanta 72:1667–1674. https://doi.org/10.1016/j.talanta.2007.03.013

Yu M, Yao J, Liang J, Zeng Z, Cui B, Zhao X, Sun C, Wang Y, Liu G, Cui H (2017) Development of functionalized abamectin poly(lactic acid) nanoparticles with regulatable adhesion to enhance foliar retention. RSC Adv 7:11271–11280. https://doi.org/10.1039/C6RA27345A

Zhang H, Chen B, Banfield JF (2009) The size dependence of the surface free energy of titania nanocrystals. Phys Chem 11:2553–2558. https://doi.org/10.1039/B819623K

Zhao YG, Shen HY, Shi JW, Chen XH, Jin MC (2011) Preparation and characterization of amino functionalized nano-composite material and its application for multi-residue analysis of pesticides in cabbage by gas chromatographyetriple quadrupole mass spectrometry. J Chromatogr A 1218:5568–5580. https://doi.org/10.1016/j.chroma.2011.06.090

Zhu H, Han J, Xiao JQ, Jin Y (2008) Uptake, translocation, and accumulation of manufactured iron oxide NPs by pumpkin plants. J Environ Monit 10:713–717. https://doi.org/10.1039/b805998e

Chapter 3
Synthesis of Nanofertilizers by Planetary Ball Milling

Chwadaka Pohshna, Damodhara Rao Mailapalli, and Tapas Laha

Abstract Plant nutrients supplied to crops as fertilizers are essential for plant growth, metabolism and production. Inappropriate application of plant nutrients induces 40–70% loss of nutrients and causes contamination of land and water systems. Nano-fertilizers provide nutrients precisely to the plant's requirement, and thus reduces the environmental loss of nutrients. Synthesis of nanoparticles is carried out by either top-down or bottom-up methods. Most nanofertilizers are synthesized by a bottom-up approach, which is a complex and requires sophisticated instruments. The top-down approach is an alternative method for large scale and low cost of production. For instance, high energy ball milling is a top-down method using planetary ball mills. To obtain optimized milling parameters in a planetary ball mill, many trials are needed. Hence optimization of the milling parameters through modeling tools is necessary to reach economically efficient and time-saving synthesis of nano-fertilizers. Here we review modeling approaches using the planetary ball milling principle for the efficient synthesis of nano-fertilizers.

Keywords Nanomaterials · Planetary ball milling · Mathematical models · Plant nutrients · Top-down method · Nano-fertilizers

C. Pohshna · D. R. Mailapalli (✉)
Agricultural and Food Engineering Department, Indian Institute of Technology Kharagpur, Kharagpur, West Bengal, India
e-mail: mailapalli@agfe.iitkgp.ernet.in

T. Laha
Metallurgical and Materials Engineering Department, Indian Institute of Technology Kharagpur, Kharagpur, West Bengal, India

© Springer Nature Switzerland AG 2020
E. Lichtfouse (ed.), *Sustainable Agriculture Reviews 40*, Sustainable Agriculture
Reviews 40, https://doi.org/10.1007/978-3-030-33281-5_3

3.1 Introduction

Plant nutrients are essential elements for plant growth and metabolism and their slight deficiency causes irregular growth of plants. Plant nutrients have various functions such as structural components of macromolecules and enzymes for reactions (Table 3.1). Plant nutrients are classified into macronutrients and micronutrients with macronutrients at a concentration of 0.1% of dry tissue weight namely: nitrogen (N), phosphorus (P), potassium (K), calcium (Ca), magnesium (Mg) and sulphur (S) and micronutrients found in the concentration of less than 0.01% of dry tissue weight are zinc (Zn), nickel (Ni), manganese (Mn), molybdenum (Mo), iron (Fe), chlorine (Cl), copper (Cu), and boron (B) (Grusak, 2001). Macronutrients are utilized from the germination stage to the ripening stage of plants growth and production and are essential for providing humans with a suitable supply of energy, nutritional value, and promotion of good health (Grusak and DellaPenna 1999). Micronutrients, though present in small amount, play a vital role in various plant processes, photosynthesis and chlorophyll formation (Monreal et al. 2015) but are also toxic to soil and plants at high concentrations (Arnon and Stout 1939).

Plant nutrients existing in the soil profile are inadequate for crops cultivation. Therefore nutrients are supplied in the form of inorganic fertilizers to suffice the nutrients requirement of crops. Application of fertilizers are inevitable but the excess application and low nutrient use efficiency of the plants lead to 40–75% (Celsia and Mala 2014) lost to the environment through leaching and volatilization causing pollution and toxicity of soil and water bodies; and wastage of fertilizers is an economic loss. The N:P:K ratio for optimal growth is 4:2:1 whereas, in India, it is practice as 10:2.7:1 (Subramanian and Tarafdar 2011); similar increased in fertilizer consumption has been observed in other countries for instances China; where the N used for rice cultivation is 90% more than global average (Guo et al. 2017). Shaviv and Mikkelsen (1993) observed that while the application of nitrogen fertilizers has increased to 15 times, there was only a 3% increase in yield. To overcome the excessive use of fertilizers, without compromising the yield, discovering new and advanced solutions are encouraged. One of the possible solutions is through the application of nanotechnology, i.e., using nano-fertilizers. Since nanomaterials are smaller and more reactive than their bulk materials, it is suggested to have the potential to revolutionize agricultural systems (Singh 2012). Numerous researchers have also reported the possible applications of nanotechnology in agriculture (Subramanian and Tarafdar 2011; Khot et al. 2012; Prasad et al. 2014; Benzon et al. 2015; Monreal et al. 2015; Dubey and Mailapalli 2016; Duhan et al. 2017; and Suppan 2017).

Table 3.1 Details of selected macro and micronutrients required for plant growth

Plant nutrient	[a]Concentration in plants Nano mol/g	[b]Abundance on earth (ppm)	Role	Deficiency
N	1000000	25	Aids in the formation of amino acids	Stunt growth, yellowing of leaves and decrease in dry weight of leaves
K	250000	21000	Adjusts water balance, and improves tolerance against high and low temperature and moisture condition	Diminish the process of photosynthesis, respiration, and translocation.
			Enhances flavor and color of plants and increases the oil content	Low yields, spotted and burned leaves,
Ca	125000	41000	Acts as a structural component of cell walls	Stunt growth stems, flowers and roots and also the presence of dark spots on crops
			Initiates enzymes for cell growth division and water movements	
Mg	80000	23000	A key element for components of the chlorophyll molecule	Its inadequate supply causes drooping and yellowing of veins leaves
			Essential for fruit and nut formation and germination of seeds	
P	60000	1000	Enhance flower and fruit quality	Purple stems and leaves, maturity and growth, are retarded and poor yields
S	30000	260	Used in the development of vitamins and enzymes	Dull green leaves, poor quality and yield of crop, low oil content of seeds and postponed maturity
			Essential for the production of chlorophyll which is responsible for the green color of crop and adds flavor to many crops	
Cl	3000	130	Aids in the movement of water or solutes in cells	Wilting, stubby roots, yellowing and bronzing of crops and small leaf area.
Fe	2000	41000	Acts as a catalyst and aids enzyme activation for the synthesis of chlorophyll	Pale-leaf color followed by yellowing of leaves and large veins

(continued)

Table 3.1 (continued)

Plant nutrient	[a]Concentration in plants Nano mol/g	[b]Abundance on earth (ppm)	Role	Deficiency
B	2000	950	Aids in at least 16 functions of plants such as flowering, pollen germination, fruiting, cell division, water relationships, hormone movements (Blevins and Lukaszewski 1998)	Spoil terminal buds, discolored and brown spot fruits; leaves become thick, curled and brittle
Mn	1000	950	Aids in enzyme activity for photosynthesis, respiration, and nitrogen metabolism	Shedding of young leaves, Brownish, black, or greyish spots may appear next to the veins
Zn	300	75	Aids in the activation of enzymes for carbohydrate metabolism, protein synthesis and stem growth	Mottled leaves with irregular yellowing areas and also cause iron deficiency
			Increased the uptake of N, Mg, and Cu	
Cu	100	50	Assists in the growth and reproduction of higher plants and enhance the activity nitrogen	Brown spots in leaves and shoot tips
Ni	1	80	Aids urease enzyme for breaking down of urea to liberate the nitrogen into a usable form for plants	Production of viable seeds and plants fail to complete their life cycle
Mo	1	1.5	Acts as an enzyme for activation of other nutrients	Thin and Yellowing leaves

[a]Average concentrations of mineral nutrients in plant shoots considered sufficient for adequate growth (Grusak 2001)
[b]Kenneth Barbalace. Periodic Table of Elements. Environmental chemistry (Barbalace 2017)

3.2 Nanofertilizers

Nanomaterials refer to any materials having a particle size of 1–100 nm (10^{-9} m). They have unique physical and chemical properties, which can be more advantageous as compared to their bulk structures (Le Brun et al. 1992). Nanotechnology is an interdisciplinary research field with many practical applications in the field of medicine, electronics, mechanical engineering; however in agriculture, it is still on its research stage (Benzon et al. 2015) where feasibility to large field areas are yet to be ascertained (Khot et al. 2012) but it is also gaining importance gradually (Prasad et al. 2014). Nanotechnology applications in agriculture as nano-fertilizers, nanosensors, nanocapsule, nano-encapsulated flavor enhancer, nanofilms in packaging to prevent spoilage and prevent oxygen absorption, nano-chip use for identification and tracking. Mastronardi et al. (2015) classified three types of nano-technologies for fertilizer inputs and plant protection, i.e., nano-fertilizer, nano-additives, nano-coatings. Nano-sensors in agriculture acts as specifying agents for the level of fertilizers and pesticides present in the soil, soil physical properties, plant health and toxicity level (Rameshaiah et al. 2015). Numerous reviews and few laboratory studies (Table 3.2), established the advantage of nano-fertilizers over conventional fertilizers in terms of low leaching rate of nutrients, higher nutrient absorption capacity, protect against fungal and bacteria growth and increasing plant biomass and yield. Hence nano-fertilizers, controlled release nano-fertilizers, and nano-pesticides can be used as a substitute for conventional ones without affecting the crop yield (Adhikari et al. 2014) while controlling other unwanted factors such as high leaching, eutrophication, and disease which may cause to humans.

Table 3.2 Effect of different types of nanoparticles and their size on crop growth (Modified after Dubey and Mailapalli 2016); The values in the brackets in column 1 represent sizes of the nanoparticles

Nanoparticles (size in nm)	Crop	Effect	References
Titanium dioxide (–[a])	Spinach (*Spinacia oleracea*)	Photosynthesis rate (~3 times), chlorophyll-a (~45%) and chlorophyll-b (~28%) was increased	Zheng et al. (2005)
Multi-walled carbon nanotubes (–[a])	Tomato (*Solanum lycopersicum*)	Water absorption of seed was increased by 58% and germination was increased by 90%	Khodakovskaya et al. (2009)
Iron) and Copper (7.5–20.5)	Potato (*Solanum tuberosum*)	The weight of sprouts was increased by 50% with Fe and it was not significant with Cu	Chalenko et al. (2010)
Multi-walled carbon nanotubes (30)	Mustard (*Brassica juncea*)	Seed germination was increased by 99%	Mondal et al. (2011)
Carbon nanotubes (10–30)	Gram (*Cicer arietinum*)	Water absorption was increased by 50% through Xylem	Tripathi et al. (2011)
Zinc Oxide (25)	Peanut (*Arachis hypogaea*)	Crop yield was increased by 25–30%	Prasad et al. (2012)
Zinc oxide (20), Iron oxide (100) and Zinc iron copper oxide (40)	Mung (*Vigna radiata*)	Root and shoot biomass ware increased by 40% and 44% with ZnO, 68% and 48% with FeO, 42% and 84% with ZnFeCu, respectively	Dhoke et al. (2013)
Hydroxyapatite (<200)	Rice (*Oryza Sativa*)	The sugar level in rice straw and rice husk was reduced by 21–41%	Dutta et al. (2014)
Iron oxide nanoparticles (20)	Peanut (*Arachis hypogaea*)	Increased root length, plant height, biomass, and chlorophyll content of peanut plants	Rui et al. (2016)

[a]– indicates 'not available'

However, efforts arise in synthesizing and stabilizing of the synthesized nano-fertilizers because it involves complex, expensive and sophisticated instruments. Some of the approved commercially available nano-fertilizers are Nano Micro Nutrient by Alert biotech in Maharashtra, India, Nano Ultra-Fertilizer by AB Industries in Taiwan, Nano-Fertilizer by Geetharam Agencies in Kerela, India, Hero Super Nano in Thailand, and Nano Calcium Magic Green-Setia Bersama in Germany (Dimkpa and Bindraban 2017). Nano-zinc, Nano-nitrogen, Nano-phosphorus, Nano-silver, Ultra bio-silver, Nano-sulphur, Nano-Nitrogen and Nano-potassium are some of the other nano-fertilizer products developed in India by Kanak Biotech, New Delhi.

3.3   Synthesis Methods for Nanoparticles

Two types of approaches are available for nanomaterial synthesis; bottom-up and top-down approaches (Fig. 3.1). Bottom-up methods are the chemical approaches for synthesizing nanomaterials with the help of elemental units to combine into larger stable structures Involving building up of the material atom by atom and cluster by cluster until it forms a nanosized material. It involves chemical synthesis,

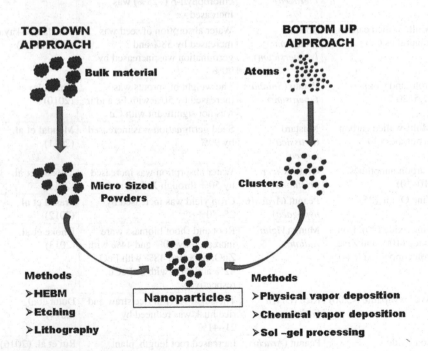

Fig. 3.1 Methods for synthesizing nanomaterials. Bulk sizes are being reduced to nano size in the top-down method, whereas, atoms aggregate and form clusters to become nanoparticles in bottom-up method. Modified after Galstyan et al. 2018

self-assembly, and positional assembly, in which suitable solvents are used for synthesizing ultrafine particles through their dissolved molecular state. Some of the methods involved in the bottom-up approach are plasma arc, rapid solidification, arc discharge, physical vapor deposition, chemical vapor deposition, sol-gel, and inert gas condensation. Bottom-up approach advantages are being able to produce uniform size, shape and distribution of the nanomaterials formed. Bottom-up approaches are common for synthesizing nanoparticles for application to plants (Chalenko 2010; Dutta et al. 2014; Tarafdar 2015; Poopathi et al. 2015; Saha and Gupta 2017). However, the process involved is highly complicated with low yield and expensive machinery are required; it is suitable for highly pure materials only, and it sometimes produces harmful by-products during the synthesis process.

3.3.1 High Energy Ball Milling

The top-down approach is a physical approach for synthesizing nanomaterials from bulk materials to nanosized materials by milling, crushing or grinding (De Castro and Mitchell 2002). Some of the top-down processes are etching technology, high energy ball milling, cold milling or cryo-milling, severe plastic deformation, mechanical polishing, nanoimprint lithography and sliding wear. The most common top-down method is the high energy ball milling; earlier, also known as mechanical milling, which involves breaking down of large-sized particles to nanosized particles through severe plastic deformation to reduce the size of materials and increase the surface area and reactivity of the particles. The high energy ball milling established during an attempt to develop homogeneous composite particles or alloys (Benjamin 1970) uses an attrition mill, followed by the use of other mills such as shaker mill, mixer mill, ball mill, and planetary ball mill. Benjamin's outcome of the study in the 1970s led to the study of the different stages that occur during the high energy ball milling process (Benjamin and Volin 1974). Subsequently, the method was used by different authors with different mills for obtaining nano-alloys, nanocrystalline and metallic-amorphous materials. The advantage of high energy ball milling is that it is simple, easy handling, the versatility of the process, ability to produce large quantities (Maurice and Courtney 1990) and the method is applicable for different types of materials and scalability of the process and its low cost. The main drawback of the high energy ball milling approach is the non-uniformity of the surface structure formed, i.e., not suitable for preparing uniformly shaped materials.

The high energy ball milling devices are of three types namely: shaker mills, attrition mills, and planetary ball mills (Suryanarayana 2001). Shaker mills have a vial where grinding media, i.e., milling balls and sample are swung vigorously to-and-fro for several times causing an impact of milling ball against each other, with the sample and the wall of the vials causing the size reduction in the end product. Attrition mill is a conventional ball mill with a fixed chamber containing centrally vertical rotating stirrer system; as the stirrer rotates the milling balls drops and grind the material. Fritsch Company in 1962 introduced the first self-developed patented Planetary Mill (Fig. 3.2a). Planetary ball mill includes a disc and vial(s) rotating on top of a disc in planet-like movement, i.e., the disc and the vials rotate in the opposite directions similar to planets rotation around the sun (Fig. 3.2b); as such the centrifugal forces act in like and opposite direction alternately. This total centrifugal force acts on the material and milling balls inside the vials, causing impact between the milling balls, the sample, and the vial and ground the material. Planetary ball milling operates in both dry and wet conditions, with easy handling and moderate costs. High energy ball milling synthesized nanofertilizer by reducing the precursor size to nanoparticle size or by milling two or more types of precursors to develop a nano-carrier or nano-sensor. However, it involves several trial runs to obtain the desired milling parameters, which consume a lot of time and energy; Paul et al. (2007) synthesized a nano-sized fly ash of 148 nm at a milling time of 60 h, Gaffet et al. (1991) synthesized a 350 nm size tungsten carbide at 180 h milling time.

Furthermore, Charkhi et al. (2010) and Guaglianoni et al. (2015) conducted several trials to optimize milling speed, milling time, ball to powder ratio and milling medium for synthesizing nanoparticles. Thus, the use of modeling comes in to picture, where optimization of the milling parameters is possible by providing basic material properties. This paper attempts to briefly study the top-down method of nanomaterial synthesis using planetary ball milling generally adopted in material science/engineering and review different mathematical models available in dealing with planetary ball milling for possible application in nano-fertilizer synthesis.

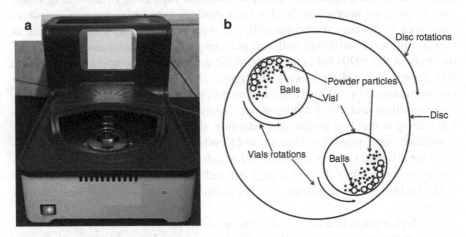

Fig. 3.2 Working principle of planetary ball milling; (a) selected planetary ball mill with front panel open: (b) schematic line diagram of rotation of disc and vials. Modified after Chen et al. 2006

3.3.2 Synthesis of Nanoparticles Using a Planetary Ball Mill

Table 3.3 presents some of the studies conducted using planetary ball mills during the past three decades. The table indicates the different input parameters involved during the High energy ball milling process in a planetary ball mill and also the final size obtained during the synthesis process. Earlier studies are more towards obtaining alloys or homogeneous amorphous products and details of the milling parameters were not explicitly mentioned, unlike the latter studies which are more size oriented. Table 3.3 highlights the importance of different milling parameters viz., milling speed, milling time, ball to powder ratio, material type, milling medium and their effect on the final size of the end product. Since planetary ball mills' s has been used for synthesizing of non-metallic materials such as fly ash (Paul et al. 2007), biochar (Peterson et al. 2012) and zeolite (Charkhi et al. 2010; Mukhtar et al. 2014), hence planetary ball mills are capable of synthesizing the nanofertilizer material. Milling parameters information gathered from previous studies (Paul et al. 2007; Rao et al., 2010; Patil and Anandhan 2012; Raghavendra et al. 2014; Patil and Anandhan 2015; Rajak et al. 2017) for fly ash material were used for studying the interaction effect of milling speed, milling time and ball to powder ratio on fly ash particle size (Figs. 3.3 and 3.4).

3.3.2.1 Milling Speed

Table 3.3 shows the vast range of milling speed from 60 to 2400 rpm used by authors for achieving the nanosized materials. Increasing the milling speed reduced the size of the bulk material as seen in Table 3.3. The higher milling speed of greater than 500 rpm evidently reduced the milling time (Canakci et al. 2013a; Feng et al. 2007; Le Brun et al. 1992; Lee et al. 2017; Wakihara et al. 2011) while the lower milling speeds have been compromised with higher milling time (Kong et al. 2000; Patil and Anandhan, 2012) to attain nanoparticles of 50–1000 nm. Figure 3.3 shows the interaction effect of milling speed and milling time on the size of the fly ash material; it also indicates rapid decreases in the size of the fly ash particle with an increase in milling speed. However, continuous increase in milling speed leads to an increase in temperature and may cause welding of nanoparticles.

3.3.2.2 Milling Time

The milling time used in most of the studies is more than 10 h (Table 3.3), while some studies have used milling time up to 180 h (Gaffet and Harmelin 1990) to obtain nano-sized products 20–350 nm. Lim et al. (2003) used a milling time of 10 min only to obtain an end product of 100 nm with a mixer mill, but it was adjusted with the higher milling speed of 920 rpm, indicating an interrelationship between milling speeds and milling time. Figure 3.3 indicates a gradual decrease in fly ash

Table 3.3 Selected planetary ball mills and the values of milling parameters used in the synthesis of nanoparticles from different bulk materials

Planetary ball mill (Model No.)	Bulk material used	Milling medium	Initial size (μm)	Speed (rpm)	Time (h)	BPR	Final size (nm)	Research highlights	References
Fritsch (N.A)	Fly Ash	Toluene	74	300	60	10:1	148	Morphological studies revealed the uneven, rough and irregular shape of the nano structured fly ash and it has become more active as compared to bulk.	Paul et al. (2007)
	Mg, MgO and Zn	Argon	44	190	10	N.A.	N.A.	The dissolution rate of Zn and Mg were controlled after high-energy ball milling	Kim et al. (2010)
Fritsch (P 4)	K_2CO_3, Na_2CO_3, Li_2CO_3, Nb_2O_5, Bi_2O_3, ZrO_2 and TiO_2	Argon	150, 1, 20, 2, 10, 5,1,	Vial-1000 Disk-2400	0.5–0.6	36:1	crystalline size :16.5–18.5	HEBM confirmed the rapid formation of nano-crystalline perovskite oxides at milling time of 40 min using the Burgio's model	Lee et al. (2017)
Fritsch (P 5) Retsch (PM400)	Ni and Zr	Argon	Ni : 1–3 Zr<177	N.A.	60	15:1	N.A.	Low milling intensity needs an extended milling time for the material to become completely amorphous	Eckert et al. (1998)
Fritsch (P 5)	PbO, TiO_2 and ZrO_2	Air	1–10	200	80	20:1	crystalline size:10	HEBM process is a promising method to synthesize Lead zirconium titanate. Also there is no further decrease in grain size with increase milling times since it was formed	Kong et al. (2000)
	PbO, ZrO_2, TiO_2, and Gd_2O_3	Toluene	45	300	35	10:1	1000	A combination of milling and sintering has synthesized nanocrystalline Gadolinium modified lead zirconate titanate	Parashar et al. (2003)
								Phase transition temperature decreases with a decrease in crystallite size.	
	WC Co	Ethanol and Argon	5.6 1	250	10	15:1	100–500	Synthesized WC nanoparticles with high dislocation density and lattice distortion	Zhang et al. (2003)
	Al	Stearic acid	63	250	12	15:1	150	Addition of PCA causes contamination of product at higher milling time.	Kleiner et al. (2005)

Fritsch (P-6)	Fe	Acetone	160	500	30	10:1	50–100	Milling time influence crystallite and particle size, porosity, magnetic properties and specific surface area of the synthesized Fe powder	Bui et al. (2013)
Fritsch (P-7)	Al	Methanol	113	500	6	10:1	crystalline size: 20	The PCA ha 84% effect on the particle size and powder morphology	Canakci et al. (2013a)
Fritsch P (7/2)	Cu	Argon	10–20 / –	200 / 200	1	7.5:1.	1000–2000	Cu and Ni show dynamic recovery and welding events during milling; Ni recrystallization temperature is slightly higher than that of Cu.	Le Brun et al. (1992)
	Ni								
	Fe		20	750				Fe particle size remains constant and the processes of fracture and welding are in equilibrium at milling speed more than 325 rpm	
Fritsch P(5/2)	Cu (W) tungsten	Argon	N.A.	N.A.	140–180	7:1	crystalline size 20–350	Milling assist the solubility of Cu to tungsten lattice and vice versa	Gaffet et al. (1991)
Fritsch P-(7/2 and 5/2)	Si	Argon	N.A.	N.A.	95 / 70 / 96	7:1	crystalline size: 8–20	Crystal-amorphous phase transition is induced with a high increase in temperature during the milling process	Gaffet and Harmelin (1990)
Fritsch P 5/2	Cu (W) tungsten	Argon	N.A.	N.A.	140–180	7:1	crystalline size 20–350	Milling assist the solubility of Cu to tungsten lattice and vice versa	Gaffet et al. (1991)
Fritsch (P-7/2 and 5/2)	Si	Argon	N.A.	N.A.	95 / 70 / 96	7:1	crystalline size: 8–20	Crystal-amorphous phase transition is induced with a high increase in temperature during the milling process	Gaffet and Harmelin (1990)
NA	Ti, V, Zr, Nb, B, C, N, Ni, Fe and Co	Argon	10–150	N.A.	16–20	10:1	100–1000	Crystallite size reduce to the saturation value of after a milling time of 16 – 20 h and there is no further reduction in crystal size when milled to 48 h	Kieback et al. (1993)
	Fe65Mo, B and TiB$_2$	Argon	74,2 and 3	500	10	10:1	N.A.	Small milling balls cause cold welding and distorted particles. For longer milling time (30 h), milling without PCA show homogeneous particles	Feng et al. (2007)

(continued)

Table 3.3 (continued)

Planetary ball mill (Model No.)	Bulk material used	Milling medium	Initial size (μm)	Milling parameters			Final size (nm)	Research highlights	References
				Speed (rpm)	Time (h)	BPR			
N.A.	Zeolite	Water	45	550	8	90:1	164.9	The optimum milling time for zeolite particles is 2 h while the particles form clusters and have irregular shapes at milling time > 2 h	Mukhtar et al. (2014)
	Fe–18Cr–8Ni–1Mo–0.5Ti–0.15Si–0.35Y$_2$O$_3$	Nitrogen	N.A.	300	30	5:1	100	Three types of oxide-dispersion-strengthened stainless steel specimens precipitates were obtained: polygonal spherical particles and extremely small scale along with aluminium contamination	Miao et al. (2015)
Retsch (PM-100)	Zeolite	Water	1000	550	0.16	1:10	<100	Dry milling as pre-treatments was conducted before wet milling of Zeolite	Charkhi et al. (2010)
		Air		500	3	4.5:1		Wet milling without any needs for dry milling pre-treatments is sufficient.	
	Fly Ash	Toluene	94	300	60	10:1	700	A smooth surface of the fly ash was converted to a rough and more reactive surface. Surfactant reduces the agglomeration of fly ash particles	Patil and Anandhan (2012)
Retsch (PM-200)	WC (Tungsten carbide)	Isopropanol	6	500	15	10:1	15	Developed an analytical equation for planetary ball milling with milling size as the function of the milling time	Gusev and Kurlov (2008)
Retsch (PM-400)	MgO and Al$_2$O$_2$	Air	N.A.	200	12	N.A.	crystalline size:100–300	HEBM enhance the reaction of magnesium oxide and aluminum oxide significantly	Kong et al. (2002)
SFM-1 (QM-3SP2)	Biochar	Water, Nacl,C$_3$H$_6$O, C$_2$H$_6$O, C$_7$H$_{16}$, C$_6$H$_{14}$	N.A.	<710	6	10:1	N.A.	The dry-milling technique can be improved with the addition of salt	Peterson et al. (2012)
						50:1		Changing the diameter of milling media does not have any effect on the end product	

N.A. Not available

BPR ball to powder ratio, *HEBM* High energy ball milling, *PCA* process control agent.

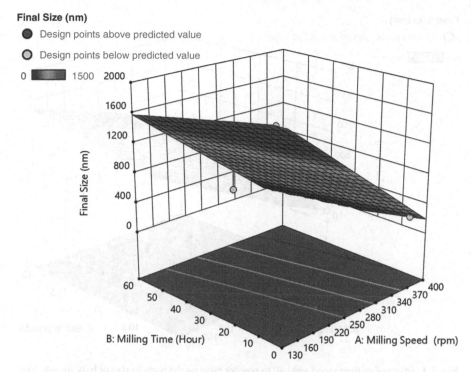

Final Size (nm)

● Design points above predicted value

○ Design points below predicted value

Fig. 3.3 Effect of milling speed and milling time on the particle size of bulk fly ash. Note the decrease in particle size with an increase in milling time and milling speed

particle size with increasing milling time. However, extended milling time decomposed or agglomerate the product and also cause contamination (Suryanarayana 2001; Burmeister and Kwade 2013; Mukhtar et al., 2014; Malayathodi et al. 2018) and an increase in temperature inside the mill and welding of particle take place. The interaction effect of milling speed and milling time on the final particle size (Fig. 3.3) indicate that milling time and milling speed parameters are interdependent hence; the interaction plays a role together to reduce the overall size of the material.

3.3.2.3   Ball to Powder Ratio

Optimum ball to powder ratio is an essential factor because with less ball to powder ratio there will not be enough impact to be able to reduce the size of the material. Table 3.3 indicates ball to powder ratio of 10:1 is the most commonly used ratio and it can go as high as 100:1 and as low as 1:10. Increase in a ball to powder ratio decreases the particle size; Guaglianoni et al. (2015) reported the crystallite size of 53.6 and 48.3 nm at the ball to powder ratio of 5:1 and 20:1 respectively. A higher ball to powder ratio generally reduces the milling time for a particular

Final Size (nm)

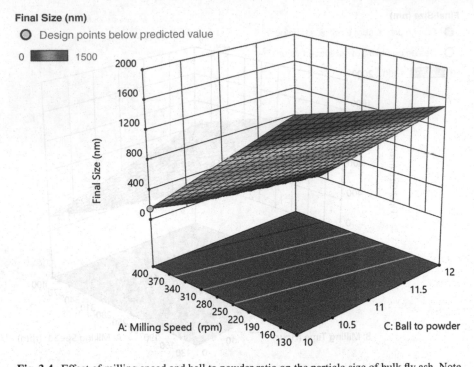

Fig. 3.4 Effect of milling speed and ball to powder ratio on the particle size of bulk fly ash. Note the decrease in the final size of fly ash particles with increasing milling speed and ball to powder ratio

material (Lee et al. 2017; Zakeri et al. 2012); but the higher ball to powder ratio also reduces the amount of initial input material. Figure 3.4 shows the slight decrease of final size of fly ash to the increase in ball to powder ratio. However increase in ball to powder ratio leads to collision of balls against each other and cause restricted movement and it also increases impurities in nanoparticles products (Li et al. 2018).

3.3.2.4 Milling Medium

Milling medium in planetary ball mills acts as surfactant or process control agent and a medium to avoid contamination during milling which may cause due to the reaction of powder materials with the surrounding particle such as the formation of oxides. Milling medium is one of the parameters governing the size of the end product it can be dry milling, wet milling or salt assisted milling. Numerous studies (Charkhi et al. 2010; Mukhtar et al. 2014; Munkhbayar et al. 2013) suggested that wet milling is efficient than dry milling, because it creates higher efficiency, lowers the enthalpy, and eliminates dust formation. Salt assisted milling is also a milling option more effective than wet milling according to Peterson et al. 2012. Argon is found to be the best milling medium (Table 3.3) since it is widely used by most of the researchers due to its inert properties.

Apart from the above milling parameters, type of milling balls, milling ball diameter, temperature and also material parameters such as the initial size of bulk material, types of material, material physical and chemical properties are also considered as factors affecting the size of the end product. An in-depth study is required to understand the effect of the milling parameters, which may be possible through mathematical models.

3.4   Modeling of Planetary Ball Milling

Generally, a number of experimental trials are conducted to determine the different milling parameters in planetary ball mills but, trial and error methods are time-consuming, inefficient and not practical. Real-time ball milling is a very slow process as compared to a discrete element method procedure which takes only a few seconds to simulate the process (Feng et al. 2004). For economical and time-saving synthesis, few models (Tables 3.4, 3.5, and 3.6) were developed to determine the optimum milling parameters in a planetary ball mill. Zhang (2004) stated that modeling and mathematical analysis of high energy ball milling process are still lacking and deserve the attention of researchers. Since large-scale production of nanomaterials is much more feasible in case of the top-down method, modeling of planetary ball milling process is essential to facilitate industrial-scale nanofertilizer production. High energy ball milling is a dynamic process, and it is a challenge to develop mathematical models to represent the description of the process. The models cannot represent the exact process but, they are still capable of providing valuable insights into the behavior of nanoscale materials. Thus, understanding the process with the help of different simulation models is an excellent deal in the synthesis of nanofertilizers, and it is urgently needed to assist effective, efficient and economical results. The model's aid in optimizing the milling parameters for nanofertilizer synthesis by providing the material properties as input parameters; based on the nanofertilizer size required, the model optimized the different milling parameters for direct synthesis in planetary ball mills without the labor of trial and error. However, not all models required the material properties as input parameters; some models required a large trials data as inputs for calibration and validation so based on the available inputs the model can be selected for nano-fertilizers synthesis.

3.4.1   Analytical Models

Table 3.4 presents some of the analytical models developed on planetary ball mills with different types of equations used and research highlights, during the last two and a half decades. The first attempt to model the underlying geometry, mechanics, and physics of the process of mechanical alloying was established by Maurice et al. (1990), using the concept of Hertzian contacts between the grinding media; the

Table 3.4 Analytical models used in the simulation of the planetary ball milling process

Governing equation	Highlights	References
$P = \left\{ -\varphi_b N_b m_b t (\Omega - \omega) \left[\dfrac{\Omega^3 (R_v - R_b)}{\Omega} + \Omega \omega R_p \right] \dfrac{(R_v - R_b)}{2\pi PW} \right.$	Reasonably correlate the input energy with the experimental results.	Burgio et al. (1991)
$\varphi_0 = \arccos\left(-\dfrac{R_v}{R_d} \left(1 - \dfrac{\omega}{\Omega}\right)^2 \right)$	Impact on end product is as follow: Rotation speed>Powder weight in grams>ball diameter. Poor agreement between model and experimental values is observed	Le Brun et al. (1993)
$P = fE$ $f = \dfrac{1}{T} = \dfrac{1}{T_1 + T_2}$	End product depends not only on kinetic energy and shock power	Abdellaoui and Gaffet (1994, 1995)
$P = K_c m_b N_b \Omega^3 R_p^2$	The model was able to predict the size of the end product with an error of 20% which is due to the oversimplification of the ball movement. Model is suitable for all types of mill	Magini et al. (1996)
$P = \dfrac{1}{2} f m_b \left(\Omega^2 R_d^2 \right) + (\Omega - \omega)^2 R_v^2 + 2\Omega(\Omega - \omega) R_d R_v \cos\varnothing$	The balls and vials properties play an essential role in determining the rate of refinement during milling	Chattopadhyay et al. (2001)
$P = C \displaystyle\int_0^\infty Eke^{-kE} dE = \dfrac{C}{k}$	The vial-to-disk speed ratio is a significant parameter in the transfer of impact energy to the powder	Alkebro et al. (2002)
$D(t) = \left. \dfrac{a_D + b_D \varepsilon(t)}{t + \left[a_D + b_D \varepsilon(t) \right]} \right/ D_{in}$	Substitution of empirical selection of milling parameters for the theoretical is possible, with a prediction error of 3% approx.	Gusev and Kurlov (2008)

Description	Equation	Reference
Speed ratio show no considerable influence on the grinding ball motion	$F = F^n + F^t + F_d^n + F_d^t$	Rosenkranz et al. (2011)
Friction is the most important factor affecting the end product		
Experimental results are matching with the prediction of the collision model	$P = P^* \dfrac{1}{2} m_b N_b \Omega^3 R_p^2$	Iasonna and Magini (1996)
Low and intermediate filling resulted in the highest energy path		
Number of balls had a minimum impact on the milling energy	$p = -\dfrac{1}{2\pi}(1-\varphi)K_c N_b m_b (\Omega-\omega)\left(\dfrac{\omega^3 (R_v - R_b)}{w_D} + \omega\Omega R_p\right)(R_v - R_b)$	Ghayour et al. (2016)
The weight loss of the ball led to a decrease in its kinetic energy and a consequent reduction in efficiency even in a high ball to powder ratio values		
At a high ball to powder ratio, both low and high ball size distributions led to a sharp decline in the milling energy		
Modeling of kinematic equations with Burgio's model.	$\Delta E_b = \dfrac{1}{2}\left[\rho_b \dfrac{\pi d_b^3}{6}\right]\Omega^2 \left[\left[\dfrac{\omega}{\Omega}\right]^2 [R_v - R_b]^2 \left[1 - 2\dfrac{\omega}{\Omega}\right] - 2R_P\left[\dfrac{\omega}{\Omega}\right][R_v - R_b] - \left[\dfrac{\omega}{\Omega}\right]^2 [R_v - R_b]^2\right]$	Lee et al. (2017)
The milling parameters were set as per the maximum energy calculated.		
Reducing the milling time and chemical losses and contamination		

P Power released by the ball to the powders

φ_b Degree of filling, $\varphi_b = 0$ (vial is completely filled with balls) & $\varphi_b = 1$ (one or few balls)

N_b, m_b, t PW— Number of balls, mass of ball, Milling time and weight of the powder respectively

Ω, ω—Absolute angular velocity of the plate of the mill and of vial respectively

R_d, R_p & R_v—Distances from the center of the mill to the center of the vial, radius of disk and radius of the vial, respectively

R_b Radius of grinding balls

(continued)

Table 3.4 (continued)

φ_0- Angle of rotation of the milling container towards the sun disc at t = 0

f Shock frequency (number of collision per second)

E energy released from the ball to the powders during a ball milling duration

T_1 time needed by the ball to go from the detachment point up to the collision point

T_2 time needed between the first collision event and the second detachment one

K_c Constant depending on gethe ometry of the mill and collisions

$\emptyset$- is the angular distance described by the ball at a

given moment during its motion

k distribution constant

C impact frequency of all energies

$D(t)$ post-milling particle size after milling time, t

a_D and b_D parameters calculated from material properties

$\varepsilon(t)$ microstrains of a material at a particular milling time

D_{In} Initial diameter of the material

$P*$ includes collision constants, number of balls and number of vials

F resulting contact force

F^n and F^t are the contact forces at normal and tangential

directions, respectively

F_d^n and F_d^t are the damping forces at normal and tangential directions, respectively

ρ_b is the density of the balls

Table 3.5 Statistical models used in simulating planetary ball milling process

Model process	Governing equations	Research highlights	References
Statistical model	$y_{ij} = \mu + \tau_i + \beta_j + (\tau\beta)_{ij} + \varepsilon_{ij}$ $\quad\begin{cases} i = 1,2,3,4... \\ j = 1,2,3,4... \end{cases}$	TV and P1/P2 variables are significant for crystallite size, lattice strain and mean particle size	Dashtbayazi and Shokuhfar (2007)
	Linear: $y_{ij} = \beta_0 + \beta_1 X_1 + \beta_2 X_2 + ... + \beta_K X_K + \epsilon$ Quadratic: $y_{ij} = \beta_0 + \sum_{i=1}^{K}\beta_i X_i + \sum_{i=1}^{K}\beta_{ii}X_i^2 + \sum\sum_{i<j}\beta_{ij}X_iX_j + \varepsilon$	There is a nonlinear response between process parameters and output quality	Hou et al. (2007)
Genetic algorithm	$y_{ij} = 279.23 - 3.14X_1 - 17.58X_2 - 0.018X_3 - 2.11X_4 - 1.63X_5$ $y_{ij} = 380.99 - 7.15X_1 - 17.58X_2 - 0.12X_3 - 9.67X_4 - 1.63X_5 + 0.4X_1^2 + 0.0000284X_3^2 + 0.91X_4^2$	Two algorithms were used MOEA and MPIGA Integrated MPIGA approach was able to determine the optimal parameters	Su and Hou (2008)
Taguchi method	$L_{16}(4^5)$ factors and levels of the orthogonal array was selected	All factors are significant except for ball to powder ratio ratio Milling medium is also a significant parameter to improve size reduction	Zhang et al. (2008)

(continued)

Table 3.5 (continued)

Model process	Governing equations	Research highlights	References
Artificial neural network	$I_i = \dfrac{x_i - x_{i\min}}{x_{i\max} - x_{i\min}}$	Optimized milling parameters i.e., milling speed 300–350 rpm and milling ball diameter range of 8–10 mm	Ma et al. (2009)
	$O_{ki} = \dfrac{1}{1+e^{-I_u}};\qquad O_{ki} = \dfrac{e^{I_u}-e^{-I_u}}{e^{I_u}+e^{-I_u}}$	Back-propagation neural network prediction is a better prediction	
	$PS = \sum\limits_{j=1}^{6} wh_j \tanh\left(\sum\limits_{i=1}^{6} y_i wi_{ij}\right)$	The model optimized milling speed: 350–400 rpm and ball to powder ratio of (20:1).	Lemine et al. (2010)
	$\varphi = f(T, V, D, P1, P2, P3)$	Artificial neural network provides a good agreement with experimental results moreover, results were confirmed by regression analysis	Dashtbayazi et al. (2007)
	$I_i = \dfrac{x_i - x_{i\min}}{x_{i\max} - x_{i\min}}$ and $O_{ki} = \dfrac{1}{1+e^{-I_u}}$	Magnetization and coercivity were optimized and observed that magnetic properties vary with the milling parameters	Hamzaoui et al. 2009

Artificial neural network	$$\frac{S}{N} = -10\log\left[\frac{\sum\left(\frac{y_{ij}^2}{}\right)}{m}\right]$$	The model obtained a mean absolute percentage error of 4.93%.	Canakci et al. (2012)
		Model also indicate the significant impact of process control agent on the particle size, apparent density and micro-hardness	
	Sequential Quadratic Programming Pattern Search Method	Assurance of preservation of nanocrystalline structure since optimized soft milling conditions are established	Dashtbayazi (2012)
	$$I_i = \sum_{i=1}^{n} x_i w_{ij} - \theta_i$$	The model obtained a mean absolute percentage error (MAPE) is 4.68%.	Canakci et al. (2013b)
	$$O_{ki} = \frac{1}{1+e^{-I_{ki}}}$$	The initial amounts of gradual PCA effectively prevent excessive cold welding during ball milling.	
	$$PS = \sum_{j=1}^{6}\left[wh_j \tanh\left(\sum_{i=1}^{6} y_i w_{ij}\right)\right]$$	The model optimized a milling time of 8.5 h and ball to powder ratio of 15.8	Lemine and Louly (2014)
		Artificial neural network model was able to predict particle size with 3.2% error	

φ crystallite size, T milling time, V milling speed, D ball diameter, $P1$ weight of balls, $P2$ weight of powders, $P3$ weight of PCA

I_i and x_i are the input formatted value and real value associated to parameter i which can be milling speed, ball diameter and ball-to-powder weight ratio

x_{imax} and x_{imin} are the maximum and minimum values associated to parameter i

k is the index of the layer i

O is the output values of the neuron indexed by i and k

PS particles size

wi and wh are respectively the input and hidden weights

$y_i$inputs: R(ball to powder ratio), S (Speed), $R*R$, $S*S$, $R*S$, and 1

w_{ij} is the connection weight from j. element to i. element, θ_i is the polarization value

n indicates the sent input signal of the artificial neuron number in the previous layer

(continued)

Table 3.5 (continued)

y_{ij} are the response data of mean particle size

m is the number of replicates

μ – Overall mean effect

τ_i – Effect of the ith level of the row factor ball to powder ratio

β_j – Effect of the jth level of column factor i.e., pra oduct of milling speed and milling time

$(\tau\beta)_{ij}$ Effect of the interaction between τ_i and β_j random error component

β_1, β_2, β_K, β_i, β_{in} β_{ij} represents the input parameter factor and K is the numbers of variables

X_1, X_2, X_k, X_i, X_j, X_{ij} represents the input parameter S/N signal to noise ratio

Table 3.6 Numerical models used to simulate planetary ball milling process

Model Type	Governing equation for contact forces/energy	Research highlights	References
Discrete element method	Kevin Model: $f_n = K_n * \delta n + C_n * v_n$	Determined the impact force for the 2-D motion of ball during ball milling	Dallimore and McCormick (1996)
	Modified Kevin model: $f_n = \hat{K}_n * \delta vol + \hat{C}_n * v_n * \delta a$ Maxwell model: $f_n = K_n * \delta n = C_n * v_n$	Increase in milling speed proportionally increase the impact force	
		Counter n rotation of the mill increase the balls impact energy	Mio et al. (2002, 2004a)
	$E_w = \sum_{j=1}^n \frac{1}{2W} mv_j^2$	Increasing rotation-to-revolution speed ratio below the critical speed ratio increases the ball impact energy	
	$F_n = K_n \Delta u_n + C_n \dfrac{\Delta u_n}{\Delta t}$	The impact energy was related to the grinding rate	Mio et al. (2004b)
	$F_t = \min\left\{ \mu F_n, K_t \Delta(u_t + r_b\varphi) + \eta_t \dfrac{\Delta(u_t + r_E\varphi)}{\Delta t} \right\}$	Effective grinding is observed during counter-rotational direction of the vial to the disk	
	$E_w = \sum_{j=1}^n \dfrac{\frac{1}{2}mv_j^2}{t}$	Powder wear increases almost linearly with milling time at the early stage and then more rapidly, there was a correlation between the impact of energy and wear rate constant.	Sato et al. (2010)
	$E_W = \sum_1^k F_n * V_n + F_s * V_s$	Increasing the milling speed increases the impact energy, number of impacts and dissipated energy	Ashrafizadeh and Ashrafizaadeh (2012)

(continued)

Table 3.6 (continued)

Model Type	Governing equation for contact forces/energy	Research highlights	References
Particle element method	$$\frac{D_{50,t}}{D_{50,0}} = \left(1 - \frac{D_{50,1}}{D_{50,0}}\right)\exp(-K_P t) + \frac{D_{50,1}}{D_{50,0}}$$	Smaller diameter balls have more effect on size reduction at higher speed	Kano and Saito (1998)
		Rate constant and rate constant of amorphization increases with a decrease in ball diameter at a high speed	
	$S = S_{max}[1 - \exp(K_t t)]$	Highest specific surface area can be expressed by the vibratory mill	Kano et al. (2000)
		The planetary mill can achieve the grinding times at which the specific surface area reaches the maximum value	
Dynamic-mechanical multi body model	$$\frac{E}{\tau V} = \frac{1}{2\tau V}\sum_{j=1}^{x}\frac{m_{1j}m_{2j}}{m_{1j}+m_{2j}}v_j^2$$	Developed an innovative jar design for the planetary ball mill with half-moon (HM) cross-section which gives a more uniform and finer end product	Broseghini et al. (2016a)
		Quick predictions of efficient milling parameters for a given material reduce experimental effort in fine-tuning the ball milling process.	Broseghini et al. (2016b)

f_n impact force; K_n & C_n Spring and dash pot coefficient respectively

v_n rethe lative velocity of approach

δn, δvol & δa are linear overlap, volume overlap, and instantaneous area of impact, respectively

E_w specific impact energy of balls

m is the mass of a ball; W weight of sample; F_n compressive force; F_t shear force

v_j relative velocity between two colliding balls or a ball colliding against the mill wall, n number of collision of a ball against other balls or the mill wall within a second

n and t subscript denote normal and tangential components.

u and φ are relative displacement and relative angular displacement, μ coefficient of friction and r_B radius of ball; t milling time

k number of particles that are having interaction with each other

$D_{50,0}$ & $D_{50,t}$ 50% passing particle size of the powder ground at time t = 0 & t = t

K_P Rate constant for size reduction

S Specific surface area of the sample; S_{max} Maximum specific surface area

V normalized over jar volume; m_i mass of the i$-$th colliding body

x the number of points sampling collisions during the

τ simulation time period

study used three types of mill viz., attrition mill, vibratory mill, and horizontal mill. In planetary ball mills, the arrangement of the vial and disc exerted centrifugal force on the balls in the milling container towards the center of the disc and the center of the milling container, resulting in frictions between the balls, the milling container, and the material. Burgio et al. (1991) originally proposed a theoretical-empirical model based on the kinematic equation of the velocity and accelerations of a ball in the vial of a planetary ball mill. The total power, P, transferred from mill to system during collisions, can be obtained from the kinetic energy (E), which is a function of velocity (V), which is a function of speed (ω) and properties (r) of the vial and disc of the mill.

$$\left.\begin{array}{c} P = C \times E \\ E = \frac{1}{2}m(V)^2 \\ V = f(r, \omega) \end{array}\right\} \tag{3.1}$$

Abdellaoui and Gaffet (1994) established a mathematical model and claimed a better geometrical description than Burgio's model and concluded that the end product depends not only on the kinetic energy but also on the shock power. Correspondingly, other authors follow the Burgio's model which is established based on the kinematics of ball movement inside the vial with few modifications (Magini et al. 1996; Iasonna and Magini 1996; Chattopadhyay et al. 2001; Radune et al. 2014; Ghayour et al. 2016; Lee et al. 2017). Further study was conducted by Abdellaoui and Gaffet (1995, 1996) for planetary ball mills and horizontal rod mill in subsequent years to follow up in detail on their previous work and to compare model results with experimental data, and concluded that the injected shock power 'f' is responsible for milling the material to nanoparticles. Chattopadhyay et al. (2001) include a detachment parameter to ensure that the detachment angle does not assume any value for a ball vial impact and calculate the dissipated energy for various conditions of milling. Gusev and Kurlov (2008) deduced an analytical expression to calculate the final size based on the milling parameters and material properties.

Subsequently, other analytical models (Le Brun et al. 1993) calculate the angle of impact of the milling balls on the vial surface but fail to obtain a good correlation between theory and in-situ observations. Alkebro et al. (2002) derived the model based on the Hertzian concept by Maurice and Courtney (1990), to describes the frequency distribution of impact energies and account the energy loss between balls in the vial. Ghayour et al. (2016) studied the effects of the vial to plate spinning rate, ball size distribution and type of balls on the performance of the mill using the Burgio's model. Most of the literature has not been able to correctly simulate the process involved in ball milling since it is a highly complex process, Rosenkranz et al. (2011) investigated the ball motion in a planetary ball mill with a high-speed video camera to recorded grinding ball motion during the process. The ball motion observed from ball trajectories, contradict earlier theoretical calculations, since previous theories do not account for friction and slip which, occur during ball milling.

The model proposes by Burgio et al. (1991) becomes the foundation for most of the latter study of planetary ball mill modeling. These models have been able to reduce the time and cost of synthesizing nanoparticle by providing insight into possible parameter values.

Furthermore, understanding and representation obtained through video recording, or transparent mills can provide in-depth knowledge of the process. Analytical models presented in Table 3.4 could be a valuable tool for synthesizing of nanofertilizers. The precursor for nano plant nutrients or nano-fertilizers are mainly ceramics and nonmetals viz., C, H, O, N, P, S; which are more brittle and soft as compared to the typical materials or metals. Hence the power required for synthesizing these nano size might be much lesser as compared to typical material. The nonmetal materials such as zeolite and biochar have lower milling time as compared to other typical metals (Table 3.3). Gusev and Kurlov (2008) derived a relationship between the milling parameters and material properties to obtain a nano-sized particle. However, most of the models do not incorporate the material properties; therefore for the synthesis of nanofertilizer; it is necessary to incorporate the material properties of the precursor.

3.4.2   Statistical Modeling

Statistical models including artificial neural network with mathematical equations are presented in Table 3.5. The statistical models cannot deliver the factors of high energy ball milling, but only provide guidelines on the reliability and validate the experimental results. They measure the error and the confidence level of the experimental results. Fixed model, artificial neural network, and regression model are some of the models used for simulating the milling process.

Regression model

$$y_{ij} = \beta_0 + \beta_1 x_1 + \beta_2 x_2 + \beta_{12} x_1 x_2 + \beta_{11} x_1^2 + \beta_{22} x_2^2 + \varepsilon \tag{3.2}$$

Where y_{ij} is the response variable including crystallite size/lattice strain, and the mean particle size of nanocomposite powders, x_1 and x_2 are the variables that represent factors/parameters, and ε represents the random error term, β_0 is the constant of the model, β_1 and β_2 represent the main effects of parameters e.g., milling time milling speed ball to powder ratio and other parameters, β_{11} and β_{22} represent the square effect of the parameters, and β_{12} represents the interaction effect of parameters.

The artificial neural network is similar to the biological neuron network. It is one of the most potent modeling techniques, in all fields of sciences, using the statistical approach. It can provide suitable and logical results with acceptable accuracy and faster prediction. Dashtbayazi et al. (2007) used artificial neural network by using two types of neural network architectures, i.e. multi-layer perceptron and radial basis function for modeling the effects of milling parameters of planetary ball mills on the characteristics of aluminium silicon nanocomposite powders; taking into consideration the crystallite size and lattice strain of the aluminum matrix. Regression analysis confirmed that the developed artificial neural network agreed with experimental data. However, Sha (2008) commented on the inappropriate extrapolation of artificial neural network modeling results due to over-emphasis on modeling. So in 2012, Dashtbayazi (2012) study the mechanical alloying process for synthesizing of aluminum silicon nanocomposite powders in a planetary ball mill through artificial neural network and established that low milling speed, low milling time and low ball to powder ratio could produce better nanocomposite structure.

Similarly, other authors (Canakci et al. 2012, 2013b; Hamzaoui et al. 2009; Lemine and Louly 2014; Ma et al. 2009) have used artificial neural network model for studying the process of milling in a planetary ball mills, by modifying the parameters, by providing weighting factor, using different sigmoid functions such as log-sigmoid and hyperbolic tangent sigmoid (Ma et al. 2009). Varol et al. (2013) suggested the artificial neural network model as a powerful tool for modeling of high energy ball milling. Canakci et al. (2012) claim that artificial neural network was successful in predicting the apparent density, particle size and microhardness of the composite powders with a mean percentage error of 4.93%. Dashtbayazi and Shokuhfar (2007) suggested a statistical approach for milling of nanocomposite powders through problem description; identifying the response variables; setting of factors, levels, and range; selection of experimental design; conducting the experiment; statistically analyzing the data and obtaining conclusions. Hou et al. (2007) integrated three methods: Taguchi model, response surface method and genetic algorithm for optimization of milling parameters of a planetary ball mill. The Taguchi method determines the proper working levels of the design factors and analyses the most significant factors of input parameters. Response surface method determines the optimized parameters that produce a maximum or minimum value of the response by developing the first and second order mathematical models. Genetic algorithm approach was applied to optimize the milling parameters using the response function of the response surface method model as the fitness function (Table 3.5). Parameter optimization using Taguchi method showed a good representation of the planetary ball mills process and provided an understanding of the most significant parameters (Su and Hou 2008; Zhang et al. 2008; Canakci et al. 2013b). The statistical models require large initial input data for modeling nanomaterial synthesis efficiently, which involved lots of trial synthesis. Combining and utilizing the results of different studies conducted for a particular material, e.g., zeolite (Charkhi et al. 2010; Mukhtar et al. 2014) can be an option, but since limited studies are available, it is difficult to conclude whether the model is suitable or not for milling parameter optimization.

3.4.3 Numerical Models

Modeling of granular particles and understanding of macroscopic particulate behavior are mostly discussed using the discrete element methods incorporating the particle properties, and interaction forces (Luding 2008). The properties of particles are calculated at every time step by integrating the translational and rotational displacements of Newton's second law of motion, while the contact forces between particles are calculated using different models, e.g., Kevin model, Maxwell model Dallimore, P.G. McCormick Linear normal contact model (Luding 2008). The expression for translational and rotational motions of a single particle is as follow (Zhao 2017);

$$m_i \frac{d^2}{dt^2} \vec{x_i} = m_i \vec{g} + \sum_{Nc} \left(\vec{f_n} + \vec{f_t} \right) + \vec{f_f} \tag{3.3}$$

$$I_i \frac{d}{dt} \vec{\omega_i} = \sum_{Nc} \left(\vec{r_c} \times \vec{f_t} + \vec{R_r} \right) \tag{3.4}$$

Where m_i is the mass of a particle i; $\vec{x_i}$ is the position of its centroid; $\vec{g}$ is the acceleration due to gravity; $\vec{f_n}$ and $\vec{f_t}$ are the normal and tangential forces exerted among the particles and the wall; Nc is the summation of the contact forces or over all the contacts; $\vec{f_f}$ is the interaction forces between fluid and particles; I_i is the moment of inertia about the grain centroid; $\vec{\omega_i}$ is the angular velocity; $\vec{r_c}$ is the vector from the particle mass centre to the contact point; $\vec{R_r}$ is the rolling resistant moment, which inhibits particle rotation over other particles.

Few studies used discrete element method to simulate the planetary ball milling process for the production of nanosized materials are presented in Table 3.6. For obtaining the tangential and normal force, different authors have used different contact models (Ashrafizadeh and Ashrafizaadeh 2012; Dallimore and McCormick 1996; Feng et al. 2004; Kano et al., 2000; Kano and Saito 1998; Mio et al. 2002, 2004a, 2004b; Sato et al. 2010). Kevin model, Modified Kevin model, the Maxwell model, and Elastic/Plastic yield (Dallimore and McCormick 1996) model simulate actual milling impacts. Kano and Saito (1998) simulate the ball movement in a planetary ball mill using the particle element method and established that the rate of size reduction and rate of amorphization increased with a decrease in ball diameter while controlling the milling speed of the mill and also conducted the same study on different types of mills (Kano et al. 2000). When rotation to revolution speed ratio increases there is an increase in the impact energy of balls as calculated from the computer simulation based on discrete element method. Mio et al. (2004b, 2002) conducted another study on a scale-up method using discrete element method and established that the impact energy is proportional to the cube of the vial diameter, the depth of the vial and the revolution radius of the disk, while the scale-up ratio of planetary ball mills is of the power of 4.87. Feng et al. (2004) stated that discrete element method simulation of the planetary ball milling dynamics is a better digital approach as compared to analytical based modeling procedures since it can easily stimulate and investigate any change of operation conditions on the dynamics of the system.

Sato et al. (2010) analyzed the abrasion mechanism of grinding media in a planetary mill using discrete element method simulation and observed the relation between impact energy and wear rate constant suggesting that the wear rate constant might be able to simulate using discrete element method. Ashrafizadeh and Ashrafizaadeh (2012) also study the simulation of planetary ball milling using discrete element method regarding effects on impact energy through the rotational speed of the disk and the ball to powder ratio. The frequency of the impacts, the abrasion of the balls and the dissipated energy, and results indicate the suitability of the method for calculating the improved efficiency of grinding operation regarding the required grinding time and reduced abrasion.

Broseghini et al. (2016a) developed a numerical dynamic-mechanical model and used an MSC ADAMS software to solve the model for planetary ball mill. The study is on the effect of milling parameters: – ball size and number, jar geometry and milling speed on the efficiency of milling in ceramic powders. Broseghini et al. (2016b) developed a new vial design for a planetary ball mill using a similar model as Broseghini et al. (2016a) to simulate the ball milling process. The new design was found to give more uniform and more reliable results. Since planetary ball mill involves a dynamic process, and numerical models are capable of handling large systems of equations and nonlinearities, it can represent the dynamic process, which is often impossible to solve analytically. The different material properties, ball motion, and impact force and other processes involve inside the planetary ball mill can be simulated iteratively to obtain actual milling parameters. Numerical models are considered as the optimal option for synthesizing nanofertilizer since the material properties can be provided as input parameters to synthesized nanofertilizer. The main issues are understanding and simulating the process inside the mill, the effect of force developed inside the mill and representing it mathematically and also obtaining the detailed material properties.

3.5 Discussion

High energy ball milling as one of the possible top-down approach for synthesizing of nano-fertilizers and can counteract the limitations of bottom-up methods; since the method is simple and easy and also suitable for large scale of production (Lam et al. 2000). However, selection of synthesis methods depend on the type of product or the kind of results desired; such as, if uniformity of product is more desired then one can opt for the bottom-up method, but the top-down method is feasible for low cost and high production. As observed in Table 3.3, most of the nanoparticles are produced using planetary ball mill; however, limited studies are available for the synthesis of nano-fertilizers. The production of nano-fertilizers in a planetary ball mill can follow the same procedure as of the commonly used materials but due to the difference in material property, retesting of the methods for a particular material type is required. Increase in milling speed and milling time decreases the material size, however prolonged milling time in some materials causes clustering of the particles (Burmeister and Kwade 2013; Mukhtar et al. 2014) or does not decrease in size any further (Kong et al. 2000; Biyika and Aydind 2014). Milling medium or process control agent aids in the milling of particles by reducing cold welding and also regulating the temperature inside the mill (Kleiner et al. 2005; Canakci et al. 2013a). Wet milling is considered more effective than salt assisted milling and dry milling (Peterson et al. 2012). Increase in a ball to powder ratio decreases the particle size (Zakeri et al. 2012) but introduce impurities (Li et al. 2018) to the end product. Milling parameters such as milling speed, milling time, milling medium and ball to powder ratio of some studies conducted for non-metals or ceramics materials viz., fly-ash (Paul et al. 2007), biochar (Peterson et al. 2012) zeolite (Charkhi et al. 2010) can be used as insights for synthesizing of nanofertilizer.

Modeling the planetary ball milling process is essential because real-time ball milling is time-consuming, inefficient and not practical. The mathematical representation of the process assists the use of models for different types of materials such as bulk fertilizer materials. Modeling tools developed for common materials such as metals can be used for modeling the synthesis of nanofertilizer. Analytical, numerical and statistical models are developed to represent the dynamic process of milling in a planetary ball mill (Tables 3.4, 3.5, and 3.6). Statistical models such as artificial neural network model, Taguchi model, regression model, response surface method and genetic algorithm are suitable methods for optimization of the milling parameters, as they provide the most and least significant parameter details; however, they required a large initial data for reliable optimization. Numerical and analytical models follow the basic principle of centrifugal force and momentum of the mill's disc and vial to predict the milling parameters regarding energy and equations involving material properties where the properties of bulk fertilizer can be incorporated to obtain desired nano-sized particles of nanofertilizer. However, the analytical model involves simplifying assumptions, due to inherent complexity combine with dynamic, non-linear behavior of the process. While numerical models can be alternative tools to simulate the planetary ball milling process efficiently, it includes multi parameters and complicated processes which are difficult to understand and incorporate mathematically. Most of the models estimate the milling energy or power as the primary output and concluded that with more power/energy more size reduction takes place because the principle of size reduction in high energy ball milling devices depends on the energy imparted to the sample during impacts between the milling media. Higher energy generates more impact and hence lead to more size reduction. Most of the models do not consider the effect of the material or precursor properties; these models have been mostly stimulated for common materials but not for the synthesis of nanofertilizer. Few models have linked the kinematic equation along with the milling energy equations (Gusev and Kurlov 2008; Choi et al. 2001). Unlike other studies whose aim is to achieved amorphous state or some stable state of the materials, the main aim of synthesizing nanofertilizer is to obtain the nanometer size these models can provide the direct relation of the impact of milling energy and material properties to the milling size of the nanofertilizer.

Hence the present models need to incorporate and link the material properties with the kinematic equation of planetary ball mill for better results. Analytical and numerical models can be used for synthesizing of nano-fertilizers. An analytical model developed by Gusev and Kurlov (2008) relates the material properties with the planetary ball milling equation, where the particle size is model as the function of milling time, can be a suitable model for synthesizing the nanofertilizer.

3.6 Conclusion

Farmers have been excessively supplying the essential nutrients to plants through the application of inorganic fertilizers to obtain the desired amount of yields. Conversely, the fertilizers applied have low nutrient use efficiency and cause pollution to the environment and affect the health of the soil. An alternative method to increase nutrient use efficiency and reduce the loss of fertilizers is through the use of nano-fertilizer (Dhewa 2015). The biological, physical, and chemical properties of nanoparticles are more enhanced than their bulk material (Dasgupta et al. 2017; Gruère 2012). Synthesis of nano-fertilizers is still in its initial stage with maximum of the methods used are of bottom-up approaches, involving sophisticated, costly instruments and low production. Simplified synthesizing methods of nano-fertilizer are essential to be able to replace conventional fertilizer and minimize the risk of environmental pollution. Nano-fertilizer can supply nutrients efficiently to plants and increase the yield while minimizing nutrient losses and controlling pollution in the environment. Understanding the milling parameters impact on the milling of nanofertilizer is essential to optimize the milling parameters to obtain desired nano sized fertilizer at low cost and less time. This review reveals a few of the laboratory studies conducted on a planetary ball mill for synthesizing of nanoparticles and also the modeling techniques developed to represent its dynamic process; analytically, numerically and statistically. It is evident from the study that planetary ball mills have been used for decades to synthesize nanoparticles and also various types of models have been successfully developed to simulate the milling process. Thus, nano-fertilizer can be economically and timely synthesized, through planetary ball mill with the help of available modeling tools. The numerical model is the best method to simulate the planetary ball milling process and material properties. However it required clear understanding and representation of the process mathematically which is a complicated process and it is not practical. Analytical models are suitable for nanofertilizer synthesis practically as it involves simplification and assumption of the process and can simulate the size up to 6–60% error (Gusev and Kurlov 2008). Since over-simplification of many of the models developed so far; further study conducted on planetary ball mill by using a transparent mill and video recorder to understand the movement and working process of the mill (Rosenkranz et al. 2011) is necessary; also further studies on linking the kinematic equation to the final size of the nanoparticle are required. More focused research is essential on the development and synthesis of nano-fertilizers using the planetary ball mill along with the modeling tools for efficient synthesis of nano plant nutrients. At the present stage of understanding, it can be speculated that the analytical models can provide an idea of the optimized milling parameters while numerical models are also an efficient method, but a better understanding of the planetary ball mill internal process is still required. Hence, models are presently best solution for understanding and simulating significant parameters of planetary ball mills by way of reducing energy, cost and time.

References

Abdellaoui M, Gaffet E (1994) Mechanical alloying in a planetary ball mill: kinematic description. Le. J Phys IV(04):C3-291-C3-296. https://doi.org/10.1051/jp4:1994340

Abdellaoui M, Gaffet E (1995) The physics of mechanical alloying in a planetary ball mill: mathematical treatment. Actametallurgicaetmaterialia 43(3):1087–1098. https://doi.org/10.1016/0956-7151(95)92625-7

Abdellaoui M, Gaffet E (1996) The physics of mechanical alloying in a modified horizontal rod mill: Mathematical treatment. Acta Mater 44:725–734. https://doi.org/10.1016/1359-6454(95)00177-8

Adhikari T, Kundu S, Meena V, Rao AS (2014) Utilization of nano rock phosphate by Maize (Zea mays L.) crop in a vertisol of Central India. J Agric Sci Technol A J Agric Sci Technol 4:384–394

Alkebro J, Bégin-Colin S, Mocellin A, Warren R (2002) Modeling high-energy ball milling in the alumina-yttria system. J Solid State Chem 164:88–97. https://doi.org/10.1006/jssc.2001.9451

Arnon DI, Stout PR (1939) The essentiality of certain elements in minute quantity for plants with special reference to copper. Plant Physiol 14(2):371–375. https://doi.org/10.1104/pp.14.2.37

Ashrafizadeh H, Ashrafizaadeh M (2012) Influence of processing parameters on grinding mechanism in planetary mill by employing discrete element method. Adv Powder Technol 23:708–716. https://doi.org/10.1016/j.apt.2011.09.002

Barbalace K (2017) Periodic table of elements. https://environmentalchemistry.com/yogi/periodic/name.html. Accessed 10/6/2017

Benjamin JS (1970) Dispersion strengthened superalloys by mechanical alloying. Met Trans 1(10):2943–2951. https://doi.org/10.1007/BF03037835

Benjamin JS, Volin TE (1974) Mechanism of mechanical alloying. Met Trans. https://doi.org/10.1007/BF02644161

Benzon HRL, Rubenecia MRU, Ultra VU Jr, Lee SC (2015) Nano-fertilizer affects the growth, development, and chemical properties of rice. Int J Agric Res 7(1):105–117. http://www.innspub.net/wp-content/uploads/2015/07/IJAAR-V7No1-p105-117.pdf

Biyika S, Aydinb M (2014) The E ect of milling speed on particle size and morphology of Cu25W composite powder. Proceedings of the 4th International Congress APMAS2014, April 24-27, 2014, Fethiye, Turkey. https://doi.org/10.12693/APhysPolA.127.1255

Blevins D, Lukaszewski K (1998) Boron in plant structure and function. Annu Rev Plant Physiol Plant Mol Biol 49:481–500. https://doi.org/10.1146/annurev.arplant.49.1.481

Broseghini M, D'Incau M, Gelisio L, Pugno NM, Scardi P (2016a) Effect of jar shape on high-energy planetary ball milling efficiency: simulations and experiments. Mater Des 110:365–374. https://doi.org/10.1016/j.matdes.2016.06.118

Broseghini M, Gelisio L, D'Incau M, Azanza Ricardo CL, Pugno NM, Scardi P (2016b) Modeling of the planetary ball-milling process: the case study of ceramic powders. J Eur Ceram Soc 36:2205–2212. https://doi.org/10.1016/j.jeurceramsoc.2015.09.032

Le Brun P, Gaffet E, Froyen L, Delaey L (1992) Structure and properties of Cu, Ni and Fe powders milled in a planetary ball mill. Scr Metall Mater. https://doi.org/10.1016/0956-716X(92)90545-P

Bui TT, Le XQ, Tommi DP (2013) Investigation of typical properties of nanocrystalline iron powders prepared by ball milling techniques. Adv Nat Sci: Nanosci Nanotechnol 4(4):045003. https://doi.org/10.1088/2043-6262/4/4/045003

Burgio N, Iasonna A, Magini M, Martelli S, Padella F (1991) Mechanical Alloying of the Fe-Zr System. Correlation between Input Energy and End Products. Nuovo Cim D 13(4):459–476. https://doi.org/10.1007/BF02452130

Burmeister CF, Kwade A (2013) Process engineering with planetary ball millss. Chem Soc Rev. https://doi.org/10.1039/c3cs35455e

Canakci A, Ozsahin S, Varol T (2012) Modeling the influence of a process control agent on the properties of metal matrix composite powders using artificial neural networks. Powder Technol 228:26–35. https://doi.org/10.1016/j.powtec.2012.04.045

Canakci A, Erdemir F, Varol T, Patir A (2013a) Determining the effect of process parameters on particle size in mechanical milling using the Taguchi method: Measurement and analysis. Measurement 46:3532–3540. https://doi.org/10.1016/j.measurement.2013.06.035

Canakci A, Varol T, Ozsahin S (2013b) Analysis of the effect of a new process control agent technique on the mechanical milling process using a neural network model: Measurement and modeling. Measurement 46:1818–1827. https://doi.org/10.1016/j.measurement.2013.02.005

De Castro CL, Mitchell BS (2002) Nanoparticles from mechanical attrition. Synth Funct Surf Treat Nanoparticles:1–15. https://pdfs.semanticscholar.org/4279/540565f0e0860db9eb9e3041d18a8c65f11c.pdf

Celsia AR, Mala R (2014) Fabrication of nanostructured slow release fertilizer system and its influence on germination and biochemical characteristics of vigna raidata. Recent Adv Chem Eng 6:4497–4503. http://www.sphinxsai.com/2014/RACE/1/(4497-4503)%20014.p

Chalenko GI, Gerasimova NG, Vasyukova NI, Ozeretskovskaya OL, Kovalenko LV, Folmanis GE, Tananaev IG (2010) Metal nanoparticles induce germination and wound healing in potato tubers. Dokl Biol Sci 435:428–430. https://doi.org/10.1134/S0012496610060165

Charkhi A, Kazemian H, Kazemeini M (2010) Optimized experimental design for natural clinoptilolite zeolite ball milling to produce nano powders. Powder Technol. https://doi.org/10.1016/j.powtec.2010.05.034

Chattopadhyay PP, Manna I, Talapatra S, Pabi SK (2001) Mathematical analysis of milling mechanics in a planetary ball mill. Mater Chem Phys 68:85–94. https://doi.org/10.1016/S0254-0584(00)00289-3

Chen Y, Li CP, Chen H, Chen Y (2006) One-dimensional nanomaterials synthesized using high-energy ball milling and annealing process. Sci Technol Adv Mater 7(8):839–846. https://doi.org/10.1016/j.stam.2006.11.014

Choi WS, Chung HY, Yoon BR, Kim SS (2001) Applications of grinding kinetics analysis to fine grinding characteristics of some inorganic materials using a composite grinding media by planetary ball mill. Powder Technol 115(3):209–214. https://doi.org/10.1016/S0032-5910(00)00341-7

Dallimore MP, McCormick PG (1996) Dynamics of planetary ball milling: A comparison of computer simulated processing parameters with CuO/Ni displacement reaction milling kinetics. Mater Trans JIM 37(5):1091–1098. https://doi.org/10.2320/matertrans1989.37.1091

Dasgupta N, Ranjan S, Ramalingam C (2017) Applications of nanotechnology in agriculture and water quality management. Environ Chem Lett. https://doi.org/10.1007/s10311-017-0648-9

Dashtbayazi MR (2012) Artificial neural network-based multiobjective optimization of mechanical alloying process for synthesizing of metal matrix nanocomposite powder. Mater Manuf Proc 27:33–42. https://doi.org/10.1080/10426914.2010.523917

Dashtbayazi MR, Shokuhfar A (2007) Statistical modeling of the mechanical alloying process for producing of Al/SiC nanocomposite powders. Comput Mater Sci 40:466–479. https://doi.org/10.1016/j.commatsci.2007.02.001

Dashtbayazi MR, Shokuhfar A, Simchi A (2007) Artificial neural network modeling of mechanical alloying process for synthesizing of metal matrix nanocomposite powders. Mater Sci Eng A 466:274–283. https://doi.org/10.1016/j.msea.2007.02.075

Dhewa T (2015) Nanotechnology applications in agriculture: an update. Octa J Environ Res 3:204–211

Dhoke SK, Mahajan P, Kamble R, Khanna A (2013) Effect of nanoparticles suspension on the growth of mung (Vigna radiata) seedlings by foliar spray method. Nanotechnol Dev 3(1):e1–e1

Dimkpa CO, Bindraban PS (2017) Nanofertilizers: new products for the industry? J Agric Food Chem 66:6462–6473. https://doi.org/10.1021/acs.jafc.7b02150

Dubey A, Mailapalli DR (2016) Sustain Agric Rev:12. https://doi.org/10.1007/978-94-007-5961-9

Duhan JS, Kumar R, Kumar N, Kaur P, Nehra K, Duhan S (2017) Nanotechnology: The new perspective in precision agriculture. Biotechnol Rep. https://doi.org/10.1016/j.btre.2017.03.002

Dutta N, Mukhopadhyay A, Dasgupta AK, Chakrabarti K (2014) Improved production of reducing sugars from rice husk and rice straw using bacterial cellulase and xylanase activated with hydroxyapatite nanoparticles. Bioresour Technol 153:269–277. https://doi.org/10.1016/j.biortech.2013.12.016

Eckert J, Schultz L, Hellstern E, Urban K (1998) Glass-forming range in mechanically alloyed Ni-Zr and the influence of the milling intensity. J Appl Phys 64(6):3224–3228. https://doi.org/10.1063/1.341540

Feng YT, Han K, Owen DRJ (2004) Discrete element simulation of the dynamics of high energy planetary ball milling processes. Mater Sci Eng A 375–377:815–819. https://doi.org/10.1016/j.msea.2003.10.162

Feng H, Jia D, Zhou Y (2007) Influence factors of ball milling process on BE powder for reaction sintering of TiB/Ti-4.0Fe-7.3Mo composite. J Mater Proc Technol. https://doi.org/10.1016/j.jmatprotec.2006.07.014

Gaffet E, Harmelin M (1990) Crystal-amorphous phase transition induced by ball-milling in silicon. J Less-Common Met 157:201–222. https://doi.org/10.1016/0022-5088(90)90176-K

Gaffet E, Louison C, Harmelin M, Faudot F (1991) Metastable phase transformations induced by ball-milling in the Cu–W system. Mater Sci Eng A 134:1380–1384. https://doi.org/10.1016/0921-5093(91)90995-Y

Galstyan V, Bhandari M, Sberveglieri V, Sberveglieri G, Comini E (2018) Metal oxide nanostructures in food applications: Quality control and packaging. Chemosensors 6(2):16–37. https://doi.org/10.3390/chemosensors6020016

Ghayour H, Abdellahi M, Bahmanpour M (2016) Optimization of the high energy ball-milling: Modeling and parametric study. Powder Technol 291:7–13. https://doi.org/10.1016/j.powtec.2015.12.004

Gruère GP (2012) Implications of nanotechnology growth in food and agriculture in OECD countries. Food Policy 37:191–198. https://doi.org/10.1016/j.foodpol.2012.01.001

Grusak MA (2001) Plant macro- and micronutrient minerals. In: Encyclopedia of life sciences. https://doi.org/10.1038/npg.els.0001306.

Grusak MA, DellaPenna D (1999) Improving the nutrient composition of plants to enhance human nutrition and health. Annu Rev Plant Physiol Plant Mol Biol. https://doi.org/10.1146/annurev.arplant.50.1.133

Guaglianoni WC, Takimi A, Vicenzi J, Bergmann CP (2015) Synthesis of wc-12wt% co nanocomposites by high energy ball milling and their morphological characterization. Tecnologia em Metalurgia, Materiais e Mineração 12(3):211–215. https://doi.org/10.4322/2176-1523.0838

Guo J, Hu X, Gao L, Xie K, Ling N, Shen Q, Hu S, Guo S (2017) The rice production practices of high yield and high nitrogen use efficiency in Jiangsu, China. Sci Rep. https://doi.org/10.1038/s41598-017-02338-3

Gusev AI, Kurlov AS (2008) Production of nanocrystalline powders by high-energy ball milling: model and experiment. Nanotechnology 19:265302. https://doi.org/10.1088/0957-4484/19/26/265302

Hamzaoui R, Cherigui M, Guessasma S, ElKedim O, Fenineche N (2009) Artificial neural network methodology: Application to predict magnetic properties of nanocrystalline alloys. Mater Sci Eng B Solid-State Mater Adv Technol 163:17–21. https://doi.org/10.1016/j.mseb.2009.04.015

Hou T-H, Su C-H, Liu W-L (2007) Parameters optimization of a nano-particle wet milling process using the Taguchi method, response surface method, and genetic algorithm. Powder Technol 173:153–162. https://doi.org/10.1016/j.powtec.2006.11.019

Iasonna A, Magini M (1996) Power measurements during mechanical milling. An experimental way to investigate the energy transfer phenomena. Acta Mater 44:1109–1117. https://doi.org/10.1016/1359-6454(95)00226-X

Kano J, Saito F (1998) Correlation of powder characteristics of talc during planetary ball milling with the impact energy of the balls simulated by the particle element method. Powder Technol 98:166–170. https://doi.org/10.1016/S0032-5910(98)00039-4

Kano J, Miyazaki M, Saito F (2000) Ball mill simulation and powder characteristics of ground talc in various types of mill. Adv Powder Technol 11:333–342. https://doi.org/10.1163/156855200750172204

Khodakovskaya M, Dervishi E, Mahmood M, Xu Y, Li Z, Watanabe F, Biris AS, Watanabe F (2009) Carbon nanotubes are able to penetrate plant seed coat and dramatically affect seed germination and plant growth. ACS Nano 3(10):3221–3227. https://doi.org/10.1021/nn900887m

Khot LR, Sankaran S, Maja JM, Ehsani R, Schuster EW (2012) Applications of nanomaterials in agricultural production and crop protection: A review. Crop Prot. https://doi.org/10.1016/j.cropro.2012.01.007

Kieback B, Kubsch H, Bunke A (1993) Synthesis and properties of nanocrystalline compounds prepared by high-energy milling. J Phys IV 3:1425–1428. https://doi.org/10.1051/jp4:19937220

Kim YH, Chung WS, Chun HH, Lee I, Kim YH, Kim DH, Park H (2010) The effect of ball milling on the ph of Mg-based metals, oxides and Zn in aqueous media. Metals Mater Int 16(2):253–258. https://doi.org/10.1007/s12540-010-0414-z

Kleiner S, Bertocco F, Khalid FA, Beffort O (2005) Decomposition of process control agent during mechanical milling and its influence on displacement reactions in the Al–TiO2 system. Mater Chem Physics 89(2-3):362–366. https://doi.org/10.1016/j.matchemphys.2004.09.014

Kong LB, Zhu W, Tan OK (2000) from high-energy ball milling powders. Mater Lett 42:232–239

Kong LB, Ma J, Huang H (2002) MgAl2O4 spinel phase derived from oxide mixture activated by a high-energy ball milling process. Mater Lett 56(3):238–243. https://doi.org/10.1016/S0167-577X(02)00447-0

Lam C, Zhang YF, Tang YH, Lee CS, Bello I, Lee ST (2000) Large-scale synthesis of ultrafine Si nanoparticles by ball milling. Journal of Crystal Growth 220(4):466–470. https://doi.org/10.1016/S0022-0248(00)00882-4

Le Brun P, Froyen L, Delaey L (1993) The modeling of the mechanical alloying process in a planetary ball mill: comparison between theory and in-situ observations. Mater Sci Eng A 161:75–82. https://doi.org/10.1016/0921-5093(93)90477-V

Lee G-J, Park E-K, Yang S-A, Park J-J, Bu S-D, Lee M-K (2017) Rapid and direct synthesis of complex perovskite oxides through a highly energetic planetary milling. Sci Rep 7:46241. https://doi.org/10.1038/srep46241

Lemine OM, Louly MA (2014) Application of neural network technique to high energy milling process for synthesizing ZnO nanopowders. J Mech Sci Technol 28:273–278. https://doi.org/10.1007/s12206-013-0960-7

Lemine OM, Louly MA, Al-Ahmari AM (2010) Planetary milling parameters optimization for the production of ZnO nanocrystalline. Int J Phys Sci 5:2721–2729

Li L, Pu S, Liu Y, Zhao L, Ma J, Li J (2018) High-purity disperse α-Al2O3 nanoparticles synthesized by high-energy ball milling. Adv Powder Technol 29(9):2194–2203. https://doi.org/10.1016/j.apt.2018.06.003

Lim HH, Gilkes RJ, McCormick PG (2003) Beneficiation of rock phosphate fertilizers by mechano-milling. Nutr Cycle Agroecosyst 67:177–186. https://doi.org/10.1023/A:1025505315247

Luding S (2008) Introduction to discrete element methods: basic of contact force models and how to perform the micro-macro transition to continuum theory. Eur J Environ Civil Eng 12(7-8):785–826. https://doi.org/10.3166/ejece.12.785-826

Ma J, Zhu SG, Wu CX, Zhang ML (2009) Application of back-propagation neural network technique to high-energy planetary ball milling process for synthesizing nanocomposite WC-MgO powders. Mater Des 30:2867–2874. https://doi.org/10.1016/j.matdes.2009.01.016

Magini M, Iasonna A, Padella F (1996) Ball milling: An experimental support to the energy transfer evaluated by the collision model. Scr Mater 34:13–19. https://doi.org/10.1016/1359-6462(95)00465-3

Malayathodi R, Sreekanth MS, Deepak A, Dev K, Surendranathan AO (2018) Effect of milling time on production of aluminium nanoparticle by high energy ball milling. Int J Mech Eng Technol 9:646–652

Mastronardi E, Tsae P, Zhang X, Monreal C, DeRosa MC (2015) Strategic role of nanotechnology in fertilizers: potential and limitations. In: Nanotechnologies in food and agriculture. Springer, Cham. https://doi.org/10.1007/978-3-319-14024-7_2

Maurice DR, Maurice DR, Courtney TH, Courtney TH (1990) The physics of mechanical alloying – a 1st. Metall Trans A-Physical Metall Mater Sci 21:289–303

Miao Y, Mo K, Zhou Z, Liu X, Lan KC, Zhang G, Miller MK, Powers KA, Almer J, Stubbins JF (2015) In situ synchrotron tensile investigations on the phase responses within an oxide-dispersion-strengthened (ODS) 304 steel. Mater Sci Eng A 625:146–152. https://doi.org/10.1016/j.msea.2014.12.01

Mio H, Kano J, Saito F, Kaneko K (2002) Effects of rotational direction and rotation-to-revolution speed ratio in planetary ball milling. Mater Sci Eng A 332:75–80. https://doi.org/10.1016/S0921-5093(01)01718-X

Mio H, Kano J, Saito F (2004a) Scale-up method of planetary ball mill. Chem Eng Sci 59:5909–5916. https://doi.org/10.1016/j.ces.2004.07.020

Mio H, Kano J, Saito F, Kaneko K (2004b) Optimum revolution and rotational directions and their speeds in planetary ball milling. Int J Miner Proc 74. https://doi.org/10.1016/j.minpro.2004.07.002

Mondal A, Basu R, Das S, Nandy P (2011) Beneficial role of carbon nanotubes on mustard plant growth: an agricultural prospect. J Nanoparticle Res 13:4519–4528. https://doi.org/10.1007/s11051-011-0406-z

Monreal CM, Derosa M, Mallubhotla SC, Bindraban PS, Dimkpa C (2015) The application of nanotechnology for micronutrients in soil-plant systems

Mukhtar NZF, Borhan MZ, Rusop M, Abdullah S (2014) Recent trends in nanotechnology and materials science. Springer, Cham, pp 41–47. https://doi.org/10.1007/978-3-319-04516-0

Munkhbayar B, Nine MJ, Jeoun J, Bat-Erdene M, Chung H, Jeong H (2013) Influence of dry and wet ball milling on dispersion characteristics of the multi-walled carbon nanotubes in aqueous solution with and without surfactant. Powder Technol 234:132–140. https://doi.org/10.1016/j.powtec.2012.09.045

Parashar SKS, Choudhary RNP, Murty BS (2003) Ferroelectric phase transition in Pb0.92Gd0.08(Zr0.53Ti0.47)0.98O3 nanoceramic synthesized by high-energy ball milling. J Appl Phys 94:6091–6096. https://doi.org/10.1063/1.1618915

Patil AG, Anandhan S (2012) Ball milling of class-F Indian fly ash obtained from a thermal power station. Int J Energy Eng 2:57–62

Patil AG, Anandhan S (2015) Influence of planetary ball milling parameters on the mechanochemical activation of fly ash. Powder Technol 281:151–158

Paul KT, Satpathy SK, Manna I, Chakraborty KK, Nando GB (2007) Preparation and characterization of nano structured materials from fly ash: A waste from thermal power stations, by high energy ball milling. Nanoscale Res Lett 2:397–404. https://doi.org/10.1007/s11671-007-9074-4

Peterson SC, Jackson MA, Kim S, Palmquist DE (2012) Increasing biochar surface area: Optimization of ball milling parameters. Powder Technol 228:115–120. https://doi.org/10.1016/j.powtec.2012.05.005

Poopathi S, De Britto LJ, Praba VL, Mani C, Praveen M (2015) Synthesis of silver nanoparticles from Azadirachta indica—a most effective method for mosquito control. Environ Sci Pollut Res 22(4):2956–2963. https://doi.org/10.1007/s11356-014-3560-x

Prasad TNVKV, Sudhakar P, Sreenivasulu Y, Latha P, Munaswamy V, Raja Reddy K, Sreeprasad TS, Sajanlal PR, Pradeep T (2012) Effect of nanoscale zinc oxide particles on the germination, growth and yield of peanut. J Plant Nutr 35:905–927. https://doi.org/10.1080/01904167.2012.663443

Prasad R, Kumar V, Prasad KS (2014) Nanotechnology in sustainable agriculture: present concerns and future aspects. African J Biotechnol 13(6):705–713. https://doi.org/10.5897/AJBX2013.13554

Radune M, Radune A, Lugovskoy S, Zinigrad M, Fuks D, Frage N (2014) Mathematical modeling of High Energy Ball Milling (HEBM) Process. Defect Diffus Forum 353:126–130. https://doi.org/10.4028/www.scientific.net/DDF.353.126

Raghavendra G, Ojha S, Acharya SK, Pal SK (2014) Fabrication and characterization of nano Fly ash by planetary ball milling. Int J Mater Sci Innovations 2:59–68

Rajak DK, Raj A, Guria C, Pathak AK (2017) Grinding of class-F fly ash using planetary ball mill: a simulation study to determine the breakage kinetics by direct-and back-calculation method. S Afr J Chem Eng 24:135–147. https://doi.org/10.1016/j.sajce.2017.08.002

Rameshaiah GN, Pallav IJ, Shabnam S (2015) Nano fertilizers and nano sensors–an attempt for developing smart agriculture. Int J Eng Res Gen Sci 3(1):314–320. http://biozarco.ir/wp-content/uploads/2015/12/40.pdf

Rosenkranz S, Breitung-Faes S, Kwade A (2011) Experimental investigations and modeling of the ball motion in planetary ball mills. Powder Technol. https://doi.org/10.1016/j.powtec.2011.05.021

Rui M, Ma C, Hao Y, Guo J, Rui Y, Tang X, Zhao Q, Fan X, Zhang Z, Hou T, Zhu S (2016) Iron oxide nanoparticles as a potential iron fertilizer for peanut (Arachis hypogaea). Front Plant Sci 7(815). https://doi.org/10.3389/fpls.2016.00815

Saha N, Gupta SD (2017) Low-dose toxicity of biogenic silver nanoparticles fabricated by Swertia chirata on root tips and flower buds of Allium cepa. J Hazard Mater 330:18–28. https://doi.org/10.1016/j.jhazmat.2017.01.021

Sato A, Kano J, Saito F (2010) Analysis of the abrasion mechanism of grinding media in a planetary mill with DEM simulation. Adv Powder Technol 21:212–216. https://doi.org/10.1016/j.apt.2010.01.005

Sha W (2008) Comment on "Artificial neural network modeling of a mechanical alloying process for synthesizing of metal matrix nanocomposite powders" by Dashtbayazi et al. [Mater Sci Eng A 466 (2007) 274]. Mater Sci Eng A https://doi.org/10.1016/j.msea.2008.03.024

Shaviv A, Mikkelsen RL (1993) Controlled-release fertilizers to increase the efficiency of nutrient use and minimize environmental degradation – a review. Fertil Res. https://doi.org/10.1007/BF00750215

Singh DS (2012) Achieving second Green Revolution through Nanotechnology in India. Agric Situat India:545–572. https://eands.dacnet.nic.in/Publication12-12-2012/1485-jan12/1485-1.pdf

Su CH, Hou TH (2008) Using multi-population intelligent genetic algorithm to find the pareto-optimal parameters for a nano-particle milling process. Expert Syst Appl 34:2502–2510. https://doi.org/10.1016/j.eswa.2007.04.017

Subramanian KS, Tarafdar JC (2011) Prospects of nanotechnology in Indian farming. Indian J Agric Sci

Suppan S (2017) By Steve Suppan applying nanotechnology to fertilizer The Institute for Agriculture and Trade Policy works locally and globally at the intersection of policy and practice to ensure fair and sustainable food, farm and trade systems

Suryanarayana C (2001) Mechanical alloying and milling Suryanarayana. Mater Sci 46:1–184. https://doi.org/10.1016/S0079-6425(99)00010-9

112 C. Pohshna et al.

Tarafdar JC (2015) Nanoparticle production, characterization and its application to horticultural crops. Winter School – Utilization of Degraded Land and Soil through Horticultural Crops for Agricultural Productivity and Environmental Quality. Ajmer, Rajasthan, pp 222–229

Tripathi S, Sonkar SK, Sarkar S (2011) Growth stimulation of gram (Cicerarietinum) plant by water-soluble carbon nanotubes. Nanoscale 3:1176–1181. https://doi.org/10.1039/c0nr00722f

Varol T, Canakci A, Ozsahin S (2013) Artificial neural network modeling to effect of reinforcement properties on the physical and mechanical properties of Al2024-B4C composites produced by powder metallurgy. Compos Part B Eng 54:224–233. https://doi.org/10.1016/j.compositesb.2013.05.015

Wakihara T, Ihara A, Inagaki S, Tatami J, Sato K, Komeya K, Meguro T, Kubota Y, Nakahira A (2011) Top-down tuning of nanosized ZSM-5 zeolite catalyst by bead milling and recrystallization. Cryst Growth Des 11:5153–5158. https://doi.org/10.1021/cg201078r

Zakeri M, Ramezani M, Nazari A (2012) Effect of ball to powder weight ratio on the mechanochemical synthesis of MoSi2-TiC nanocomposite powder. Mater Res. https://doi.org/10.1590/S1516-14392012005000111

Zhang DL (2004) Processing of advanced materials using high-energy mechanical milling. Prog Mater Sci 49:537–560. https://doi.org/10.1016/S0079-6425(03)00034-3

Zhang FL, Wang CY, Zhu M (2003) Nanostructured WC/Co composite powder prepared by high energy ball milling. Scr Mater 49:1123–1128. https://doi.org/10.1016/j.scriptamat.2003.08.009

Zhang FL, Zhu M, Wang CY (2008) Parameters optimization in the planetary ball milling of nanostructured tungsten carbide/cobalt powder. Int J Refract Met Hard Mater 26:329–333. https://doi.org/10.1016/j.ijrmhm.2007.08.005

Zhao T (2017) Coupled DEM-CFD analyses of landslide-induced debris flows. Springer/Science Press, Singapore/Beijing, pp 1–220. https://doi.org/10.1007/978-981-10-4627-8

Zheng L, Hong F, Lu S, Liu C (2005) Effect of nano-TiO2 on strength of naturally aged seeds and growth of spinach. Biol Trace Elem Res 104:83–91. https://doi.org/10.1385/BTER:104:1:083

Chapter 4
Materials and Technologies for the Removal of Chromium from Aqueous Systems

Fayyaz Salih Hussain and Najma Memon

Abstract Chromium (Cr) enters into the environment through activities related to mining, various industrial and certain geological processes. Globally more than 170,000 tons of chromium - containing wastewater is discharged annually. Chromium is toxic for all living organisms, therefore, its removal is crucial to save plants and aquatic life. Due to its toxic and carcinogenic nature, efficient removal technologies and materials are continuously investigated and widely documented. This chapter focuses on recent technologies and materials employed to remove chromium from water bodies. Techniques such as membrane filtration, electro-chemical, and phytoremediation have emerged over the years. Sorptive removal of chromium is another approach that is investigated in detail for the development of efficient sorbent materials. Many natural and synthetic materials, such as carbon nanotubes, composite of nanomaterials, zeolites, biochar and others are reported as sorbents. Electrochemical precipitation has superseded conventional coagulation technology in terms of efficiency. However, it still requires investigation for industrial scale utilization and also, like all precipitation techniques, it produces chromium-containing sludge. On the other hand, functionalized materials with high surface area are promising candidates as sorbents.

Keywords Chromium removal · Wastewater treatment · Removal techniques · Adsorbents · Membrane technology · Electrocoagulation · Polypyrrole · Carbon nanotubes · Biocarbon · Activated carbon · Bio sorbents

Acronyms

CNTs	Carbon Nanotubes
APTS	3-aminopropyl-triethoxysilane
PVDF	Poly (vinylidene fluoride)

F. S. Hussain · N. Memon (✉)
National Centre of Excellence in Analytical Chemistry, University of Sindh,
Jamshoro, Sindh, Pakistan

© Springer Nature Switzerland AG 2020   113
E. Lichtfouse (ed.), *Sustainable Agriculture Reviews 40*, Sustainable Agriculture
Reviews 40, https://doi.org/10.1007/978-3-030-33281-5_4

DMAc	Dimethylacetamide
PVP	Polyvinyl pyrrolidone
ECP	Electrochemical precipitation
COD	Chemical oxygen demand
TSS	Total suspended solids
SWCNT	Single walled carbon nanotubes
PEUF	Polymer-enhanced ultrafiltration
MWCNT	Multi walled carbon nanotubes
BET	Brunauer, Emmett and Teller
PDMAEMA	poly(2-dimethylaminoethyl methacrylate)
TA	Tataric acid
GAC	Green Macroalgae Cladophore
AC	Activated carbon
PPy	Polypyrrole
GO	Graphene oxide
LDHs	Layered Double Hydroxides
CRIS	Constructed rapid infiltration systems
PEG	Polyethylene glycol
SLM	Supported liquid membrane
NF	Nano-filtration
CPC	Cetylpyridinium chloride
NAD(P)H	Nicotinamide adenine dinucleotide phosphate

4.1 Introduction

The industrial revolution has led to exponential growth of economic development but simultaneously caused severe damage to environment. Uncontrolled expansion produced enormous waste material from all the segments of industrialization, which ranges from small cottage to heavy industries. Waste products were generated in the form of chemicals, process waste and discarded material. Untreated waste material from those industries continuously polluted all spheres of environment i.e. hydrosphere, lithosphere and atmosphere, when that wastes cross the assimilation capacity of the environment, it created pollution. Control of industrial waste is a first-rate challenge, especially in developing countries. Large quantities of industrial wastes are being dumped on the soil floor, resulting in eco-unfavorable consequences and due the excessive cost of treatment, little or no attention is being given to proper disposal of built-up waste (Abbas et al. 2016; Alamgir and Kanwal 2018).

Among huge number of pollutants (metals) that industries generate, one of them is chromium. It is the 7th maximum abundant element in the world and the twenty-first within the crustal rocks. Chromium naturally occurs in different form of crustal rocks, predominantly as ferro chromite ($Fe_2Cr_2O_4$) however, it does not contribute to chromium contamination but main source of pollution is various industrial processing units. It is used in different industries like refractory, pigments, electroplating, chemical, metallurgical, and tanning; whereas major contributor to chromium pollution is tanning industry. Disposal of chromium contaminated sludge is one of the major problem of tannery industry besides huge amount of untreated wastewater (Alamgir and Kanwal 2018). Chromium containing sludge results from coagulation treatment process, which is currently employed in the industry to treat chromium waste. Out of the total world production of 31,000 metric tons which is marketable gross weight of chromite ore, more than 60–70% is used in alloy preparation and stainless steel industry The Fig. 4.1 shows the different industrial processes, which use chromium more than 15% for industrial scale products (Papp 1999).

Depending upon the type of activities, industries can be classified based on toxicity and quantity of waste released into environment. A metallurgy and heavy industry use major fraction of chromium in solid forms whereas chemical industries utilize chromium in soluble form therefore later generate solvated chromium that can easily travel and contaminate the environment. At present, more than 4000 tanning industries are using chromium for tanning process. Effluents from tannery industry in the form of Cr(VI) and Cr(III) salts is loaded with about 40% of the spent chromium (Sundaramoorthy et al. 2010). Besides natural rocks, major sources of chromium are effluents from various industries, ferrochromium slag, and solid wastes containing chromium as by products, leachates and dust particles where chromium concentration is found strikingly above permissible limits.

Solvated chromium exists mainly in two oxidation states; Cr(III) and Cr(VI), where former is positively charged species and later remain as oxoanion, $Cr_2O_7^{2-}$ and others. Migration and toxicity of chromium in soil and water depends upon its oxidation state (Mubarak et al. 2016). In its oxidized Cr(VI) form, chromium is quite soluble in water and therefore can easily penetrate cell wall, whereas the

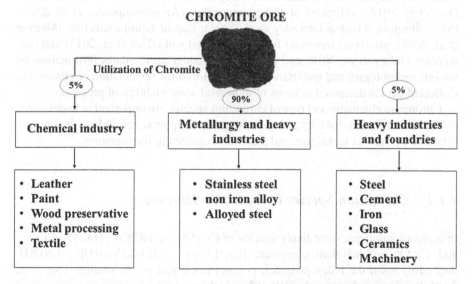

Fig. 4.1 Chromite is the major ore of chromium. Chromium has a wide range of industrial uses. Chromium is utilized in in various industries during different industrial processes

reduced Cr(III) form is less soluble in water thus its movement in environment is somewhat restricted (Oze et al. 2007). Cr (III) concentration of 5–3000 mg per gram of soil is also reported (Polti et al. 2011) which proves the contamination of soils with chromium. Uptake of chromium from soil to plant is also reported (Ertani et al. 2017; Shanker et al. 2005). Therefore, Cr enters in food chain through eating and from drinking of contaminated water. Intake of chromium has enormous effects on living organism and plants due to their toxicity and bioaccumulation (Kozlowski and Walkowiak 2005; Polti et al. 2011). Details on adverse health effects are given in another section of this chapter.

Main ecological toxic burden of chromium is anthropogenic and concerned with industrial operations (Alloway 2013) and due to its toxic nature, its release in industrial effluents is regulated. World health organization (WHO) recommends maximum permissible limits for the discharge of Cr(VI) into inland surface and drinking water is 0.05 mg/L. To achieve these limits of chromium in discharge waters, flocculation is most commonly adopted method for treatment of chromium contaminated wastewater. Therefore, new strategies are always sought to tackle chromium issues; in industrial wastewaters at point-of discharge, in contaminated surface/ground water bodies, and contaminated soil.

Last two decades can be labeled as revolution in engineering of new materials and techniques. A huge number of novel approaches have emerged in literature for removal of chromium as well. This chapter is an effort to evaluate the existing literature and to come up with solutions for environmentally sustainable options for treatment of chromium from industrial wastewaters. There are many separate reviews on chromium removal using silica based sorbents (Dinker and Kulkarni 2015), natural minerals (Dimos et al. 2012), other adsorbents (Jung et al. 2013), treatment of chromium in tannery effluents (Kumar and Pandey 2006), reduction techniques (Barrera-Díaz et al. 2012), utilization of carbon nanotubes (Anagnostopoulos et al. 2017), Phyto-filtration (Gardea-Torresdey et al. 2004), Liquid-liquid extraction (Memon et al. 2004), microbial treatment for contaminated soil (Dhal et al. 2013) and biosorption (Jobby et al. 2018) and more. This chapter accumulates information on various technologies and materials that are explored for remediation of chromium. Collected data is discussed in terms of industrial scale viability of processes.

Chromium chemistry and type of chromium species are important in understanding of accumulation, toxicity and mobility; this chapter starts with such concepts followed by various techniques and materials reported in the literature.

4.1.1 Chromium Species in Aqueous Solution

In aqueous solutions, most likely species of Cr (VI) are $HCrO_4^-$, H_2CrO_4, $Cr_2O_7^{2-}$ and CrO_4^{2-} and trivalent chromium $[Cr(H_2O)^6]^{3+}$, $[(H_2O)_5Cr(OH)]^{2+}$, $Cr(OH)_3$ depending upon the redox potential, concentration and pH of solution (Ng et al. 2010; Sundarapandiyan et al. 2010). Chemical properties, reactivity and biological role of chromium depends on its oxidation state; Cr(VI)/Cr(III) which are interchangeable. The electric potential difference between Cr(III) and Cr(VI) indicates the strong oxidizing potential of Cr(VI) and considerable amount of energy (1.33 eV) is needed to convert Cr(VI) to Cr(III). Cr(III) does not get oxidized in living system while reduction of Cr(VI) occur spontaneously. In blood Cr(VI) is quickly reduced to Cr(III) but once its reduced to Cr(III) it is unable to leave cell because it is bound to cell components (Dayan and Paine 2001). Cr(III) hydroxy species would be dominated in aquatic environment and their mobility greatly depends upon type of species. Cr(III) is reported to form following hydroxy compounds in an aqueous solution; Cr^{3+}, $Cr(OH)_2^+$, $Cr(OH)_3$, $Cr(OH)_4^-$ and $Cr_3(OH)_4^{5+}$ as well (Latimer 1964; Rai et al. 1987). Correlation of various species as functions of pH is explained A plot of pH and log[Cr] is reproduced in Fig. 4.2. It represents

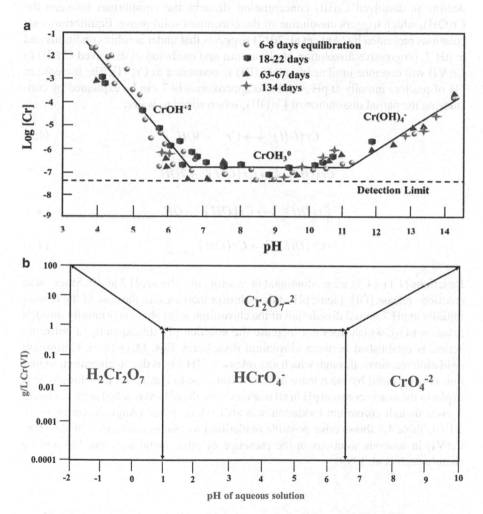

Fig. 4.2 (a) Graph of pH versus Log[Cr], showing dominancy of Cr(III) hydroxyl species as function of pH. (b) Relative distribution of Cr(VI) species in water as function of pH and Cr(VI) concentration. Adapted from Pourbaix 1974 with permission

$Cr(OH)^{2+}$, $Cr(OH)_3(s)$ and $Cr(OH)_4^-$ at pH 3.5–6.4, 6.4–11.5 and > 11.5 exists respectively. This shows that Cr(III) hydroxyl occurs as amphoteric form in aqueous solutions. Fig.4.2(a) also explains the stability of Cr(III) in aqueous solution with pH below 5 upto 1 week, however, Cr(VI) is stable in whole pH range (Fig.4.2b).

Soluble Cr(III) in any form is considered unstable due oxidation to Cr(VI). Apte et al. studied conversion rate of Cr(III) to Cr(VI) as function of pH (Apte et al. 2017). These reactions were carried out to ascertain the effect of initial pH on conversion rate in aqueous suspensions containing $Cr(OH)_3$ under aerobic conditions. The dissolved Cr(III) concentration was monitored in all reactors, initially at pH values of 3, 5 and 7, respectively. Dissolution of Cr(III) was found in equilibrium with the $Cr(OH)_3$ solid phase. When Cr(III) is oxidized to Cr(VI), the resultant decline in dissolved Cr(III) concentration disturbs the equilibrium between the $Cr(OH)_3$ which triggers dissolution of the chromium solid phase. Equilibrium calculations presented by (Apte et al. 2017) suggests that under aerobic conditions and at pH 7, progressive dissolution of chromium and oxidation of dissolved Cr(III) to Cr(VI) will continue until nearly all Cr(III) is converted to Cr(VI). The increase in pH of reactors initially at pH 3 and 5 to approximately 7 can be explained by considering the partial dissolution of $Cr(OH)_3$ when added to water,

$$Cr(OH)_3 \downarrow \rightarrow Cr^{3+} + 3OH^- \tag{4.1}$$

$$Cr(OH)_3 \downarrow \rightarrow Cr(OH)^{2+} + 2OH^- \tag{4.2}$$

$$Cr(OH)_3 \downarrow \rightarrow Cr(OH)^{2+} + OH^- \tag{4.3}$$

$$Cr(OH)_3 \downarrow \rightarrow Cr(OH)_{3(aqueous)} \tag{4.4}$$

Reactions (4.1)–(4.3) are predominant in reactors initially at pH 3 and 5. Since these reactions release [OH⁻] ions, pH of the solution increases. In the case of the reactor initially at pH 7, initial dissolution of the chromium solid phase is primarily through reaction (4.4), which does not increase the solution pH subsequently, a buffering action is established between chromium dissolution Eqs. (4.1)–(4.4). Chromium oxidation reactions, through which any release of [H⁺] ions due to chromium oxidation is neutralized by the release of [OH⁻] ions due to chromium dissolution. This explains the nearly constant pH in all reactors once the pH has reached approximately 7 even though chromium oxidation was still taking place (Augustynowicz et al. 2010). Table 4.1 shows other possible oxidation process for conversion of Cr(III) to Cr(VI) in aqueous solutions in the presence of other metal ions like Mn and Fe (Kimbrough et al. 1999).

Table 4.1 Various reaction of chromium in an aqueous solution which shows the conversion of Cr(III) to Cr(VI) in the presence of different metal ions

Oxidation	E (V)
$3MnO_2 + 2Cr(OH)_3 \Leftrightarrow 3Mn^{2+} + 2CrO4^{2-} + 2H_2O + 2OH^-$	1.328
$2Cr^{3+} + 5H_2O + 3O_3 \Leftrightarrow 2CrO4^{2-} + 10H^+ + 3O_2$	0.87
$2Cr^{3+} + 2H_2O + 3H_2O_2 \Leftrightarrow 2CrO4^{2-} + 10H^+$	0.58
$2Cr^{3+} + 3H_2O + 2MnO4^- \Leftrightarrow Cr_2O7^{2-} + 6H^+ + 2MnO_2$	0.35
$2Cr^{3+} + 7H_2O + 6Mn^{3+} \Leftrightarrow Cr_2O7^{2-} + 14H^+ + 6Mn^{2+}$	0.18
$2Cr^{3+} + H_2O + 3PbO_2 \Leftrightarrow Cr_2O7^{-2} + 2H^+ + Pb^{2+}$	0.13
Reduction	E (V)
$2HCrO4^- + 5H^+ + 3HSO_3^- \Leftrightarrow 2Cr^{3+} + 5H_2O + 3SO_4^{2-}$	2.115
$HCrO4^- + 3\ V^{2+} + 7H^+ \Leftrightarrow Cr^{3+} + 3V^{3+} + 4H_2O$	1.45
$2HCrO4^- + 3H_2S + 8H^+ \Leftrightarrow 2Cr^{3+} + 5H_2O + 3S$	1.18
$HCrO4^- + 3Fe^{2+} + 7H+ \Leftrightarrow Cr^{3+} + 3Fe^{3+} + 4H_2O$	0.56
$2HCrO4^- + 5H^+ + 3HNO_2 \Leftrightarrow 2Cr^{3+} + 5H_2O + 3NO_3^-$	0.35

4.1.2   Chromium Toxicity and Legislation

Due to its high mobility chromium can simply infiltrate the wall of cell and its toxic nature cause numerous cancer diseases (Janssen and Koene 2002). It is not only toxic to human but also for animal, plants, bacteria, algae and fungi. The health hazards of exposure to Cr(VI) and Cr(III) are well documented by the World Health Organization and the Agency for Toxic Substances and Disease Registry (Tao and Michael Bolger 1999). Cr(VI) is listed by the United States Environmental Protection Agency (USEPA) among seventeen chemicals posing greatest threat to humans (Cheung and Gu 2007). Owing to a very high positive redox potential, Cr crosses cell membranes damaging the cellular and molecular components of the cell leading to membrane disruption, protein degradation and DNA alterations in humans, animals and plants (da Silva Pereira et al. 2010).

Hexavalent chromium induces mutation by interfering with DNA protein cross-links and causes single-strand breakage (Shanker et al. 2005). It damages kidney and liver functions and may cause epigastric pain, nausea, vomiting, allergic reactions, stomach ulcers, and hemorrhage (Mancuso 1997). In plants and many other organisms, reducing agents such as NAD(P)H, FADH2, several pentoses and glutathione in the cell pool, reduce Cr(VI) to Cr(III) (Hossain et al. 2012). During this conversion, transient formation of chromium unstable states occurs leading to free radical formation, which induces oxidative stress conditions in plants. In adult human subjects, the lethal oral dose is 50–70 mg soluble chromates per kilogram body weight. The clinical features of acute poisoning are vomiting, diarrhea, hemorrhage and blood loss into the gastrointestinal tract, causing cardiovascular shock. If the patient survives for more than about 8 days, the major effects resulting from oral ingestion of toxic doses of chromium are liver and kidney necrosis. Although parenteral administration of chromium to experimental animals can lead to teratogenic effects, birth defects have not been associated with human exposure to chromium (Sundaramoorthy et al. 2010). Globally more than 170,000 tons of chromium containing wastewater is discharged into environment annually (Gadd and White 1993). WHO and Canadian legislation recommends that chromium concentration should not exceed 0.05 mg/L in drinking water whereas US-EPA recommends 0.01 mg/L (Edition 2011) (Davies and Mazumder 2003; Sutton 2010).

4.2 Techniques for Chromium Clean-up from Contaminated Water

There are two type of techniques for chromium removal; (1) conversion of metal from one oxidation to another (redox systems) and (2) accumulation of metal ions through adsorption, ion-exchange or membrane processes (adsorptive systems) or amalgam formation (precipitation systems). Most of the reported systems are discussed in following sections.

4.2.1 Coagulation/Flocculation and Electrochemical Precipitation

Coagulation or flocculation is used to destabilize the charged particles of suspended solids. Industries utilize inorganic coagulants i.e. ferric chloride ($FeCl_3$), aluminum sulphate ($AlSO_4$) etc. to reduce the organic load, suspended solids to eliminate toxic heavy metals i.e. chromium (Lofrano et al. 2006). Fe(II) sulfate converts Cr(VI) to Cr(III) to nearly 100% by forming of precipitate of $Cr(OH)_3$ (Faust and Ali 1998; Lee and Hering 2003; Qin et al. 2005). Coagulation/flocculation can be carried by other chemicals also or may be assisted through electrochemical process. Table 4.2 shows chemical and electrochemical process reported for removal of chromium, respectively.

Song et al. used ferric chloride and aluminum sulphate for the treatment of organic carbon and toxic heavy metals. In this case, each coagulant works effectively at specific and favorable pH. Alum is used as coagulant, which removes 30–37% of Chemical oxygen demand (COD), 38–46% of suspended solids (SS) and 74–99% of chromium by using optimum alum dose i.e. 800 mg/L from tannery wastewater. During the treatment process, ferric chloride gives better results than aluminum sulphate as coagulant. The coagulation study was investigated by Ayoub et al. by using lime and bittern monitored by activated carbon for the treatment of tannery effluent (Ayoub et al. 2011). The pH was maintained by using lime slurry up to 11.3. This results in remediation of total suspended solids 97%, total chromium 99.7% and phosphorous 87%. The chemical oxygen demand and biological oxygen demand (BOD) removal obtained up to 71% and 57%. Kabdasli at el reported greater than 99% removal of chromium by using alum, $FeCl_3$ and $FeSO_4$

Table 4.2 Coagulation and flocculation methods including redox assisted, sulfide precipitation, conventional coagulation with alum and Fe(III) and electrochemical reduction, their advantages and disadvantages for removal of chromium

Technology	Process	Advantage	Disadvantages	References
Coagulation with Fe(II) (redox assisted)	Fe(II) reduced Cr (VI) to Cr(III) and Fe(II) oxidized to Fe(III) than Cr(III) adsorbed or co precipitated as Fe(III) hydroxides	Effective removal 99% with use of excess Fe(II)	Performance effected by floc and filterability of solid precipitated. Additional treatment required for floc removal	Lee and Hering (2003) and Qin et al. (2005)
Precipitation with sulfide	Reduction of Cr(VI) with sulfide and formation Cr(III) precipitate as chromium sulfide	Cr(VI) reduced and chromium precipitate in single step	Relatively expensive and not suitable for drinking water treatment application	Calder (1988)
		Less metal residue than hydroxide faster settlement of metal		
Coagulation precipitation (conventional)	Coagulation with alum and Fe(III) salts precipitation with lime, caustic soda and sodium carbonate (to increase the pH) two-stage process for Cr(VI) removal: Reduction of Cr (VI) to Cr (III) and precipitation/ filtration of Cr(III) reduction proceeds rapidly at low pH reducing agents most commonly employed are sulfur dioxide, sodium sulfite and Fe(II) sulfate	Low capital and O&M costs low pretreatment requirement effective for Cr(III) removal recovery of chromium for recycling reaction time is short	Two-stage process for Cr(VI) removal precipitation is often ineffective if metals are complexed or if they are present as anions. Removal of the microflocs formed is often difficult and critical for process efficiency. Produces high volume of sludge	Beszedits (1988) and Calder (1988)
Electrochemical reduction	Uses consumable iron electrodes and an electric current to generate Fe(II) ions which reacts with Cr(VI) to give Cr(III)	Reaction occurs rapidly and requires minimum retention time	Increased quantity of sludge due to additional precipitation of iron hydroxide	Calder, (1988) and Mukhopadhyay et al. (2007)

(Kabdaşlı et al. 1999). Removal of chromium using electrocoagulation process where Fe(II) is either added or electrogenerated to reduce Cr(VI) to Cr(III) is reported. Then pH of solution is increased which favors the formation of precipitates of $Cr(OH)_3$. Electrochemical reduction using iron sacrificial electrode in various operating conditions shows excellent removal of Cr(VI). Un et al. (2017) achieved 100% removal using current density 20 mA/cm^2, pH 2.4 and 0.05 M NaCl electrolyte, initial Cr(VI) concentration of 1000 mg/L, time 25 min and energy cost 2.68 kWh/m^3. Similar reports with good efficiency for removal of Cr(VI) are carried out using iron as electrodes (Akbal and Camcı 2010; Cheballah et al. 2015; El-Taweel et al. 2015; Esmaeili et al. 2015; Espinoza-Quiñones et al. 2012; Hamdan and El-Naas 2014a; Taa et al. 2016; Verma et al. 2013; Zewail and Yousef 2014).

Iron electrodes in continuous flow operating conditions were reported and found equally efficient in terms of removal where final concentration of 0.2 mg/L was achieved (Hamdan and El-Naas 2014b; Kongsricharoern and Polprasert 1995).

Electrodes other than iron have emerged in the literature for electro-precipitation of chromium. Selvaraj et al. (2018) recovered of Cr(III) without oxidizing into Cr(VI) from spent liquor tannery effluent by using two compartment electro-floation reactor. In a two compartment, membrane electro-chemical reactor RuO_2/TiO_2-Ti cell was used as anode and Ti as cathode while 0.01 N H_2SO_4 was used as electrolyte. Cell was separated by Nafion 117 membrane where more than 98% of chromium was recovered in the form of $Cr(OH)_3$. An electrochemical cell consist of three electrode, Ag/AgCl and graphite as a counter electrode while aluminum copper foil was established to remove chromium from real sample of tannery wastewater having concentration of 2654 mg/L and 2775 mg/L. Cr(VI) removal was more than 99% in the form of insoluble $Cr(OH)_3$ using cyclic voltametry at 50 mV/s at temperature 25 °C (Ramírez-Estrada et al. 2018). Enhancement in removal of Cr(VI) ions was observed significantly when carbon aerogel electrode was used at high charge conditions and low pH. It was observed that on decreasing pH the removal of metal ion was reduced up to 98.5% at acidic conditions (Rana et al. 2004). It was observed that at high charge density, treatment time and energy consumption was reduced after treatment, residual concentration of Cr(VI) reached to 0.5 mg/L.

Table 4.3 shows few more applications for the remediation of Cr(VI) from effluent by using electrochemical precipitation. Electrodes other than iron in electrocoagulation are not comparable in removal efficiency for chromium at higher concentration. On the other hand iron electrodes are consumable and generate waste in the form of chromium sludge.

Chemical or electrocoagulation despite being efficient process requires removal and proper disposal of sludge containing chromium. Precipitates also need to be removed quickly from aqueous solution as Cr(III) is unstable and gets converted to Cr(VI) especially at basic pH (refer to Fig. 4.2). However, regeneration of sludge back to chromium and iron into the compounds useful for industrial applications would be of interest.

Table 4.3 Electrocoagulation of chromium using electrodes including carbon nanotube electrode, modified glass electrode, silver nanoparticle coated electrode, graphene polyaniline, mild steel electrode, iron aluminum electrode, polypyrrole coated aluminum, carbon aerogel, stainless steel nets coated with single wall carbon nanotubes, carbon aerogel, titanium electrodes, microbial fuel cell graphite and Iron rotary

Electrodes	Initial conc. (mg/L)	Time (h)	pH	Removal (%)	References
Carbon nanotube poly vinyl alcohol ultrafiltration membranes	1.0	4.0	6.6	99.0	Duan et al. (2017)
Binder-free carbon nanotube electrode	12	1.5	3.0	97.0	Na and Wang (2016)
Liquid crystal coated polaroid glass electrode	100.0	24	2.0	99.8	Gangadharan et al. (2015)
Fe-Fe	887.2	1	4.0	100.0	Bazrafshan et al. (2015)
Au NPs TiO$_2$ NTs	20.0	3	11.0	100.0	Jin et al. (2014)
Polyaniline graphene polyaniline	50.0	2	0.5	95.0	Yang et al. (2014b)
RuO$_2$/Ti	5.0	2	4.4	99,0	Dharnaik and Ghosh (2014)
Carbon nanotubes	12.0	1.9	3.0	96.0	Wang and Na (2014)
Electrocoagulation with an Fe-Al electrode	44.5	0.3	3.0	100.0	Akbal and Camcı (2011)
Mild steel electrode	50.0	0.58	2.0	100.0	Golder et al. (2011)
Stainless steel nets coated with single wall carbon nanotubes	10.0	4.2	4.0	99.0	Liu et al. (2011)
Reticulated vitreous carbon	100.0	2	2.0	100.0	Rodriguez-Valadez et al. (2005)
Polypyrrole coated aluminum	350.0	3	0.7	100.0	Conroy and Breslin (2004)
Carbon felt	3–50.0	13	2.0	100.0	Frenzel et al. (2006)
Carbon aerogel	2.0	–	2.0	99.6	Rana-Madaria et al. (2005)
Iron rotary	130.0	–	8.5	99.6	Chen (2004)
Stainless steel electrodes	147.0	1.16	1.8	100.0	Sanjay et al. (2003)
Titanium electrodes	405.6	–	<4.0	87.0	Herrmann (1999)
Microbial fuel cell graphite	100.0	150	2–6	100.0	Yan Li et al. (2009)
Microbial fuel cell (MFC),graphite plate anode, rutile coated graphite plate cathode	26.0	26	2.0	97.0	Yan Li et al. (2009)

4.2.2 Membrane Filtration

Semipermeable membrane with very narrow pores is involved in membrane filtration. Retention of particles, compounds or ions depends upon pore size as well as on selective binding sites in the pores or on the surface on membrane. Mechanical and chemical stability are also important factors for successful operation of membrane-based systems. Therefore, membrane treatment process has to be explored in the terms of efficiency, cost and energy-effectiveness (Chelme-Ayala et al. 2009). Membrane has capability to remediate toxic metals i.e. chromium, arsenic and others. Several types of membrane techniques are employed i.e. microfiltration, polymeric, inorganic and liquid membrane for the removal of chromium. Membrane filtration is very popular but disposal of concentrate from RO and NF after water treatment create environmental problem as concentrate contain organic and inorganic contaminant and considered as disadvantage (Van der Bruggen et al. 2003). Membranes classification is based on porosity i.e. micro, nano and type i.e. supported or Inorganic and state i.e. liquid. Membranes employed for chromium removal are discussed below.

4.2.2.1 Micro and Nano Filtration

Microfiltration is a process used for the treatment of industrial wastewater. The pore size of microfiltration membrane ranges from 0.1 μm–10 μm. The separation through microfiltration process is usually carried at low pressure to overcome the barrier of pressure (Dutta 2007). This technique can be used with polymers, polyelectrolytes, biomass and surfactants which improve the efficiency of chromium removal by complex formation or adsorption (Aroua et al. 2007; Daniş and Keskinler 2009; Ge et al. 2013; Ghosh and Bhattacharya 2006; Ramrakhiani et al. 2011; Zeng et al. 2014).

Suresh M. Doke used surfactant enhanced Titania membrane prepared from polymeric solgel method. Micellar Enhanced Micro Filtration (MEMF) was used with a cationic surfactant cetylpyridinium chloride (CPC). A porous ceramic membrane was prepared from nano-crystalline titanium oxide powder (TiO_2) to obtain a stable structure. Membrane was sintered for 1 h at 450 °C and was characterized through Scanning Electron Microscope (SEM), X-Ray Diffraction(XRD) and mercury porosimetry. Results show pore size of 0.58 μm, porosity 0.32 and water permeability of membrane 1049 L/m².h.bar. A 99% removal was found under optimized conditions, surfactant to chromate ratio (CPC/Cr) 2.5, initial chromate concentration 100 mg/L, pressure 1 bar (Doke and Yadav 2014).

Bao et al. (2015) synthesized modified ultrafiltration membrane functionalized with amine using Mobil Composition of Matter No.41 (MCM-41). Equilibrium for chromium was achieved in 5 min which shows faster kinetics, however, membrane was good at initial concentration of 0.5 mg/L. Vasanth et al. (2012), used Baker's yeast ceramic biomass assisted microfiltration for the removal of Cr(VI). Membrane was prepared by mixing of kaolin, quartz and calcium carbonate 2:1:1 respectively.

Aroua et al. (2007) used polymer-enhanced ultrafiltration (PEUF) for the treatment of Cr(VI) from wastewater. Water-soluble polymers i.e. pectin, chitosan, poly-ethyleneimine (PEI) were used. The ultrafiltration process equipped with polysulfone hollow fiber membrane was also used. It was found that the pH is the important factor for the elimination of chromium. Bohdziewicz utilized polyacrilonitrile membrane for the treatment of chromium from industrial wastewater. The results obtained shows that 17.5% polymer eliminate 98% of chromium ions and the remaining ions were passed from membrane and present in water (Bohdziewicz 2000). Muthukrishnan investigated the different nano-filtration membranes for the remediation of chromium ions by changing its pH. Two types of nano-filtration membranes are employed i.e. high rejection membrane (NFI) and a low rejection membrane (NFII). The elimination of chromium was augmented by increasing the pH of solution.(Muthukrishnan and Guha 2008) Also, aromatic polyamide thin film membrane was utilized for the treatment of chromium from aqueous system (Hafiane et al. 2000). The results obtained shows that the elimination rate depended on pH and ionic strength. The elimination of chromium was decreased by increasing the ionic strength and at basic pH the better results were obtained.

The non-interpenetrating modified ultrafiltration carbon membrane by gas nitration using NO_x with hydrazine hydrate is reported (Pugazhenthi et al. 2005). The supported membrane is used for the removal of Cr(VI) ions. The pore radius of aminated carbon, nitrated and unmodified membrane is observed 3.3, 2.8 and 2.0 nm. The water flux of modified membrane is increased two times as compared with unmodified membrane. The removal efficiency of Cr(VI) ions by using aminated carbon (88%), nitrated (84%). Some other reports on removal of chromium using membrane filtration are given in Table 4.4.

4.2.2.2 Liquid Membranes

These membranes consist of liquid phase a thin oil film which exists either in supported or non-supported form that serves as membrane barrier between two phases of aqueous solution. There are two types of liquid membrane emulsion liquid membrane (ELM) and immobilized liquid membrane (ILM) which is also called a supported liquid membrane (SLM). An emulsion liquid membrane is consisting of "bubble within bubble". The inner bubble is receiving phase and outer bubble is separation phase, which contains the carriers. The emulsion liquid membrane can be affected by ionic strength and pH. An immobilized membrane is simpler to visualize. It is made-up of rigid polymer membrane with several microscopic pores, which is filled with organic liquid through which separation occurs. Liquid membrane is used as effective and inexpensive method for the treatment of toxic

Table 4.4 Removal of chromium using various membranes

Membrane	Conditions	Removal (%)	References
Photocatalytic couple nanoparticle multilayer membrane	The water flux 39.7 L/m^2, initial conc. 10 mg/g, pH 3.0, pressure 5 bar and temperature 25 °C.	99.1	Kazemi et al. (2018)
Polymeric membrane	Time 50 min, pH 5.0, applied current 0.6 amp, temperature 25 °C, donor phase: 2×10^{-4} M K$_2$Cr$_2$O$_7$ in 0.1 M HCl, acceptor phase pH 5.0 acetic acid/ammonium acetate buffer solution.	99.7	Onac et al. (2018)
Polyvinylidene fluoride/2-Amino-4-thiazoleacetic acid ultrafiltration membrane	Water flux 318.11 L/m^2.h, contact angle was 81.2°, adsorption capacity 165 µg/cm^2.	–	Zhou et al. (2018)
Poly(2-dimethylaminoethyl methacrylate) PDMAEMA polymer-enhanced ultrafiltration	Adsorbent capacity 165 mg/g polymer 25 mg L^{-1} in the feed, pH 4.0, polymer Cr molar ratio of 40:1, pressure 1.0 bar	100.0	Sánchez et al. (2018)
CuO/hydroxyethyl cellulose composite ceramic membrane	Water flux (22.19 L m^{-2} h^{-1} at 0.5 bar TMP, time 120 min, pressure 3 bar	91.44	Choudhury et al. (2018)
Chitosan/polyvinyl alcohol(PVA)/ polyethersulfone (PES) dual layers nanofibrous membrane	The maximum adsorption capacity 509.7 mg/g initial conc. 100 mg/L, pH 2.0, adsorbent dosage of 0.5 g/L, contact time 60 min and temperature 30 °C.	–	Koushkbaghi et al. (2018)
1-octanol/polyvinyl chloride in polymer inclusion membrane	P 43.38 µm·s^{-1}, Cr(VI) conc. Feed phase 1 mg/L, current 30 V, pH 2.0, initial current density 0.1 A.	>90.0	Meng et al. (2018)
Chitosan thin films	BET surface area 88.47 m^2/g, time 40 min, pH 4.0, initial conc. 10 mg/L	100.0	Nayak et al. (2015)
Modified ultrafiltration membrane NH$_2$-MCM-41	BET surface area 437.2 m^2/g, adsorption capacity of 2.8 mg/g, time 5 min, initial conc. 0.5 mg/L, pH 3.5	–	Bao et al. (2015)
Supported liquid membrane using trioctylphosphine oxide	Initial conc. 19.2×10^{-4} mol L^{-1}, feed phase 1.5 mol L^{-1} H$_2$O$_2$, 0.1 mol L^{-1}, conc. Membrane phase and 0.001 mol L^{-1} DPC and 1.5 mol L^{-1} H$_2$SO$_4$ as stripping phase. Time 180 min, stable up to 10 days.	80	Nawaz et al. (2016)
Cationic hydrophilic polymers coupled to ultrafiltration membranes	pH 9.0, maximum retention capacity 164 mg Cr(VI)/g polymer, initial conc. 30 mg/L, pressure 1 bar,	>90.0	Sánchez and Rivas (2011)

(continued)

Table 4.4 (continued)

Membrane	Conditions	Removal (%)	References
Supported ionic liquid membrane containing CYPHOS IL101.	Initial conc. 7 mg L^{-1}, 0.01 mol/L HCl, time 5 h, permeability of 7.4 × 10^{-5} m s^{-1}, initial strip permeability of 2.0 × 10^{-5} m s^{-1}, initial fluxes of 5.5 × 10^{-6}, mol m^{-2} s^{-1} for the feed and 2.3 × 10^{-6} mol m^{-2} s^{-1} for the strip solutions.	90.0	de San Miguel et al. (2014)
Thiol-modified cellulose nanofibrous composite membranes	Adsorption capacity 87.7 mg/g, initial conc. 50 mg/g, pH 5.0, time 30 min, temperature 25 °C	>93.0	Yang et al. (2014a)
Baker's yeast ceramic membrane	Membrane flux (2.07 × 10^{-5} m^3/m^2 s, pressure 207 kPa, pH 1.0, initial conc. 100 mg/g	94.0	Vasanth et al. (2012)
Polymer inclusion membrane	Time 40 min, pH 5.0, applied voltage 70 V, applied current 0.6A	98.3	Kaya et al. (2016)
Peroxyacyl nitrates (PAN) based ultrafiltration membrane	Initial conc. 25 mg/g, pH 8.06, cross flow velocity 0.72 ms^{-1} and 0.05 ms^{-1} and transmembrane pressure 25 kPa & 200 kPa, temperature 25 °C	90.0	Muthumareeswaran et al. (2017)

heavy metals. Venkatesan and Meera Sheriffa Begum (2009) synthesized Emulsion Liquid Membrane (ELM) was water/oil/water or (w/o/w). The process of Emulsion Liquid Membrane is couple transport mechanism in which metal ion and the carrier present in membrane coupled and carrier exchanges metal ion with appropriate ion present in the stripping phase (Babcock et al. 1980; Kobya et al. 1997; Kunungo and Mohapatra 1995). Table 4.5 lists some of the processes reported for removal chromium using liquid membranes. Liquid membrane based reactors for removal of chromium seems an interesting option as compared to coagulation. Unlike precipitation based process liquid membranes can selectively preconcentrate and extract chromium from wastewaters which can be reused after little treatment.

4.2.3   Ion Exchange Methods

Ion Exchanger is the solid resins having capacity to exchange cations or anions from surrounding medium. These resins are natural solid or synthetic organic materials (Bashir et al. 2018) which exchange charged ions positively or negatively from an electrolyte solution (Barakat 2011).

Disadvantages of ion exchange process are high cost, not suitable for concentrated solution and incomplete removal of certain ions. In spite of this, there are many attempts reported in literature as shown in Table 4.6 regarding removal of

Table 4.5 Removal of chromium using liquid membrane with different membrane and strip phases

Membrane phase	Membrane/method	Strip phase	References
Tri-n-octylamine(TOA)/o--nitrophenyl pentyl(ONPPE) ether/dichloromethane	Polymer inclusion membrane	NaOH	Kozłowski (2007)
Aliquat 336/tertiary amines	Polymer inclusion membrane	NaOH	Kozlowski and Walkowiak (2005)
Aliquat 336/kerosene	Polytetrafluoroethylene (PTFE) membrane	HNO_3 or $NaNO_3$	Soko et al. (2002)
Alamine 336/kerosene	Emulsion liquid membrane	NaOH	Chakraborty et al. (2005)
Methylcholate/toluene	Polyvinylidene difluoride (PVDF)	HNO_3	Benjjar et al. (2012)
Cyanex 921/Solvesso 100	Polyvinylidene difluoride (PVDF)	H_2N-NH_2	Alguacil et al. (2003)
Alamine 336	Artificial neutral networks	NaOH	Eyupoglu et al. (2010)
Trioctylamine	Supported liquid membrane (SLM)	NaOH	Nawaz et al. (2016)
Alamine 336	Supported liquid membrane (SLM)	NaOH	Eyupoglu and Tutkun (2011)
Dicyclohexano-18-crown-6/dichloromethane	Bulk liquid membrane	KOH	Zouhri et al. (1999)
Aliquat 336, NPOE/THF	Polyvinyl chloride based SLM	NaCl	Güell et al. (2008)
Aliquat 336/dodecane	Hollow fiber supported liquid membrane (HFSLM)	HNO_3	Choi and Moon (2005)
CYPHOS IL101/toluene	Millipore GVHPO4700	NaOH	de San Miguel et al. (2014)
Hostarex A327/cumene	Microporous Polyvinylidene difluoride (PVDF)	NaOH	Guo et al. (2012)

a feed phase was Cr(VI) in all cases

chromium using ion exchange process. Koujalagi et al. 2018 studied removal of Cr(VI) from water as well as organic solvent by using weak base anion exchanger Tulsion A-2X (MP), and found 70% removal at pH 5.0–5.5 and contact time of 15 min and when contact time was increased up to 225 min removal to increased 97%. Value of Gibbs free energy showed that adsorption process is spontaneous. Cavaco et al. 2009 studied two chelating resins based on diphosphonic (Diphonix) and sulfonic groups containing iminodiacetic acid group (Diaion CR 11 and Amberlite IRC 748) results shows sorption capacity for Cr(III) was found 3.6 and 3.4 mEq./g dry resin Amberlite and Diaion and Diphonix respectively. Shan 2018, prepared Nanoscale zero-valent iron particles (NZVI), which was immobilized in Poly vinylidene fluoride (PVDF) hybrid film (cation-exchange) for removal of Cr(VI) at initial concentration 2 mg/L, pH 3.88, time 120 min showed removal

Table 4.6 Removal of chromium using ion-exchange materials under different optimized conditions

Membrane	Optimized conditions	Initial conc. (mg/L)	Efficiency (%)	References
D301 poly-epichlorohydrin-dimethylamine (EPIDMA)	BET surface area 32.356 m^2/g adsorption capacity 194 mg /g, temperature 25 °C, time 24 h, stirring rate 150 rpm, pH 2.0	50	93.76	Zang et al. (2018)
Polypyrrole multi-walled carbon nanotubes on carbon cloth (CC-MWCNTS-Ppy) nanocomposite	Time 80 min, pH 2.0–2.2, applied potential −2.5 V, time 24 h	50	80	Xing et al. (2018)
Gel-type anion exchanger	Sorption capacity 85.5 mg/g, pH 1.6, temperature 25 °C, flow rate 8.6 mL/min, time 90 min	830	–	Xiao et al. (2016)
Macroreticular anion exchange resin (Amberlite IRA900)	Initial conc. 100 mg/L, pH 4.5, dosage 1 g/L, capacity 116 mg Cr(VI) per gram of resin, energy consumption 0.07 kw h/ m^3	100	98.5	Alpaydin et al. (2011)
Acrylic anion exchanger (tertiary amine, quaternary ammonium and ketone groups)	Initial conc. 100 mg/L, pH 3.5, time 72 h	100	80.0	Wójcik et al. (2011)
By ion exchange resins containing carboxylic acid and sulphonic acid groups	Removal of Cr(III) through P(AAGA-co-APSA), P(AAGA-co-ESS), P(AAm-co-ESS), and P(APSA-co-AAc), at pH 3.62	–	89.4, 88.3, 86.8, 89.3 Respectively	Rivas et al. (2018)
Metal organic Resin-1 (MOR-1) and alginic acid (HA). MOR-1-HA	MOR-1-HA BET surface area 1000 m^2/g, pH 3.0, sorption capacity 242 mg/g. MOR-1, 252 mg/g	21.6	97.5	Rapti et al. (2016)
Amberlite XAD-4 (MAX-4)	pH 6.9, temperature 25 °C, time 60 min,	1.0 × 10^4 mol/L	98.7	Bhatti et al. (2017)
Amberlite 200 resin	Removal of Cr(III), pH 3.0, temperature 25 °C and time 6 min	100.0	–	Alguacil et al. (2012)

(continued)

Table 4.6 (continued)

Membrane	Optimized conditions	Initial conc. (mg/L)	Efficiency (%)	References
Anion exchanger (chitosan/poly vinyl amine cryogels)	The sorption capacity 200–320 mg/g, temperature 25 °C, pH 6.0. The sorption process was spontaneous and endothermic.	100.0	98.0	Dragan et al. (2017)
EIX cell (RuO$_2$/Ti)	Time 2 h, adsorption capacity 71.42 mg/g, applied voltage 10 V, temperature 22 °C, pH 4.4–4.6	5.0	99.0	Dharnaik and Ghosh (2014)
D314 resin (weak base ion exchange resin)	pH 6.45, temperature 25 °C, time 60 min	2.21 g/L	99.2	Fan et al. (2013)

efficiency of 44%. Alvarado et al. (2013) used a new hybrid technology by combining ion exchange and electro-deionization and studied systemically (Amberlite IRA900) a strong basic macroreticular anion exchange resin. The result showed that resin have high capacity 116 mg Cr(VI)/gram of resin for ion exchange with Cr(VI) and removed 97.7% chromium. Fan et al. (2013) used D314 resin (weak base ion exchange resin) containing matrix of microporous acrylic acid copolymer, and separate vanadium and chromium from V/Cr mixture, at initial concentration 3.89 g/L of Cr$_2$O$_3$ optimized pH 6.45, temperature 25 °C and contact time for 60 min, D314 resin gives 99.2% recovery of chromium and vanadium. Table 4.6 shows few more applications for removal of chromium using ion-exchange method.

4.2.4 Biological Treatment

Removal of chromium from wastewater using biological treatment is achieved by reorganization, accumulation and sorption (Barrera-Díaz et al. 2012; Pan et al. 2014). Bacteria can remove chromium into three stages, in first stage chromium binds to bacterial cell surface, in second stage chromium enters into cell and lastly it reduced form chromium (VI) to chromium(III) (Singh et al. 2011). Diverse aerobic or anaerobic bacteria like *Ochrobactrum, Bacillus, Enterobacter Pseudomonas, Exiguobacterium, Arthrobacter, Pannonibacter* and *Acinetobacter* have ability to reduce Cr(VI) to Cr(III), which is less toxic from of chromium (Bhattacharya and Gupta 2013; Das et al. 2014; Xu et al. 2009). However, it still need to be removed from water systems, therefore, standalone biological processes may not be suitable for industrial applications. Table 4.7 represents chromium reducing bacteria, on other hand nonliving biomass such as *Clodophora crispate, Chlorella vulgaris, Saccharomyces cerevisiae* and *Rhizopus arrhizus* have been reported for good

Table 4.7 Removal of chromium using different species of microorganisms, their mechanism of removal and optimized conditions

Bacterial type	Mechanisms	Conditions	References
Halomonas species isolated from Cr contaminated soil	Microbial reduction of Cr(VI) to Cr(III) through electron donor	Initial pH 9.0, temperature 30 °C, time 25 days	(Lara et al. (2017)
Alcaligenes faecalis and Pseudochrobactrum	Reduction	Initial conc. 100 mg/L, time 72 h, temperature 30 °C, Luria Bertani culture medium in 300 ml, at 180 rpm; efficiency 100%	Carlos et al. (2016)
Arthrobacter viscosus, biomass	Cr(III) is bonded by biomass functional groups through an ion-exchange mechanism	150 rpm, star shaped column 17 mm external diameter and height of 10 mm flow rate-10 mL/min up flow method for 120 h and pH 2.0, temperature 20 °C, efficiency 100% Cr uptake 20.37 mg/g	Hlihor et al. (2017)
Pannonibacter phragmitetus	Alkaline conditions, six batch cycles	Removal 100%, initial conc. 100–1000 mg, time 9–24 h, 37 °C and pH 9.0.	Xu et al. (2011a)
Chromium-reducing, sulfate-reducing, iron-reducing bacteria	Sulfate and iron Cr(VI) reduction.	Removal 100%, 20 mg L⁻¹ of Cr(VI), time 500–648 h	Somasundaram et al. (2011)
Bacillus subtillis	Constitutive membrane bound enzymes, decrease of pH and growth bacterium	Removal efficiency 100% at pH 9	Mangaiyarkarasi et al. (2011)
Bacillus cereus isolated from soil Sample	Reduction of chromium	Tannery effluent, total chromium 2.4 mg/L, temperature 35 °C, 120 rpm, time 48 h, efficiency 92.0%	Kumari et al. (2016)
Lysinibacillus fusiformis ZC1	Large numbers of NADH-dependent chromate reductase genes:	Initial conc. 1 mm K₂CrO₄, time 12 h, efficiency 100%	He et al. (2011)
Pseudomonas genus isolated from circulating cooling system of iron and steel plan	Reduction	pH 7.0–9.0, initial conc. 3 mmol/L, and inoculating dose-10%(v/v) for both Growing cells and free cells, removal efficiency 100%	Zhang et al. (2016)

(continued)

Table 4.7 (continued)

Bacterial type	Mechanisms	Conditions	References
Staphylococcus capitis and Bacillus sp. JDM-2-1	Induced protein of molecular weight around 25 kda	Removal efficiency 86% and 89%, respectively, time 144 h	Zahoor and Rehman (2009)
Rhodococcus erythopolis isolated From coal mine area	Used lactate as preferable carbon sources	Synthetic $K_2Cr_2O_7$ solution, 1–100 mg/L, pH 5.0–7.0, temperature 20 °C to 35 °C; efficiency 89%	Banerjee et al. (2017)
Escherichia coli	Glucose as electron donor to promote the reduction process	Initial conc. 50–250 mg/L, time 4 h, removal efficacy 97.5%	Liu et al. (2010a)
Hansenula polymorpha cells	Reduction by cytochrome c-oxidoreductase (flavocytochrome b2, FC b2) in the presence of l-lactate	pH 6.3, 0.5 mm chromate, time 30 min removal efficiency 39–53%,	Smutok et al. (2011)
Bacillus sp. immobilized in calcium alginate	Electron donors such as glucose, fructose, sucrose and bagasse extract	pH 7 and 37 °C	Kathiravan et al. (2011)
Pannonibacter phragmitetus	Electron donors such as lactose, fructose, glucose, pyruvate, citrate, formate, lactate, NADPH and NADH	Removal efficiency 100%, time 24 h initial conc. 1917 mg/L, with the maximum reduction rate of 562.8 mg L^{-1} h^{-1}.	Shi et al. (2012)
Pseudomonas putida and SL14 Serratia proteamaculans	A bacterium carrying reductive enzyme(s)	Removal efficiency 93% Cr(VI) and 100% color of reactive black-5 azo dye in 24 h at pH 7.2 and 35 °C	Mahmood et al. (2013)

removal of chromium under optimum pH (1.0–2.0), temperature 25–35 °C, and concentrations of 200, 200, 100, 125 mg/L respectively (Saranraj and Sujitha 2013). However disadvantage is higher concentration of pollutant inhibit the biological processes, take longer time to achieved desired results and hard to separate bacteria from treated water (Stasinakis et al. 2002). That's why it is necessary to develop new biotechnological approaches for rapid separation of bacteria from wastewater post treatment (Alvarez et al. 2010).

Among various techniques mentioned above, chemical or iron electrode based precipitation technique was able to remove chromium efficiently at higher concentrations but produce sludge. Other techniques like bacterial, ion-exchange or membrane filtration are useful for removal of chromium at low concentrations for example contaminated surface waters. In addition, biological treatment is not feasible due to accumulation of chromium in microorganism that may destroy organism and it is also difficult to recover chromium from environment.

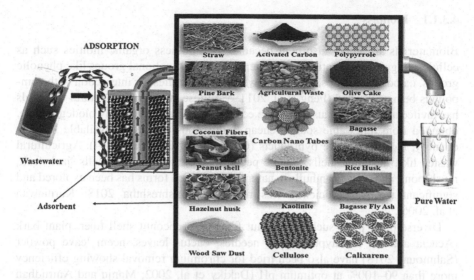

Fig. 4.3 Sorbent materials, including natural, modified sorbents and synthetic materials having good capacity to adsorb chromium form wastewater

Another approach, which has been continuously investigated for liquid waste treatment, is adsorptive removal of contaminants. Therefore, a section is dedicated to adsorption and adsorbents reported for removal of chromium. Figure 4.3 shows multitude of sorbents or raw materials used to prepare sorbents for removal of chromium.

4.3 Sorptive Removal Methods and Sorbents

Sorption processes is the retention of species, elements, compounds or ions by multitude of surface phenomenon and can also be used in treatment strategies for chromium removal. Due to the tunability and versatility in preparing sorbents, there is plethora of materials reported in literature including new generation of nano and carbon based materials with diverse chemistries. This section of chapter is dedicated to discover potential of natural and synthetic sorbents reported in literature towards developing industrial scale reactors for chromium reclamation.

4.3.1 Natural Sorbents

Natural sorbents are those materials, which have tendency to adsorb dissolved component form the contaminant media such as aqueous solutions with or without modifying them chemically. Biomass and clays with or without modification are widely reported in this category of sorbents.

4.3.1.1 Biomass

Biomaterials are good adsorbent because they possess organic moities such as cellulose, lignin and keratin based materials having functional groups like phenolic groups, carboxylic groups, amide etc. which provide sites for interaction with compounds being adsorbed (Feng et al. 2011; Varghese et al. 2018). Natural materials has invited attention because these are economically feasible, easily biodegradable, obtained from renewable source, cheap, indigenous and easily available in large quantity, mostly they are byproducts (Cutillas-Barreiro et al. 2014). Agricultural wastes like, hazelnut shells, orange peels, maize cobs, peanut shells, jack fruit, mushrooms and soya bean hulls in natural or modified forms has been explored and significant removal efficiency was reported (Kulshreshtha 2018; Kurniawan et al. 2006).

Diverse plant parts such as coconut fiber pith, coconut shell fiber, plant bark (Acacia arabica, Eucalyptus), pine needles, cactus leaves, neem leave powder (Sahmoune 2018) have also been tried for chromium removal showing efficiency more than 90–100% at optimum pH (Dakiky et al. 2002; Manju and Anirudhan 1997; Mohan and Pittman Jr. 2006; Sarin and Pant 2006). Utilization of rice bran and wheat bran as an adsorbent are found to be less effective as only 50% removal efficiency was reported (Farajzadeh and Monji 2004; Venkateswarlu et al. 2007). Gardea-Torresdey et al. (2004) reported Avena monida (whole plant biomass) showed 90% removal efficiency of Cr(VI) at optimum pH 6.0. Rice husk in natural form as well as activated rice husk carbon was used for the removal of chromium (VI) and comparable results were observed with commercial activated carbon and other adsorbents (Bishnoi et al. 2004; Mehrotra and Dwivedi 1988).

Saw dust of Indian rose wood prepared by treatment with formaldehyde and sulphuric acid showed efficient removal of Cr(VI) (Garg et al. 2004). Beech saw dust and rubber wood saw dust was also tried for chromium removal (Acar and Malkoc 2004; Karthikeyan et al. 2005). Sugarcane bagasse was used in natural as well as modified form and efficiency for both the forms was compared for the removal of Chromium (V. K. Gupta and Ali 2004; Rao et al. 2002). Utilization of mustard oil cake has been reported with significant removal efficiency and the results of activated carbon of sugar industry waste and commercial granular activated carbon for sequestering of heavy metal ions from aqueous solutions were compared (Fahim et al. 2006; Sud et al. 2008). Recently sugar cane baggase, maize corn cob and jatropha oil cake as such were used for removal of chromium under optimized conditions (Garg et al. 2007).

Agricultural waste peels, biomass based activated carbon and industrial byproducts have been used as low-cost adsorbents for the removal of pollutants from water (Ahmad and Danish 2018; Dehghani et al. 2016; Okoli et al. 2017; Yagub et al. 2014). Dula et al. 2014 used bamboo waste for removing of chromium, batch adsorption was used and found 98.28% removing at pH 2, initial conc. 100 mg/L, contact time 3 min, adsorbent dose 0.25 g. Owalude and Tella (2016) used unmodified groundnut shell and modified groundnut shell and found at pH 2, contact time 60 min, initial concentration 8 mg/L gives 82% and 96% removing efficiency.

Effectiveness of various adsorbents has been compared with commercial activated carbon where chitosan is found to be one of the most important materials in adsorption applications. Amino and hydroxyl groups present in the molecules contribute to possible adsorption interactions between chitosan and pollutants (dyes, metals, ions, phenols, pharmaceuticals/drugs, pesticides, herbicides, *etc.*) (Kyzas and Bikiaris 2015).

Gonçalves et al. 2018, used Endocarps of the *acai berry* as an absorbent for the removal of Cr(III) and other metals. Ravikumar et al. (2018) used biomass of sulfate reducing bacteria and polymer-nano zerovalent iron (nZVI) composite under anaerobic environment. Schwantes et al. 2016 used chemically modified cassava roots or peel (*Manihot esculenta* Crantz) for the removal of Cr(III) and other metal ions.

Other reports on chromium removal using natural biosorbents are given in Table 4.8 and closer look reveals that most of natural sorbents are not satisfactory to fulfill industrial needs. However, aminated wheat straw showed highest adsorption capacity of 454 mg/g. It is also capable of removing chromium at concentration levels of 500 mg/L. Corn bract also showed comparable adsorption of 438 mg/g with removal efficiency of nearly 100% at 24 h of contact time. Mostly biosorption occurs in acidic medium particularly at pH 2.0.

4.3.1.2 Clays

Clays and their minerals are small particles, found on earth surface and mainly composed of alumina, silica, weathered rock and water. It is fine grained raw material which is inexpensive and abundant material used for decades for the effective removal of heavy material form aqueous solution (Gu et al. 2018). The surface of clay can be modified by different methods to improve the adsorption capacities. Adsorption of heavy metals on clay involved a complex mechanism such as complexation, bonding with cations or ion exchange (Bergaya and Lagaly 2006). Treatment of clay can increase pore volume, surface area, binding sites that can enhance the uptake of metals (Ismadji et al. 2015). Some other studies reported for the removal of chromium by using clay are given blow in Table 4.9.

Clays were found efficient in acidic pH and all reports employed Cr(VI) as feeding phase. Highest adsorption capacity of 308 mg/g was reported for polyaniline/ Montmorillonite composite (Chen et al. 2014) which is still lower than aminated wheat straw (Yao et al. 2016). Therefore, it may be stated that modified biomass based natural sorbents are better sorbents among various studies reported so far.

4.3.2 Silica Based Material for the Removal of Chromium

Silica and its composites possess high thermal, mechanical stabilities, economic feasibility, large surface area, high porosity and great number of functionalities. (Morin-Crini et al. 2018) Many modified silica based material such as polyethylene

Table 4.8 Natural and modified biomass-based adsorbents for the removal of chromium

Material	Adsorption. capacity (mg/g)	Concentration range (mg/L)	Contact time	Sorption (%)	References
Sargassum oligocystum biomass (modified by CaCl$_2$)	34.46	10–100	100 min	93.2	Foroutan et al. (2018)
Acidically prepared rice husk carbon (APRHC)	–	80.0	120 min	99.9	Khan et al. (2016)
Raw Macadamia nutshell powder (RMN)	45.2	100.0	10 h	100.0	Pakade et al. (2017)
Acid-treated Macadamia nutshell (ATMN)	44.8	100.0	10 h	100.0	Pakade et al. (2017)
Base-treated Macadamia nutshell (BTMN)	42.4	100.0	10 h	100.0	Pakade et al. (2017)
Chenopodium album and Eclipta prostrate plant	–	10–50	2.5 h	93.0	Babu et al. (2016)
Chitosan	41.5	10–100	24 h	80.0	Zuo and Balasubramanian (2013)
Melaleuca diosmifolia leaf	62.5	250	2 h	99.9	Kuppusamy et al. (2016)
Eggshell membrane	–	5–25	2 h	81.4	Daraei et al. (2014)
Hemp fibers	6.16	13–26	80 min	90.0	Tofan et al. (2015)
Fine rice straw	7.9	50–200	12 h	85.0	Elmolla et al. (2016)
Rice straw carbon	18.8	50–200	12 h	95.0	Elmolla et al. (2016)
Rice straw activated carbon	40.3	50–200	12 h	97.0	Elmolla et al. (2016)
Corn bract (Polyethyleneimine functionalized)	438.0	100.0	24 h	100.0	Luo et al. (2017)
Populus fiber	180.5	1000	30 min	98.2	Li et al. (2016)
Mango kernel activated with H$_3$PO$_4$	7.8	20.0	150 min	100.0	Rai et al. (2016)
Banana peels	–	400.0	120 min	96.0	Ali et al. (2016)
Almond green hull	10.1	60.0	60 min	99.0	Nasseh et al. (2017b)

(continued)

Table 4.8 (continued)

Material	Adsorption. capacity (mg/g)	Concentration range (mg/L)	Contact time	Sorption (%)	References
Garlic stem (GS)-Allium sativum L.	103.0	3000.0	–	72.0	Parlayıcı and Pehlivan (2015)
Horse chesnut shell (HCS)-Aesculus hippocastanum	142.8	3000.0	–	95.0	Parlayıcı and Pehlivan (2015)
Grapefruit peelings (treated with H₂O₂)	39.0	35.0	–	100.0	Rosales et al. (2016)
Almond green hull	–	80.0	60 min	99.9	Negin Nasseh et al. (2017a)
Aminated wheat straw	454.0	500.0	24 h	99.0	Yao et al. (2016)
Dew melon peel biochar	198.7	100.0	8 h	98.6	Ahmadi et al. (2016)
Peganum harmala	9.4	100.0	30 min	100.0	Khosravi et al. (2014)
Mosambi (Citrus limetta) peel	250.0	200–300	120 min	–	Saha et al. (2013)
Ash gourd peel powder	18.7	75–350	60 min	91.0	Sreenivas et al. (2014)
Colocasia esculenta leaves	47.6	20.0	120 min	97.7	Nakkeeran et al. (2016)
Rye husk	0.435 mmol/g	5.0 mM	140 min	80.0	Altun et al. (2016)
Pinus sylvestris bark	9.77	5–20	24 h	90.0	Alves et al. (1993)
Eucalyptus bark	45	250	–	–	Sarin and Pant (2006)
Bentonite	0.512–6.0	–	–	–	Khan and Khan (1995)
Kaolinite	0.108 meq/g	0.62–8.27	–	–	Tavani et al. (1997)
Hazelnut shell	3.99 g/kg	0.1–2.0	5 h	97.8	Cimino et al. (2000)
Alkali treated straw	3.91	–	60 min	–	Kumar et al. (2000)
Insoluble straw	1.88	–	60 min	–	Kumar et al. (2000)
Sawdust	4.44	50	45–75 min	–	Zarraa (1995)
Polyacrylamide grafted	12.4	100–1000	4 h	91.0	Raji and Anirudhan (1998)

(continued)

Table 4.8 (continued)

Material	Adsorption. capacity (mg/g)	Concentration range (mg/L)	Contact time	Sorption (%)	References
Wool, olive cake, sawdust, pine needles, almond shells and cactus leaves	41.15, 33.44, 15.82, 21.50, 10.62, 7.08, and 6.78 respectively	20–1000	2 h	81.0	Dakiky et al. (2002)
Coconut shell fibers (acid-treated and activated)	12.23	1–100	48 h	–	Mohan et al. (2006)
Coconut shell charcoal oxidized with nitric acid, sulfuric acid	10.88, 4.05	–	3 h	–	Babel and Kurniawan (2004)
Groundnut husk carbon	7.01	–	5 h	–	Dubey and Gopal (2007)
Red mud	35.66	9.60×10^{-4} to 9.60×10^{-3} M	6–8 h	60–100	Gupta et al. (2001)

imine silica nanospheres, polyaniline silica gel composite, polyacrylamide-silica microspheres and aniline formaldehyde condensate coated silica gel have been extensively used for removal of chromium form wastewater. Highly branched and numerous functional groups like aromatic amines, aliphatic amine and other nitrogen containing groups attached to these polymers provide selective capacity to adsorb chromium. The functional groups have two cites per molecule like diamines can increased the chromium adsorption. J. Lee et al. (2018) prepared amino contained functional group silica and studied the chromium removal mechanism, the synthesized (3-aminopropyl) tri-methoxy silane mesoporous functionalized silica studied under batch experiments, material was characterized and found BET surface area of 402.6 m^2/g and uptake of chromium was a chemisorption process, endothermic in nature and maximum Cr(VI) adsorption was 84.90 mg/g in 60 min of time. Janik et al. (2018) worked on removal of preconcentrated Cr(VI), he modified graphene oxide with different amino silane having one(GO-1 N), 3-aminopropyltriethoxysilane(APTES), two (GO-2 N), N-(3-trimethoxysilylpropyl)ethylenediamine(TMSPEDA) and three nitrogen(GO-3 N), N1-(3-trimethoxysilylpropyl)diethylenetriamine. The results show detection limits of 0.17 ng/mL with maximum adsorption capacities 15.1 mg/g (APTES > TMSPEDA > TMSPDETA) and recovery of chromium is 99.7% at pH 3.5, temperature 25 °C, initial concentration 0.025 mg/L and time 180 min. (El-Mehalmey et al. 2018) used silica as a porous support and formed a composite on it with Zr-based MOF (UiO-66-NH2) "Zr carboxylate amino derivative of Metal Organic Framework (MOF)" and used it for column adsorption of chromium. UiO-66-NH2@ silica has high surface area found 687 m^2/g. Cr(VI) uptake was reported as 277.4 mg/g at pH 5.0 and time 2 h. Huang et al. (2018) used surface ion imprinting technique for the synthesis of chromium (VI) Ion Imprinting Polymer Cr(VI)IIP,

Table 4.9 Removal of chromium by natural clay-based sorbents

Type of clay	Adsorbate	Conditions	Adsorption capacity (mg/g)	References
Vesicular basalt	Cr(VI)	pH 2, initial conc. 5.0 mg/L, adsorbent dose 50 g/L, mono layer adsorption.	79.20 mg/kg	Alemu et al. (2018)
Amino-functionalized alkaline clay	Cr(VI)	Temperature 30 °C, pH 4.0	137.9	Pan et al. (2016)
Natural Kaolinite	Cr(VI)	BET surface area 48.75 m²/g	2.94	Hezil et al. (2018)
Modified Kaolinite	Cr(VI)	BET surface area 63.72 m²/g	4.01	Hezil et al. (2018)
Alkyl ammonium surfactant bentonite	Cr(VI)	BET surface area 28 m²/g	8.36	Stanković et al. (2011)
Cellulose-clay composite	Cr(VI)	BET surface area 87.09 m²/g, initial conc. 100 mg/L, pH 5.5, adsorption 99.5%.	22.2	Kumar et al. (2011)
Bentonite	Cr(VI)	BET surface area 119.8 m²/g, pH 2.0, time 120 min, initial conc. 400 mg/L, spontaneous and endothermic in nature.	48.83	Wanees et al. (2012)
Sepiolite	Cr(III)	Mono layer adsorption, temp 20 °C Initial conc. 100 mgL/L	27.07	Kocaoba (2009)
Bentonite	Cr(III)	BET surface area 62.56 m²/g, porosity 16.7%, initial conc. 200 mg/L, temperature 20 °C, time 180 min	13.79	Al-Jlil (2015)
Polyaniline/ montmorillonite composite	Cr(VI)	Temperature 25 °C, pH 2.0, dose 1 g/L	308.60	Chen et al. (2014)
Modified Na-montmorillonite	Cr(VI)	BET surface area 23.18 m²/g, initial conc. 10 mg/L, pH 2.5, adsorption 99.1%	23.69	Kumar et al. (2012)
Brazilian smectite	Cr(VI)	Contact time 20 min, BET surface area 787.3 m²/g pH 4.0.	97.23	Guerra et al. (2010)
Turkish vermiculite	Cr(VI)	pH 1.5, adsorbent dosage 10 g/L and 20 °C, monolayer adsorption, contact time 120 min.	87.70	Sari and Tuzen (2008)
Gaomiaozi bentonite	Cr(III)	pH 7.0, temperature 20 °C, contact time 120 min, monolayer adsorption	4.68	Chen et al. (2012)
Fe²⁺ modified vermiculite	Cr(VI)	pH 1.0, time 60 min, removal 95% desorption 80%, initial concentration 50 mg/L mono layers adsorption.	87.72	Liu et al. (2010b)

(continued)

Table 4.9 (continued)

Type of clay	Adsorbate	Conditions	Adsorption capacity (mg/g)	References
Natural sepiolite	Cr(VI)	pH 2, time 120 min, 25 °C	37.00	Marjanović et al. (2013)
Acid-activated sepiolite	Cr(VI)	pH 2, time 120 min, 25 °C, multi layers adsorption	60.00	Marjanović et al. (2013)

(GO-MS) on graphene oxide mesoporous silica by using functional monomer, 3-(2-amino ethyl amino) propyltrimethoxysilane and successfully removed Cr(VI) form aqueous media, with a maximum adsorption capacity of 438.1 mg/g with a good reusability of more than 5 cycles. Some other studies reported for the removal of chromium by using silica based materials are given blow in Table 4.10.

4.3.3 Cellulose-Based Sorbents for the Removal of Chromium

Cellulose is considered renewable and most abundant polymer worldwide, More than 1000 tons of cellulose is synthesized every year in a highly pure form by photosynthesis. Cellulose has been used for years in the form of cotton and serves as clothing material, energy source and building materials. It is also used for the removal of heavy metals by directly or by modifying it to have better adsorption sites for the attachments of metal ions. For example, modification using different functional groups onto cellulose backbone or grafting of monomers have shown affinity towards metals ions.

Yang et al. (2018) used bacterial cellulose for the adsorption of Cr(VI) by using poly m-phenylenediamine (BC/PmPD), which imparted high adsorption capacity to BC/PmPD 434.78 mg/g and very good removal was reported among various cellulosic materials. Adsorptive interactions of Cr(VI) was attributed to protonated NH_2. Jiale Wang et al. (2018) used starch and Na-carboxymethyl cellulose coated with Fe and Fe/Ni (SS-nZVI-Ni) nanoparticles for the removal of Cr(VI) and found maximum removal at lower pH 2.0. Velempini et al. (2017) synthesized ion imprinted polymer (IIP) from sodium carboxymethyl cellulose for the removal of Cr(VI), under optimized condition 20 mg adsorbent was used and the maximum adsorption capacity of Cr(VI) 177.62 mg/g which is comparatively high than non-imprinted polymer (NIP) 149.93 mg/g. Synthesized ion imprinted polymer gave very good removal of chromium, the synthesized ion imprinted polymer can be recycled more than 5 time with desorption capacity of more than 98% by using NaOH (0.1 M) as a leachate. In an acidic medium, protonated amino functionalities attract the chromium (Table 4.11).

Table 4.10 Removal of chromium using silica and modified silica-based materials trimethoxysilylpropyl diethy-lenetriamine, polyaniline, imidazole modified silica, silica magnetite nanoparticles, nano-hydrotalcite, aminopropyl and with different activated carbon

Material	Conditions/ Characteristics	Adsorption capacity (mg/g)	Removal (%)	References
SBA-15 N-(3-trimethoxysilylpropyl diethylenetriamine (DAEAPTS)- grafted mesoporous silica	BET surface are of SBA-15891.3 m²/g, optimized parameters pH 3, Cr(VI) concentration 328.7 mg/L, temperature 30 °C, pH 3.0, adsorbent dose 1 g/L, initial Cr(VI) concentration 10-1000 mg/L, time 4 h, removal efficiency 100% at a dose of 3.0 g/L.	330.88	100	Kim et al. (2018)
nZVI@MCM-41 silica	BET surface area 609 m²/g, temperature 25 °C, dosage 180 mg/L, pH 3.0, time 3 h,	–	100	Petala et al. (2013)
Silica composite with polyaniline (PANI)	Surface area 63.25 m²/g, pH 2, adsorbent dose 8 mg, initial conc. 100 mg/L, 450 min, temperature 35 °C	193.85	95.0	Ahalya et al. (2003)
Imidazole-modified silica (SilprIm-cl)	BET surface area 260.13 m²/g, initial conc. 150 mg/L, pH of 2.0, temperature, 30 °C, adsorbent, 30 mg, time 60 min, reusability up to eight times. Desorption 99.3%.	47.79	97.4	Z. Wang et al. (2013b)
Functionalized silica mesoporous magnetite nanoparticles	BET surface area 241.68 m²/g, adsorbent 80 mg, pH 2.0, contact time 15 min	185.2	90.0	Shariati et al. (2017)
Silica magnetite nanoparticles	pH 2.0, time 30 min, temperature 25 °C, desorption 97%	30.2	90.0	Araghi et al. (2015)
Nano-hydrotalcite/SiO₂ (Nano-HT)	Initial conc. 30 mg/L, adsorbent dose 1 g/L	–	94.6	Pérez et al. (2015)
Magnetic iron oxide/ mesoporous silica nanocomposites (MCM-41)	BET surface area 1032 m²/g pH 5.4, temperature 25 °C, time 2 h	2.08 mmol/g	–	Egodawatte et al. (2015)
Aminopropyl-functionalized mesoporous silica	BET surface area 834 m²/g, pH 2.2, temperature 25 °C, time 15 min	87.1	95	Fellenz et al. (2015)

(continued)

Table 4.10 (continued)

Material	Conditions/ Characteristics	Adsorption capacity (mg/g)	Removal (%)	References
Gelatin–silica-based hybrid materials	BET surface area 427.79 m²/g, pH 4.0, temperature 30 °C, contact time 12 h, adsorbent dose 2 g/L, concentration of adsorbate 50 mg/g,	94.47	89.7	Thakur and Chauhan (2014)
Silica-based adsorbent grafting dimethyl aminoethyl methacrylate (DMAEMA)	BET surface area 303.8 m²/g, ion exchange capacity (IEC) 1.30 mmol/g, pH 4.0, initial conc. 100 mg/g and time 40 min	68	80.0	Qiu et al. (2009)
NH₂-functionalized cellulose acetate/silica composite FCA/SiO₂	BET surface area 126.49 m²/g, pH 1.0, initial conc. 100 mg/L and time 60 min,	19.64	97.0	Taha et al. (2012)
Activated corban derived from algal bloom residue	Initial conc. 100 mg/L, pH 1, absorbent dosage 1 g/L, temperature 30 °C and contact time 120 min	155.52	99.9	Zhang et al. (2010)
Prawn shell activated carbon	Adsorbent dose 0.04 g initial conc. 100.6 mg/L and time 31 min and removing efficiency 99%	100	99.9	Arulkumar et al. (2012)
Modified activated corban form corn cob	BET surface area 282 m²/g, pH 2.0, adsorbent. Dosage 0.05 g/10 mL, temperature 27 °C and time 24 h and removing efficiency 3.5 g of MCCAC was able to treat 1.38 L of effluent containing 25 mg/L of Cr(VI),	57.37	86.9	Nethaji et al. (2013)
Activated corban from longan seed	BET surface area, 1511.8 m²/g, pH 3.0, initial conc. 100 mg/L, adsorbent dose 0.1 g, temperature 27 °C.	–	–	Yang et al. (2015)
Activated corban fiber felt (ACFF)	BET surface area 935.1 m²/g, initial conc. 70 mg/L, pH 5.0, dosage 0.545 g and temperature 25 °C.	–	80.2	Huang et al. (2014)

Table 4.11 Removal of chromium using cellulose-based sorbents including natural cellulose and modified with polyethyleneimine, amino-functionalized, aerogels, glycidyl methacrylate grafted densified cellulose and with different cellulose based bio-composites

Adsorbent	Conditions	Adsorption Capacity (mg/g)	Removal (%)	References
Polyethyleneimine grafted magnetic cellulose	pH 4.8, time 10 min, adsorbent dosage, 1.0 g, initial conc. 108.0 mg/L, contact time 120 min, temperature 30 °C reusability 96% after six cycles.	198.8	100.0	Li et al. (2018)
Amino-functionalized magnetic cellulose	pH 2.0, temperature 25 °C, time 10 min, desorption efficiency 98%.	171.5	–	Sun et al. (2014a)
Cellulose aerogels/zeolitic imidazolate framework (ZIF-8)	Time 120 min, initial conc. 20 mg/g	41.8	99.7	Bo et al. (2018)
Ficus carica fiber cellulose grafted acrylic acid	pH 3.5, time 3 h, sorbent dose 0.5 g temperature 30 °C, conc. 400 mg/L	28.90	–	Gupta et al. (2013b)
Cellulose grafted-(HEMA-co-AAm) hydrogel	pH 7.0, time 2 h, sorbent dose 0.1 g, temperature 25 °C, conc. 5.0 mg/L	–	19.67	Sharma and Chauhan (2009)
Glycidyl methacrylate grafted densified cellulose with quaternary ammonium groups	pH 4.5, time 1 h, sorbent dose 0.1 g temperature 30 °C, conc. 100 mg/L	123.6	100.0	Anirudhan et al. (2013)
Quaternary ammonium functionalized cellulose nanofibers	pH 3.0, time 50 min, sorbent dose 1 g, temperature 25 °C, initial conc. 1.0 mg/L	–	100.0	He et al. (2014)
Cellulose grafted-p(MA) with poly hydroxamic acid ligand	pH 6.0, time 2 h, sorbent dose 0.15 g temperature 30 °C initial conc. 0.1 M.	280	99.0	Rahman et al. (2016)
Poly(amidoamine)-grafted cellulose	pH 2.0, contact time 10 h.	377.4	–	Zhao et al. (2015)
Polyethylenimine functionalized cellulose aerogel beads	pH 2.0, initial conc. 100 mg/L, time 24 h, temperature 25 °C	229.1	–	Guo et al. 2017)
Polyethylenimine facilitated ethyl cellulose	pH 3.0, initial conc. 4.0 mg/L adsorbent dose 3.0 g/L, time 5 min	36.8	100.0	Qiu et al. (2014)
Cellulose based bio-composites	pH 7.7, sorbent dose 0.1 g, temperature 25 °C, time 50 min	–	100.0	Periyasamy et al. (2017)
Spherical cellulose-based adsorbent	pH 2.0, sorbent dose 0.01 g, conc. 150 mg/L, time 24 h.	209.6	100.0	Dong et al. (2016)

4.3.4 Carbonized Biomass-Based Sorbent

One of the most attractive and efficient adsorbent for the removal of chromium is activated carbon and its derived materials from different sources (Gopinath and Kadirvelu 2018; Mohan and Pittman Jr. 2006; Mudhoo et al. 2018; Pan et al. 2014). Activated carbon adsorption seems to be an attractive choice for chromium removal both for its exceptionally high surface areas which range from 500 to 1500 m^2/g, well-developed internal microporous structure as well as the presence of a wide spectrum of surface functional groups like carboxylic group (Owlad et al. 2010). For these reasons, activated carbon adsorption has been widely used for the treatment of chromium containing wastewaters. Based on its size and shape, activated carbon is classified into four types: powder-activated carbon (PAC), granular-activated carbon (GAC), activated carbon fibrous (ACF), and activated carbon clothe (ACC). Due to the different sources of raw materials, extent of chemical activation and physicochemical characteristics; each type of activated carbon has its specific application as well as inherent advantages and disadvantages in wastewater treatment (Kurniawan and Babel 2003).

Sharma et al. (2018), used NaOH and NaClO treated activated carbon obtained from stem of *cornulaca-Monacantha* and used it for the removal of Cr(VI). Results show that adsorption is highly pH dependent, maximum adsorption was recorded at pH 2. Monolayer adsorption capacity was found 68 mg/g and thermodynamically reaction was spontaneous and endothermic. *Cornulaca-Monacantha* activated carbon can be used for more than five cycle with Cr(VI) uptake of 89.19%. Pérez-Candela et al. (1995) used different types of powder-activated carbon prepared from different raw materials to remove Cr(VI). It was found that the adsorption process depends on the pretreatment of activated carbon and that the highest removal performance was obtained with those prepared by physical activation. It was also reported that at pH of 1.0, the retention of Cr(VI) was affected by its reduction to Cr(III) (Pérez-Candela et al. 1995). Sharma (1996) studied the removal of Cr(VI) from aqueous solution using GAC type Filtrasorb 400. It was reported that an adsorption capacity of 145 mg/g was achieved at a pH range of 2.5–3.0 (Sharma et al. 2018). This result is not in agreement with that obtained in the latter study (Fu et al. 2013) conducting a similar comparative study using activated carbon LB 830 and Filtrasorb 400. It was reported that the maximum adsorption capacity of Filtrasorb 400 in the latter study is only 0.18 mg of Cr(VI)/g. Hamadi et al. (2001) studied the removal of Cr(VI) from aqueous solution using granular-activated carbon type Filtrasorb 400. It was found that reduction in particle size of adsorbents increases its surface area for metal adsorption, and it results in higher removal efficiency on Cr(VI). It was also indicated that the adsorption of Cr(VI) was more favorable at higher temperature. Park and Jung (2001) removed Cr(VI) by activated carbon fibrous (ACFs) plated with copper metal. It was reported that the introduction of Cu(II) on activated carbon fibrous lead to an increase in the adsorption capacity of Cr(VI) from an aqueous solution. It was pointed out that the adsorption of chromium ions was essentially dependent on surface properties, rather than by surface area and porosity of ACFs (Park and Jung 2001).

Many investigations are carried out to study the adsorption of Cr(VI) using activated carbon prepared from different raw material. Natale et al. (2007) used activated carbon produced by Sutcliffe Carbon starting from a bituminous coal to uptake Cr(VI). It was found that the adsorption capacity for the activated carbon strongly depends on solution pH and salinity, with maximum values around 7 mg/g at neutral pH and low salinity levels. (Di Natale et al. 2007) Selomulya et al. (1999) used different types of activated carbons, produced from coconut shell, wood and dust coal to remove Cr(VI) from synthetic wastewater. The coconut shell and dust coal activated carbons have protonated hydroxyl groups on the surface (H-type carbons), while the surface of the wood-based activated carbon has ionized hydroxyl groups (L-type carbons). It was found that the optimum pH to remove total chromium was 2.0 for wood-based activated carbon, while for coconut shell and dust coal activated carbons, the optimum pH was around 3–4. The difference in the optimum pH for different activated carbons to remove Cr(VI) from water can be explained by the different surface characteristics and capacity of the activated carbons to reduce Cr(VI) to Cr(III) (Selomulya et al. 1999). Kobya 2004 used hazelnut shell-activated carbon for the adsorption of Cr(VI) from aqueous solution. It was reported that the adsorption of Cr(VI) was pH dependent. The adsorption capacity as calculated from the Langmuir isotherm was 170 mg/g at an initial pH of 1.0 for a Cr(VI) solution of 1000 mg/L concentration. Thermodynamic parameters were evaluated, indicating that the adsorption was endothermic and involved monolayer adsorption of Cr(VI) (Kobya 2004). Mohanty et al. (2005) prepared several activated carbons from *Terminalia arjuna* nuts, an agricultural waste, by chemical activation with zinc chloride and then tested for aqueous Cr(VI) removal. The isotherm equilibrium data were well fitted by the Langmuir and Freundlich models. The maximum removal of chromium was obtained at pH 1.0. (Mohanty et al. 2005) Karthikeyan et al. (2005) used rubber wood sawdust-activated carbon for removal of Cr(VI) in a batch system. It was found that Cr(VI) removal is pH dependent and is maximum at pH 2.0. Many commercial activated carbons have been used for Cr(VI) adsorption. The mechanism for Cr(VI) removal in most of the studies is surface reduction of Cr(VI) to Cr(III) followed by adsorption of Cr(III) (Kalavathy et al. 2005). Huang and Wu (1975) showed that Cr(VI) adsorption by activated carbon, filtrasorb 400 (Calgon), occurred by two major interfacial reactions: adsorption and reduction. (Huang and Wu 1975). Cr(VI) adsorption reached a peak value at pH 5.0–6.0. Carbon particle size and the presence of cyanide do not change the magnitude of chromium removal. Hu et al. (2003) used three commercial activated carbons FS-100, GA-3, and SHT to remove Cr(VI) from aqueous solution (Hu et al. 2005). Physiochemical factors such as equilibrium time, temperature, and solution pH that affect the magnitude of Cr(VI) adsorption were studied. It was found that both micropores and mesopores have important contribution on the adsorption. However, desorption is more dependent on the mesoporosity of activated carbons. Therefore, regeneration is easier for the carbon with high mesoporosity. Hamadi et al. (2001) used commercial activated carbon F-400 for batch removal of Cr(VI) from wastewater under different experimental conditions (Hamadi et al. 2001). It was reported that the adsorption capacity was calculated from the Langmuir isotherm was 48.5 mg/g at an initial pH of 2.0 for a Cr(VI) solution of 60 mg/L concentration. Mohan et al. (2006) used commercially available activated carbon fabric cloth for the removal of Cr(VI) from tannery wastewater (Mohan and Pittman Jr. 2006). The results indicated that the Langmuir adsorption isotherm model fitted the data better than the Freundlich adsorption isotherm model. It was found that the activated carbon fabric cloth performed better than the other tested adsorbents with adsorption capacity of 22.29 mg/g at an initial pH of 2.0.

Studies are reported where activated carbons are modified to improve the efficiency of sorbents for removal of chromium (Bello and Raman 2018). Ranjbar et al. (2018) studied the efficiency of the bone char -ZnO composite by using sol–gel method. They studied that at pH 3, absorbent dose 0.9 g/L, temperature 48 °C, concentration of chromium 55 mg/L, contact time 3 h was able to remove 84% chromium (VI). Derdour et al. (2018) have studied the removal of chromium (VI) using iron catalysts derived activated carbon form walnut shell by impregnation process at pyrolysis at 900 °C and heating under nitrogen flow, this activated carbon was used as supporting material for iron oxides. Activated carbon was permeated with 5% iron and under nitrogen flow calcined at 400 °C, the adsorptive capacity of this activated carbon was found 10.11 mg/g at pH 2. (Mohammad et al. 2017) synthesized activated carbon to remove Cr(VI) form hull of *jatropha curcas* seeds and used $ZnCl_2$ for chemical activation and carbonized at 800 °C and found that adsorption is monolayer on absorbent surface with maximum capacity of 25.189 mg/g. Some other studies reported for the removal of chromium by using activated carbon are given blow in Table 4.12.

4.3.5 Nanomaterials

Nanomaterials refer to materials on the nanoscale level between approximately 1 nm and 100 nm (Miretzky and Cirelli 2010). Generally, nanomaterials can be categorized into carbon-based nanomaterials such as carbon nanotubes and graphene or polypyrrole, and inorganic nanomaterials including the ones based on metal oxides and metals. Combinations of different nanomaterials are also developed (Shi et al. 2011). Nanomaterials hold great promise in reducing contamination of heavy metals.(Kumar et al. 2018) Among two types of organic nanomaterials; carbon nanotubes and polypyrrole nanomaterials widely reported in metal ion removal specifically for chromium removal. Both are discussed here separately.

4.3.5.1 Carbon Nanotubes

Carbon nanotubes (CNTs), mainly including single-walled nanotubes (SWCNTs) and multi-walled nanotubes (MWCNTs) (Pyrzynska 2008). These have been widely studied regarding their potential in environmental application as superior adsorbents

Table 4.12 Removal of chromium using activated carbons and modified activated carbons with different functional groups

Type of activated carbon	Optimized conditions	Adsorption Capacity (mg/g)	Removal (%)	References
Activated carbon modified with micro-sized geothite (mFeOOH)	Cr(VI) initial conc. 50 mg/L, pH: 5.6, temperature 30 °C	27.2	90.4	Su et al. (2019)
Sludge based magnetic carbon mFeOOH@AC,	Carbonization temperature 800 °C, time 60 min, BET surface area 131.1 m²/g, pH 1.0, removing capacity of Cr(VI) 200 mg/L at initial conc. 800 mg/L, time 10 min	203	95.0	Gong et al. (2018)
Powder-activated carbon counter current two-stage adsorption (CTA)	Initial Cr(VI) conc. 50 mg/L, PAC dose 1.250 g/L, pH 3.0, temperature 20 °C.	–	80.0	Wang (2018)
Polyacrylonitrile-based porous carbon (PPC-0.8-800)	BET surface are 2151.42 m²/g, pH 1.0, initial Cr(VI) conc. 427.3 mg/L, contact time 24 h and adsorbent dose 1.0 g/L, reusability five cycles.	374.90	82.0	Feng et al. (2018)
Magnetic biochar Melia azedarach wood (MMABC)	BET surface area 5.219 m²/g, super-paramagnetic magnetization 17.3, adsorbent of dosage 5 g/L, pH 3.0, initial Cr(VI) conc. 10 mg/L, time 540 min.	–	99.8	Zhang et al. (2018)
Hydro thermal carbon obtained from microalgae Chlorococcum sp. AC-KOH-750 °C and AC-NH3-900	BET surface area 1784 m²/g, (AC-KOH-750) and 487 m²/g (AC-NH3–900), time 24 h temperature 25 °C, initial conc. 500 mg/L	370.37 and 95.60	–	Sun et al. (2018)
Fe₂O₃-carbon foam	BET surface are e 458.59 m²/g, time 30 min, initial conc. 38.6 mg/L	25.96	95.0	Lee et al. (2017)
Activated carbon from Fox nutshell (FNAC-700-1.5)	BET surface area of 2636 m²/g, initial conc. of Cr(VI) 35 mg/L, pH of 2.0, temperature 45 °C and contact time of 3 h. Adsorption process endothermic and spontaneous, column parameters 4 cm bed height, and 5 mL/min flow rate.	74.95	71.86	Kumar and Jena (2017)

(continued)

Table 4.12 (continued)

Type of activated carbon	Optimized conditions	Adsorption Capacity (mg/g)	Removal (%)	References
Activated carbon from molasses MP2(500)	BET surface area 1400 m²/g, initial conc. 100 mg/L, temperature 30 °C, contact time 120 min, pH 2.0	--		Legrouri et al. (2017)
Magnetic porous carbonaceous (MPC-300)	pH 5.0, initial conc, Cr(VI) 100 mg/g, temperature 25 °C.	21.23	81.8	Wen et al. (2017)
Activated carbon prepared from apple (ACAP)	Initial conc. Cr(VI) 50 mg/L, pH of 2, adsorbent 0.05 g/50 mL, contact time 4 h, temperature of 28 °C, spontaneous and endothermic.	36.01	95.0	Ahalya et al. (2003)
Activated carbon developed from P. Capillacea (CRA)	pH 1.0, time 120 min, temperature 27 °C, initial conc. 125 mg/L	66	85.0	El Nemr et al. (2015)
Activated carbon (microwave-assisted H₃PO₄ mixed with Fe/Al/Mn activation)	BET surface area 1332 m²/g, initial conc. 23.3 mg/L, time 24 h, dosage 0.1 g/100 mL, temperature 20 °C, pH 5.38	–	85.0	Sun et al. (2014b)
AC-TA (tartaric acid employed Zizania caduciflora activated carbon)	BET surface area 1270 m²/g, pH 2.0–3.0, dosage 40 mg/50 mL, initial conc. 20 mg/L, time 48 h, temperature 22 °C	–	100.0	Liu et al. (2014)
Activated carbon Corn Cob	BET surface area 924.9 m²/g, temperature 25 °C, initial conc. of 10 mg/L pH of 2.5, time 4 h.	34.48	78.9–100	Tang et al. (2016)
Activated carbon / nanoscale zero-valent iron (C–Fe⁰) composite	pH 2.0, initial conc. 10 mg/L, time 10 min.	–	99.0	Wu et al. (2013)
Activated carbon waste rubber tire	Surface area 465 m²/g, initial Cr(III) conc. 100 mg/L, adsorbent dosage 0.5 mg, pH 5.0, temperature 25 °C	–	90.0	Gupta et al. (2013a)
Activated carbon derived from Macadamia nutshells	BET surface are 0.518 m²/g, pH 5.0, contact time 120 min, and sorbent mass 0.10 g, initial conc. 100 mg/L.	145.5	98.0	Pakade et al. (2016)
Activated carbon African biomass residues Jatropha wood, peanut shells (JKV2T700 and CHV100T400)	BET surface areas 1305 m²/g and 751 m²/g, pH 2.0, initial conc. 60 mg/L, temperature 40 °C.	106.4 and 140.8	98.0	Gueye et al. (2014)

(continued)

Table 4.12 (continued)

Type of activated carbon	Optimized conditions	Adsorption Capacity (mg/g)	Removal (%)	References
Activated carbon nitrogen-enriched (based bamboo)	BET surface area 1511 m²/g, initial conc. of 100 mg/ L, pH 2.0 time 24 h,	–	89.0	Zhang et al. (2015)
Activated carbon carbonized rice husk by ozone activation	BET surface area 380 m²/g, initial conc. 100 mg/L, pH 2.0, temperature 32 °C adsorbent dosage of 0.2 g, time of 2.5 min the adsorption is spontaneous and exothermic.	–	94.0	Sugashini and Begum (2015)

for heavy metals (Chen et al. 2010; Gupta et al. 2011; Sitko et al. 2012) and organic compounds (Fu and Wang 2011; Li et al. 2008; Rao et al. 2007; Yang and Chen 2008) including solid-phase extraction and wastewater treatment. Currently research is focused on the adsorption of single solute by Carbon nanotubes in aqueous solution and ignores the potential interactions between mixtures of metal ions and organic substances that may affect adsorption (V. Gupta et al. 2011; Nguyen et al. 2013; Yu et al. 2012). To improve the adsorption performance and avoid the disadvantage of Carbon nanotubes in adsorption process (e.g. easy aggregation and inherent insolubility), various carbon nanotubes based composites have been synthesized to explore the effectiveness of metal ions removal under different circumstances. Specifically, combining magnetic properties of iron oxide with adsorption properties of carbon nanotubes is of increasingly environmental concern as a rapid, effective and promising technology for removing hazardous pollutants in water and has been proposed for widespread environmental applications in wastewater treatment and potentially in situ remediation (Fu et al. 2013). Oxidized carbon nanotubes have high adsorption capacity for metal ions with fast kinetics. The surface functional groups (e.g., carboxyl, hydroxyl, and phenol) of carbon nanotubes are the major adsorption sites for metal ions, mainly through electrostatic attraction and chemical bonding (Venkateswarlu et al. 2007). As a result, surface oxidation can significantly enhance the adsorption capacity of carbon nanotubes. Overall, carbon nanotubes may not be a good alternative for activated carbon as wide-spectrum adsorbents. Rather, as their surface chemistry can be tuned to target specific contaminants, they may have unique applications in polishing steps to remove recalcitrant compounds or in pre-concentration of trace organic contaminants for analytical purposes. These applications require small quantity of materials and hence are less sensitive to the material cost.

Carbon nanotubes have been extensively studied during recent years for the treatment of water. Carbon nanotubes are rolled up graphene sheets which form a tube like structure with a small diameter usually one nanometer (Dai 2002). Based on number of layers of graphene sheets, carbon nanotubes are classified into single walled carbon nanotubes consist of single layer graphene shell or multi walled carbon nanotubes consist of multiple layers graphene shell (Pan and Xing 2008). Easily and eco-friendly synthesis gave great importance to carbon nanotubes for used as adsorbent. Various factors controls adsorption properties of carbon nanotubes including surface area, functional groups, adsorption site, open or close ended carbon nanotubes (Abbas et al. 2016; Agnihotri et al. 2006). As compared to activated carbons, carbon nanotubes have definite adsorption sites (Ren et al. 2011). Simple and modified carbon nanotubes were studied for the removing of Cr(III) from aqueous solution, successful modification of carbon nanotubes with COOH increased the adsorption. Atieh et al. (2010) synthesized florine doped carbon nanotubes and investigated different parameters for removal of Cr(III). at 28 °C and 20 min of contact time 50% and 65% removal was observed, respectively. ΔH_0 was -160.9 J/mol indicates the exothermic adsorption process, sticking probability S^* 0.719 J/mol, Ea (activation energy) -162 J/mol K and ΔG_0 values was -0.770 J/mol K indicates physisorption mechanism (Osikoya et al. 2014).

Gholami and Aghaie (2015) synthesized MnO_2 Multi-walled carbon nanotubes by coprecipitation and used this nanocomposite for the removal of Cr(III) from aqueous solution and optimized the condition for Cr(III) uptake, highest adsorption was observed at pH 5, absorbent dose 0.005 g, temperature 25 °C, initial concentration 10 mg/L, contact time 40 min. Adsorption phenomenon was found spontaneous and exothermic. Gupta et al. (2011), synthesized composite of multiwalled carbon nanotubes with iron oxides (MWCNT/nano-iron oxide) for the removal of Cr(III) from aqueous solution. Results from characterization through XRD, SEM, BET indicates magnetite and maghemite (magnet phase) and revealed nano-iron-oxide clusters with surface area of 92 m^2/g, optimized parameters for uptake of Cr(III) were found highest adsorption at pH 5–6, contact time 60 min, dosage 100 mg and agitation speed 150 rpm. Mubarak et al. (2016) used microwave heated fabricated carbon nanotubes for removal of Cr (III). It was found that at pH 8.0, contact time 60 min, agitation 150 rpm, initial concentration 2 mg/L and carbon nanotubes dosage .0.09 g removal efficiency of Cr(III) was 95.5% whereas maximum adsorption capacity was 24.45 mg/g. Moosa et al. (2015) used oxidized multiwalled carbon nanotubes with (1:3 by volume) (HNO_3:H_2SO_4) for adsorption of Cr(III) form wastewater. BET specific area was found 63.17 m^2/g, optimized adsorption parameters were pH 6, temperature 45 °C, and adsorption dosage 25 mg shows maximum adsorption 99.83%. Thermodynamic parameters shows that process is spontaneous, endothermic and increased randomness at solid/solution interface.

Peng et al. (2018) used polyethyleneimine of two different molecular weight 1800 and 1700 multi walled CNTs-18 and multi walled CNTs-70 respectively. These materials were characterized through TGA, XRD and SEM. TGA showed the loss of mass 46.5% and 51.9% in multi walled CNTs-70 and multi walled CNTs-18 respectively. Entirely different form original multiwalled carbon nanotubes XRD showed no significance difference between original and functionalized multi walled carbon nanotubes. Optimized parameters were initial concentration of 0.4 mmol. L^{-1}, pH 6, adsorption capacity of multiwalled CNTs-18 0.27 mmol/g and Multi Walled CNTs-70 was 0.33 mmol/g and adsorption phenomenon was spontaneous. Barakat et al. (2016) synthesized nanocomposite of binary metal oxyhydroxide AlOOH and FeOOH decorated oxidized-multiwalled carbon nanotubes (FeOOH/AlOOH/MWCNTs) and characterized through XRD, BET. Surface area of prepared nanocomposite was found 340.105 m^2/g which is greater than multiwalled Carbon nanotubes (289.33 m^2g^{-1}), maximum adsorption was found at pH 3, contact time 240 min and monolayer adsorption capacity found 60.6 mg/g. Desorption of 39% showed the strong interaction of Cr(VI) and synthesized nanocomposite.

Ghasemi et al. (2016) used functionalized carbon nanotubes with carboxylic groups and magnetite nanoparticles (Fe_3O_4) were used to prepare multiwalled CNTs/Fe_3O_4 nanocomposite to remove Cr(VI) from aqueous solution. TEM images of multiwalled carbon nanotubes and nanocomposite confirmed twisted network of carbon nanotubes with Fe_3O_4 clusters. XRD analyses show crystallite size of the nanocomposite is 26.3 nm, optimized conditions were pH 2, contact time 45 min, absorbent dosage 0.035 g for the maximum removal of Cr(VI) upto 90%. used activated carbon functionalized multiwalled carbon nanotubes (AC/f-MWCNTs) and activated carbon functionalized carbon nanospheres (AC/f-CNSs) with surface area of AC/fMWCNTs, 982 m^2/g as compared to AC/f-CNSs, 875.01 m^2/g. The activated carbon functionalized multiwalled carbon nanotubes have more active sites for metal binding which leads to higher adsorption capacity 113.29 mg/g, and adsorption process is mono layer at pH 2, contact time 50 min showed greater adsorption. Some other studies reported for the removal of chromium by using carbon nanotubes are given blow in Table 4.13.

4.3.5.2 Polypyrrole-Based Adsorbents

Polypyrrole is polymer obtained from pyrrole monomers using oxidants to trigger the reaction. It can also be synthesized electrochemically. Polyrpyrrole (PPy) synthesized in solutions with small dopants, such as Cl^-, ClO_4^-, and NO_3^-, mainly exhibits anion-exchanging behavior due to the high mobility of these ions in the polymer matrix. Polypyrole has also exhibited good prospects in adsorption applications because of the nitrogen atoms present in the polymer chains (Muhammad et al. 2016). Polypyrole graphene oxide (PPy-GO) nanocomposite have superior efficiency of 625 mg/g among all the reported sorbents for chromium removal at pH 2.0. The composite is positively charged at this pH and have BET surface area of 21.15 m^2/g. Enhanced adsorption was seen using composite as compared to Polypyrrole only. The composite is reported equally efficient in batch and flow modes of adsorption (Setshedi et al. 2015).

Polypyrole nanoparticles hybrid nanocomposite fixed on nanosheets of graphene/silica with greater surface area were obtained polypyrrole-graphene/silica (GS-PPy) by in-situ polymerization. At pH 2, temperature 25 °C, maximum Cr(VI) was adsorbed 429.2 mg/g and mechanism of removal was ion exchange, electrostatic attraction and reduction process and adsorption process was endothermic (Fang et al. 2018). Tahar et al. (2018) used pure maghemite (Magh) and crystalline magnetite rich (Magn) nanoparticles. Due to high active surface chemistry of these synthesized particles,100% removal of chromium using 20 mg/L and the dose of 3 g/L was reproted (Tahar et al. 2018). Some other studies reported for the removal of chromium by using different polypyrrole are given blow in Table 4.14.

Table 4.13 Removal of chromium using carbon nanotubes and functionalized carbon nanotubes with various functional groups for the increase adsorption capacity

Adsorbents	Adsorption capacity Q(mg/g)	Conditions	Interaction mechanism	References
Oxidized carbon nanotubes, bio sorbents (multiwalled carbon nanotubes rhizobium)	24.8	BET surface area 69.81 m²/g, pH 2.0, initial conc. 5 mg/L, time 180 min	–	Sathvika et al. (2018)
Carbon nano fiberss-carbon nanotubes (CNF's-CNT) grown on carbon nanofibers by plasma-enhanced chemical vapor deposition (PECVD)	234.9	BET surface area 398.127 m²/g removal efficiency 98%, initial conc. 60 mg/L, pH 3.0, time 20 min, Cr(VI) removal	–	Chen et al. 2018
Raw multiwalled carbon nanotubes	3.1	pH 3, initial conc. 1 mg/L, adsorbent dosage 75 mg	Electrostatic interactions	Ihsanullah et al. (2016)
Oxidized multiwalled carbon nanotubes	85.83	Temperature 25 °C, pH 2.5–4.0, adsorbent dose 0.15 g, removal efficiency 95%	Electrostatic interaction	Kumar et al. (2015)
Raw multiwalled carbon nanotubes	4.2	pH 2, initial conc. 3 mg/L, time 20, temperature 20 °C	Surface complexation	Hu et al. (2009)
Acid modified multiwalled carbon nanotubes	1.3	pH 3, initial conc. 1 mg/L, adsorbent dosage 75 mg	Electrostatic interactions	Ihsanullah et al. (2016)
Oxidized carbon nanotubes bio sorbents CNTY multiwalled carbon nanotubes-yeast	31.6	BET surface area, 37.029 m²/g, pH 2.0, initial conc. 5 mg /L, time 180 min,	–	Sathvika et al. (2018)
Single walled carbon nanotubes	20.3	pH 4.0, initial conc. 500 µg/L adsorbent dose 100 mg/L, contact Time 12 h, removal efficiency 72.9%	–	Jung et al. (2013)
Multiwalled carbon nanotubes	13.2	pH 2.5–4.0, initial conc. 20 mg/L	Electrostatic, cation–p interaction, anion–p interaction	Kumar et al. (2015)

(continued)

Table 4.13 (continued)

Adsorbents	Adsorption capacity Q(mg/g)	Conditions	Interaction mechanism	References
Multiwalled carbon nanotubes	2.48	pH 4.0, initial conc. 500 µg/L, adsorbent dose 100 mg/L, contact Time 12 h removal efficiency 51.9%	–	Jung et al. (2013)
zCeria nanoparticles supported on aligned carbon nanotubes (CeO2/ACNTs)	30.2	pH 7, initial conc. 3 mg/L	Ion exchange	Di et al. (2006)
Raw MWCNTs	0.37	pH 7, initial conc. 1 mg/L	Electrostatic interactions	Atieh et al. (2010)
Acid modified multiwalled carbon nanotubes	0.5	pH 7, initial conc. 1 mg/L	Electrostatic interactions	Atieh et al. (2010)
Nitrogen doped magnetic carbon nanotubes	12.28 mmol/g	pH 8	Chemical adsorption	Shin et al. (2011)
Raw carbon nanotubes	20.56	pH 7.5, initial conc. 33.28 mg/L	Ion exchange	Di et al. (2004)
Carbon nanotubes activated carbon	9	pH 2, initial conc. 0.5 mg/L	–	Atieh (2011)
Raw multiwalled carbon nanotubes	1.02	pH 3, initial conc. 1 mg/L, adsorbent dosage 75 mg	Electrostatic interactions	Ihsanullah et al. (2016)
Acid modified multiwalled carbon nanotubes	0.96	pH 3, initial conc. 1 mg/L, adsorbent dosage 75 mg	Electrostatic interactions	Ihsanullah et al. (2016)
Multiwalled carbon nanotubes	–	pH 2.5–4.0, initial conc. 20 mg/ L	Electrostatic, cation- π interaction, anion- π interaction	Kumar et al. (2015)
Pristine carbon nanotubes	–	pH 7.8	–	Xu et al. (2011b)
Acid treated carbon nanotubes	–	pH 7.8	Ion exchange	Xu et al. (2011b)
Oxidized multiwalled carbon nanotubes	1.0	pH 5.0, initial conc. 3 mg/L, temperature 20 °C	Surface complexation	Hu et al. (2009)

Table 4.14 Removal of chromium using polypyrrole based sorbents

Adsorbents	Optimized parameters	Adsorption (mg/g)	References
Polypyrrole graphene oxide (PPy–GO NC)	Temperature 25 °C, pH 2, time 50 min, 200 rpm and dose 0.025 g,	625	Setshedi et al. (2015)
Polypyrrole graphene oxide nanosheets (GO-αCD-PPy)	Temperature 25 °C, contact time 24 h, pH 3.0	497.1	Alvarado et al. (2013)
Polypyrrole/maghemite (PPy/γ-Fe$_2$O$_3$)	Temperature 25 °C, equilibrium time 15 min, pH 5.0 and pH 2.0	209	Chávez-Guajardo et al. (2015)
Hierarchical porous Polypyrrole nano clusters	Temperature 20 °C, batch adsorption equilibrium time 20 min, and pH 2	3.47 mmol/g	Yao et al. (2011)
Polypyrrole–titanium(IV) phosphate nanocomposite (PPy–TP)	Temperature 25–45 °C, pH 2, batch adsorption equilibrium time 19 min, initial conc. 200 mg/L and dose 0.2 g	31.64	Baig et al. (2015)
Graphene/Fe3O4@PPy nanocomposites	Temperature 25 °C, pH 2.0, magnetic separation and batch adsorption	348.4	Yao et al. (2014)
Polypyrrole coated Fe$_3$O$_4$ nanocomposites	Temperature 25 °C, residence time 30 min, and continuous flow rate 0.2 L/min with 20 mg/L		Muliwa et al. 2016)
Exfoliated polypyrrole-organically modified montmorillonite clay nanocomposite (PPy-OMMT-NC)	Temperature 20 °C, pH 2.0, contact time 24 h and dose 0.15 g	119.34	Setshedi et al. 2013)
Polypyrrole decorated reduced graphene oxide–Fe$_3$O$_4$ magnetic composites (Ppy–Fe3O4/Rgo)	Temperature 30–45 °C pH 3, batch adsorption equilibrium time 720 min	293.3	Wang et al. (2015)
Polypyrrole functionalized chitin	Temperature 30–50 °C, pH 4.8, contact time 60 min, initial conc. 50 mg/L dose 0.1 g, 250 rpm	28.92–35.22	Karthik and Meenakshi (2014)
Fe$_3$O$_4$/Polypyrrole composites microspheres	Temperature 25 °C, pH 2.0, contact time 30-180 min	209.2	Wang et al. (2012)
Peroxyacyl nitrates/ Polypyrrole core shell nanofiber mat	Temperature 25 °C, equilibrium time 30-90 min, pH 2.0	62	Jianqiang et al. (2013a)
Polypyrrole glycine doped composites	Temperature 25 °C, pH 2.0–5.0 and adsorption contact time 3 h.	217	Bhaumik et al. (2012)

(continued)

Table 4.14 (continued)

Adsorbents	Optimized parameters	Adsorption (mg/g)	References
Fe_3O_4@glycine-doped Polypyrrole magnetic nanocomposites	Temperature 25 °C, pH 2.0, adsorption contact time 3 hours, initial concentration. 200 mg/L and dose 0.1 g	238	Ballav et al. (2014)
Bamboo-like Polypyrrole nanotubes	Temperature 25 °C and pH 2.0	482.6	Li et al. (2012)
Polypyrrole/Fe_3O_4 magnetic nanocomposites	Temperature 25 °C, pH 2.0 adsorption contact time 12 h	169.4	Bhaumik et al. (2011)
Polypyrrole saw dust	Temperature 25 °C, time 15 min, dose 0.5 g, initial conc. 100 mg/L, pH 5.0	3.4	Ansari and Fahim (2007)

4.4 Conclusion

Reuse of metals in industrial process, especially those involve solvated metals, can reduce the burden of mining which eventually reduce the contamination of environment. Precipitation methods like coagulation and flocculation which are currently employed techniques in tannery waste generates huge amount of sludge containing chromium that require effective treatment again for safe disposal. Even after removal, metals remain in one or another form in nature; therefore, reclamation and reuse of spent metals would help reduce the burden of metal contamination in the environment. Therefore, removal methods that can efficiently remove chromium (or other metal ions) from environment and pose little treatment of the waste generated during treatment processes would be of interest in solving metal decontamination issue. It would also be of interest to focus on the development of technologies where reclamation of chromium is feasible. This would result in reuse of waste chromium, reduction in wastewater treatment cost and savings from reusing the chromium in place of using fresh chromium.

Sorptive removal process is seen as viable technology for removal and reclamation of chromium. Among various sorbents compiled in this chapter, aminated wheat straw showed highest adsorption capacity of 454 mg/g in the category of natural sorbents (Yao et al. 2016)while Polypyrrole graphene oxide nanocomposite showed 625 mg/g in synthetic sorbents (Setshedi et al. 2015). It may be concluded that removal of chromium is efficient when sorbents possess high density of amine functional groups. Efficiency can be further enhanced by improving accessibility of active sites of sorbents. Morphological changes in Polypyrrole and its composite would be materials of interest in this regard.

Continuous flow processes should be investigated using highly efficient adsorbent materials keeping in view industrial conditions. These studies should focus on the methods, which can reclaim chromium from industrial wastewaters. Membrane based processes have gained more attention over fixed bed reactors in recent years for chromium removal. Aminated-Fe_3O_4 nanoparticles filled chitosan/PVA/PES dual layers nanofibrous membrane is a good effort in this direction (Koushkbaghi et al. 2018). Certain electro driven membrane-based processes are also reported, however more studies are needed to develop an industrially feasible process.

Acknowledgement Financial support from Pak-US Science and Technology Cooperation Program, Phase VI (Project No. 6/6/PAK-US/HEC/2015/06) is highly acknowledged.

References

Abbas A, Al-Amer AM, Laoui T, Al-Marri MJ, Nasser MS, Khraisheh M, Atieh MA (2016) Heavy metal removal from aqueous solution by advanced carbon nanotubes: critical review of adsorption applications. Sep Purif Technol 157:141–161. https://doi.org/10.1016/j.seppur.2015.11.039

Acar F, Malkoc E (2004) The removal of chromium (VI) from aqueous solutions by Fagus orientalis L. Bioresour Technol 94(1):13–15. https://doi.org/10.1016/j.biortech.2003.10.032

Agnihotri S, Mota JP, Rostam-Abadi M, Rood MJ (2006) Theoretical and experimental investigation of morphology and temperature effects on adsorption of organic vapors in single-walled carbon nanotubes. J Phys Chem B 110(15):7640–7647. https://doi.org/10.1021/jp060040a

Ahalya N, Ramachandra T, Kanamadi R (2003) Biosorption of heavy metals. Res J Chem Environ 7(4):71–79

Ahmad T, Danish M (2018) Prospects of banana waste utilization in wastewater treatment: A review. J Environ Manag 206:330–348. https://doi.org/10.1016/j.jenvman.2017.10.061

Ahmadi M, Kouhgardi E, Ramavandi B (2016) Physico-chemical study of dew melon peel biochar for chromium attenuation from simulated and actual wastewaters. Korean J Chem Eng 33(9):2589–2601. https://doi.org/10.1007/s11814-016-0135-1

Akbal F, Camcı S (2010) Comparison of electrocoagulation and chemical coagulation for heavy metal removal. Chem Eng Technol 33(10):1655–1664. https://doi.org/10.1002/ceat.201000091

Akbal F, Camcı S (2011) Copper, chromium and nickel removal from metal plating wastewater by electrocoagulation. Desalination 269(1–3):214–222. https://doi.org/10.1016/j.desal.2010.11.001

Alamgir A, Kanwal SA (2018) Quantification and composition of solid waste abundance on the beaches of Karachi, Pakistan. Curr World Environ. https://doi.org/10.12944/CWE.13.2.08

Alemu A, Lemma B, Gabbiye N, Alula MT, Desta MT (2018) Removal of chromium (VI) from aqueous solution using vesicular basalt: a potential low cost wastewater treatment system. Heliyon 4(7):e00682. https://doi.org/10.1016/j.heliyon.2018.e00682

Alguacil FJ, Caravaca C, Martín MI (2003) Transport of chromium (VI) through a Cyanex 921-supported liquid membrane from HCl solutions. J Chem Technol Biotechnol 78(10):1048–1053. https://doi.org/10.1002/jctb.903

Alguacil FJ, Garcia-Diaz I, Lopez F (2012) The removal of chromium (III) from aqueous solution by ion exchange on Amberlite 200 resin: batch and continuous ion exchange modelling. Desalin Water Treat 45(1–3):55–60. https://doi.org/10.1080/19443994.2012.692009

Ali A, Saeed K, Mabood F (2016) Removal of chromium (VI) from aqueous medium using chemically modified banana peels as efficient low-cost adsorbent. Alexandria Eng J 55(3):2933–2942. https://doi.org/10.1016/j.aej.2016.05.011

Al-Jlil SA (2015) Kinetic study of adsorption of chromium and lead ions on bentonite clay using novel internal series model. Trends Appl Sci Res 10(1):88–58. https://doi.org/10.3923/tasr.2015.38.53

Alloway BJ (2013) Sources of heavy metals and metalloids in soils. In: Heavy metals in soils. Springer, pp 11–50. https://doi.org/10.1007/978-94-007-4470-7_2

Alpaydin S, Saf AÖ, Bozkurt S, Sirit A (2011) Kinetic study on removal of toxic metal Cr (VI) through a bulk liquid membrane containing p-tert-butylcalix [4] arene derivative. Desalination 275(1–3):166–171. https://doi.org/10.1016/j.desal.2011.02.048

Altun T, Parlayıcı Ş, Pehlivan E (2016) Hexavalent chromium removal using agricultural waste "rye husk". Desalin Water Treat 57(38):17748–17756. https://doi.org/10.1080/19443994.2015.1085914

Alvarado L, Torres IR, Chen A (2013) Integration of ion exchange and electrodeionization as a new approach for the continuous treatment of hexavalent chromium wastewater. Sep Purif Technol 105:55–62. https://doi.org/10.1016/j.seppur.2012.12.007

Alvarez L, Perez-Cruz M, Rangel-Mendez J, Cervantes F (2010) Immobilized redox mediator on metal-oxides nanoparticles and its catalytic effect in a reductive decolorization process. J Hazard Mater 184(1–3):268–272. https://doi.org/10.1016/j.jhazmat.2010.08.032

Alves MM, Beca CG, De Carvalho RG, Castanheira J, Pereira MS, Vasconcelos L (1993) Chromium removal in tannery wastewaters "polishing" by Pinus sylvestris bark. Water Res 27(8):1333–1338. https://doi.org/10.1016/0043-1354(93)90220-C

Anagnostopoulos VA, Bhatnagar A, Mitropoulos AC, Kyzas GZ (2017) A review for chromium removal by carbon nanotubes. Chem Ecol 33(6):572–588. https://doi.org/10.1080/02757540.2017.1328503

Anirudhan T, Nima J, Divya P (2013) Adsorption of chromium (VI) from aqueous solutions by glycidylmethacrylate-grafted-densified cellulose with quaternary ammonium groups. Appl Surf Sci 279:441–449. https://doi.org/10.1016/j.apsusc.2013.04.134

Ansari R, Fahim NK (2007) Application of polypyrrole coated on wood sawdust for removal of Cr (VI) ion from aqueous solutions. React Funct Polym 67(4):367–374. https://doi.org/10.1016/j.reactfunctpolym.2007.02.001

Apte JS, Messier KP, Gani S, Brauer M, Kirchstetter TW, Lunden MM, Hamburg SP (2017) High-resolution air pollution mapping with google street view cars: exploiting big data. Environ Sci Technol 51(12):6999–7008. https://doi.org/10.1021/acs.est.7b00891

Araghi SH, Entezari MH, Chamsaz M (2015) Modification of mesoporous silica magnetite nanoparticles by 3-aminopropyltriethoxysilane for the removal of Cr (VI) from aqueous solution. Microporous Mesoporous Mater 218:101–111. https://doi.org/10.1016/j.micromeso.2015.07.008

Aroua MK, Zuki FM, Sulaiman NM (2007) Removal of chromium ions from aqueous solutions by polymer-enhanced ultrafiltration. J Hazard Mater 147(3):752–758. https://doi.org/10.1016/j.jhazmat.2007.01.120

Arulkumar M, Thirumalai K, Sathishkumar P, Palvannan T (2012) Rapid removal of chromium from aqueous solution using novel prawn shell activated carbon. Chem Eng J 185:178–186. https://doi.org/10.1016/j.cej.2012.01.071

Atieh MA (2011) Removal of chromium (VI) from polluted water using carbon nanotubes supported with activated carbon. Procedia Environ Sci 4:281–293. https://doi.org/10.1016/j.proenv.2011.03.033

Atieh, M. A., Bakather, O. Y., Tawabini, B. S., Bukhari, A. A., Khaled, M., Alharthi, M., . . . Abuilaiwi, F. A. (2010). Removal of chromium (III) from water by using modified and non-modified carbon nanotubes. J Nanomater, 2010, 17. https://doi.org/10.1155/2010/232378

Augustynowicz J, Grosicki M, Hanus-Fajerska E, Lekka M, Waloszek A, Kołoczek H (2010) Chromium (VI) bioremediation by aquatic macrophyte Callitriche cophocarpa Sendtn. Chemosphere 79(11):1077–1083. https://doi.org/10.1016/j.chemosphere.2010.03.019

Ayoub G, Hamzeh A, Semerjian L (2011) Post treatment of tannery wastewater using lime/bittern coagulation and activated carbon adsorption. Desalination 273(2):359–365. https://doi.org/10.1016/j.desal.2011.01.045

Babcock W, Baker R, Lachapelle E, Smith K (1980) Coupled transport membranes II: the mechanism of uranium transport with a tertiary amine. J Membr Sci 7(1):71–87. https://doi.org/10.1016/S0376-7388(00)83186-5

Babel S, Kurniawan TA (2004) Cr (VI) removal from synthetic wastewater using coconut shell charcoal and commercial activated carbon modified with oxidizing agents and/or chitosan. Chemosphere 54(7):951–967. https://doi.org/10.1016/j.chemosphere.2003.10.001

Babu AN, Mohan GK, Ravindhranath K (2016) Removal of chromium (VI) from polluted waters using adsorbents derived from Chenopodium album and Eclipta prostrate plant materials. Int J ChemTech Res 9(03):506–516

Baig U, Rao RAK, Khan AA, Sanagi MM, Gondal MA (2015) Removal of carcinogenic hexavalent chromium from aqueous solutions using newly synthesized and characterized polypyrrole–titanium (IV) phosphate nanocomposite. Chem Eng J 280:494–504. https://doi.org/10.1016/j.cej.2015.06.031

Ballav N, Choi H, Mishra S, Maity A (2014) Synthesis, characterization of Fe3O4@ glycine doped polypyrrole magnetic nanocomposites and their potential performance to remove toxic Cr (VI). J Ind Eng Chem 20(6):4085–4093. https://doi.org/10.1016/j.jiec.2014.01.007

Banerjee S, Joshi S, Mandal T, Halder G (2017) Insight into Cr^{6+} reduction efficiency of Rhodococcus erythropolis isolated from coalmine waste water. Chemosphere 167:269–281. https://doi.org/10.1016/j.chemosphere.2016.10.012

Bao Y, Yan X, Du W, Xie X, Pan Z, Zhou J, Li L (2015) Application of amine-functionalized MCM-41 modified ultrafiltration membrane to remove chromium (VI) and copper (II). Chem Eng J 281:460–467. https://doi.org/10.1016/j.cej.2015.06.094

Barakat M (2011) New trends in removing heavy metals from industrial wastewater. Arab J Chem 4(4):361–377. https://doi.org/10.1016/j.arabjc.2010.07.019

Barakat M, Al-Ansari A, Kumar R (2016) Synthesis and characterization of Fe– Al binary oxyhydroxides/MWCNTs nanocomposite for the removal of Cr (VI) from aqueous solution. J Taiwan Inst Chem Eng 63:303–311. https://doi.org/10.1016/j.jtice.2016.03.019

Barrera-Díaz CE, Lugo-Lugo V, Bilyeu B (2012) A review of chemical, electrochemical and biological methods for aqueous Cr (VI) reduction. J Hazard Mater 223:1–12. https://doi.org/10.1016/j.jhazmat.2012.04.054

Bashir A, Malik LA, Ahad S, Manzoor T, Bhat MA, Dar G, Pandith AH (2018) Removal of heavy metal ions from aqueous system by ion-exchange and biosorption methods. Environ Chem Lett:1–26. https://doi.org/10.1007/s10311-018-00828-y

Bazrafshan E, Mohammadi L, Ansari-Moghaddam A, Mahvi AH (2015) Heavy metals removal from aqueous environments by electrocoagulation process–a systematic review. J Environ Health Sci Eng 13(1):74

Bello MM, Raman AAA (2018) Synergy of adsorption and advanced oxidation processes in recalcitrant wastewater treatment. Environ Chem Lett:1–18. https://doi.org/10.1007/s10311-018-00842-0

Benjjar A, Hor M, Riri M, Eljaddi T, Kamal O, Lebrun L, Hlaïbi M (2012) A new supported liquid membrane (SLM) with methyl cholate for facilitated transport of dichromate ions from mineral acids: parameters and mechanism relating to the transport. J Mater Environ Sci 3(5):826–839

Bergaya F, Lagaly G (2006) General introduction: clays, clay minerals, and clay science. Dev Clay Sci 1:1–18. https://doi.org/10.1016/S1572-4352(05)01001-9

Beszedits S (1988) Chromium removal from industrial wastewaters. In: Chromium in the natural and human environments. John Wiley, New York, pp 232–263

Bhattacharya A, Gupta A (2013) Evaluation of Acinetobacter sp. B9 for Cr (VI) resistance and detoxification with potential application in bioremediation of heavy-metals-rich industrial wastewater. Environ Sci Pollut Res 20(9):6628–6637. https://doi.org/10.1007/s11356-013-1728-4

Bhatti AA, Memon S, Memon N, Bhatti AA, Solangi IB (2017) Evaluation of chromium (VI) sorption efficiency of modified Amberlite XAD-4 resin. Arab J Chem 10:S1111–S1118. https://doi.org/10.1016/j.arabjc.2013.01.020

Bhaumik M, Maity A, Srinivasu V, Onyango MS (2011) Enhanced removal of Cr (VI) from aqueous solution using polypyrrole/Fe3O4 magnetic nanocomposite. J Hazard Mater 190(1–3):381–390. https://doi.org/10.1016/j.jhazmat.2011.03.062

Bhaumik M, Maity A, Srinivasu V, Onyango MS (2012) Removal of hexavalent chromium from aqueous solution using polypyrrole-polyaniline nanofibers. Chem Eng J 181:323–333. https://doi.org/10.1016/j.cej.2011.11.088

Bishnoi NR, Bajaj M, Sharma N, Gupta A (2004) Adsorption of Cr (VI) on activated rice husk carbon and activated alumina. Bioresour Technol 91(3):305–307. https://doi.org/10.1016/S0960-8524(03)00204-9

Bo S, Ren W, Lei C, Xie Y, Cai Y, Wang S et al (2018) Flexible and porous cellulose aerogels/zeolitic imidazolate framework (ZIF-8) hybrids for adsorption removal of Cr (IV) from water. J Solid State Chem 262:135–141. https://doi.org/10.1016/j.jssc.2018.02.022

Bohdziewicz J (2000) Removal of chromium ions (VI) from underground water in the hybrid complexation-ultrafiltration process. Desalination 129(3):227–235. https://doi.org/10.1016/S0011-9164(00)00063-1

Calder L (1988) Chromium contamination of groundwater. Advances in environmental science and technology (USA)

Carlos FS, Giovanella P, Bavaresco J, de Souza Borges C, de Oliveira Camargo FA (2016) A comparison of microbial bioaugmentation and biostimulation for hexavalent chromium removal from wastewater. Water Air Soil Pollut 227(6):175. https://doi.org/10.1007/s11270-016-2872-5

Cavaco SA, Fernandes S, Augusto CM, Quina MJ, Gando-Ferreira LM (2009) Evaluation of chelating ion-exchange resins for separating Cr (III) from industrial effluents. J Hazard Mater 169(1):516–523. https://doi.org/10.1016/j.jhazmat.2009.03.129

Chakraborty S, Datta S, Bhattacharya P (2005) Studies on extraction of chromium (VI) from acidic solution by emulsion liquid membrane. J Membr Sci 325:460–466. https://doi.org/10.1016/j.memsci.2008.08.009

Chávez-Guajardo AE, Medina-Llamas JC, Maqueira L, Andrade CA, Alves KG, de Melo CP (2015) Efficient removal of Cr (VI) and cu (II) ions from aqueous media by use of polypyrrole/maghemite and polyaniline/maghemite magnetic nanocomposites. Chem Eng J 281:826–836. https://doi.org/10.1016/j.cej.2015.07.008

Cheballah K, Sahmoune A, Messaoudi K, Drouiche N, Lounici H (2015) Simultaneous removal of hexavalent chromium and COD from industrial wastewater by bipolar electrocoagulation. Chem Eng Process Process Intensif 96:94–99. https://doi.org/10.1016/j.cep.2015.08.007

Chelme-Ayala P, Smith DW, El-Din MG (2009) Membrane concentrate management options: a comprehensive critical review A paper submitted to the journal of environmental engineering and science. Can J Civ Eng 36(6):1107–1119. https://doi.org/10.1139/L09-042

Chen G (2004) Electrochemical technologies in wastewater treatment. Sep Purif Technol 38(1):11–41. https://doi.org/10.1016/j.seppur.2003.10.006

Chen J-C, Wang K-S, Chen H, Lu C-Y, Huang L-C, Li H-C, Chang S-H (2010) Phytoremediation of Cr (III) by Ipomonea aquatica (water spinach) from water in the presence of EDTA and chloride: effects of Cr speciation. Bioresour Technol 101(9):3033–3039. https://doi.org/10.1016/j.biortech.2009.12.041

Chen Y-G, He Y, Ye W-M, Lin C-h, Zhang X-F, Ye B (2012) Removal of chromium (III) from aqueous solutions by adsorption on bentonite from Gaomiaozi, China. Environ Earth Sci 67(5):1261–1268. https://doi.org/10.1007/s12665-012-1569-3

Chen J, Hong X, Zhao Y, Zhang Q (2014) Removal of hexavalent chromium from aqueous solution using exfoliated polyaniline/montmorillonite composite. Water Sci Technol 70(4). https://doi.org/10.2166/wst.2014.277

Chen L, Chen N, Wu H, Li W, Fang Z, Xu Z, Qian X (2018) Flexible design of carbon nanotubes grown on carbon nanofibers by PECVD for enhanced Cr (VI) adsorption capacity. Sep Purif Technol 207:406–415. https://doi.org/10.1016/j.seppur.2018.06.065

Cheung K, Gu J-D (2007) Mechanism of hexavalent chromium detoxification by microorganisms and bioremediation application potential: a review. Int Biodeterior Biodegrad 59(1):8–15. https://doi.org/10.1016/j.ibiod.2006.05.002

Choi YW, Moon SH (2005) A study on supported liquid membrane for selective separation of Cr (VI). Sep Sci Technol 39(7):1663–1680. https://doi.org/10.1081/SS-120030797

Choudhury PR, Majumdar S, Sahoo GC, Saha S, Mondal P (2018) High pressure ultrafiltration CuO/hydroxyethyl cellulose composite ceramic membrane for separation of Cr (VI) and Pb (II) from contaminated water. Chem Eng J 336:570–578. https://doi.org/10.1016/j.cej.2017.12.062

Cimino G, Passerini A, Toscano G (2000) Removal of toxic cations and Cr (VI) from aqueous solution by hazelnut shell. Water Res 34(11):2955–2962. https://doi.org/10.1016/S0043-1354(00)00048-8

Conroy KG, Breslin CB (2004) Reduction of hexavalent chromium at a polypyrrole-coated aluminium electrode: synergistic interactions. J Appl Electrochem 34(2):191–195. https://doi.org/10.1023/B:JACH.0000009924.52188.f6

Cutillas-Barreiro L, Ansias-Manso L, Fernández-Calviño D, Arias-Estévez M, Nóvoa-Muñoz J, Fernández-Sanjurjo M et al (2014) Pine bark as bio-adsorbent for cd, cu, Ni, Pb and Zn: batch-type and stirred flow chamber experiments. J Environ Manag 144:258–264. https://doi.org/10.1016/j.jenvman.2014.06.008

da Silva Pereira A, Vieira CBL, Barbosa RMS, Soares EA, Lanzillotti HS (2010) Análise comparativa do estado nutricional de pré-escolares. Revista Paulista de Pediatria 28(2):176–180

Dai H (2002) Carbon nanotubes: opportunities and challenges. Surf Sci 500(1):218–241. https://doi.org/10.1016/S0039-6028(01)01558-8

Dakiky M, Khamis M, Manassra A, Mer'Eb M (2002) Selective adsorption of chromium (VI) in industrial wastewater using low-cost abundantly available adsorbents. Adv Environ Res 6(4):533–540. https://doi.org/10.1016/S1093-0191(01)00079-X

Daniş Ü, Keskinler B (2009) Chromate removal from wastewater using micellar enhanced crossflow filtration: effect of transmembrane pressure and crossflow velocity. Desalination 249(3):1356–1364. https://doi.org/10.1016/j.desal.2009.06.023

Daraei H, Mittal A, Mittal J, Kamali H (2014) Optimization of Cr (VI) removal onto biosorbent eggshell membrane: experimental & theoretical approaches. Desalin Water Treat 52(7–9):1307–1315. https://doi.org/10.1080/19443994.2013.787374

Das S, Mishra J, Das SK, Pandey S, Rao DS, Chakraborty A et al (2014) Investigation on mechanism of Cr (VI) reduction and removal by Bacillus amyloliquefaciens, a novel chromate tolerant bacterium isolated from chromite mine soil. Chemosphere 96:112–121. https://doi.org/10.1016/j.chemosphere.2013.08.080

Davies J-M, Mazumder A (2003) Health and environmental policy issues in Canada: the role of watershed management in sustaining clean drinking water quality at surface sources. J Environ Manag 68(3):273–286. https://doi.org/10.1016/S0301-4797(03)00070-7

Dayan A, Paine A (2001) Mechanisms of chromium toxicity, carcinogenicity and allergenicity: review of the literature from 1985 to 2000. Hum Exp Toxicol 20(9):439–451. https://doi.org/10.1191/096032701682693062

de San Miguel ER, Vital X, de Gyves J (2014) Cr (VI) transport via a supported ionic liquid membrane containing CYPHOS IL101 as carrier: system analysis and optimization through experimental design strategies. J Hazard Mater 273:253–262. https://doi.org/10.1016/j.jhazmat.2014.03.052

Dehghani MH, Sanaei D, Ali I, Bhatnagar A (2016) Removal of chromium (VI) from aqueous solution using treated waste newspaper as a low-cost adsorbent: kinetic modeling and isotherm studies. J Mol Liq 215:671–679. https://doi.org/10.1016/j.molliq.2015.12.057

Derdour K, Bouchelta C, Khorief Naser-Eddine A, Medjram MS, Magri P (2018) Removal of Cr (VI) from aqueous solution using activated carbon supported iron catalysts as efficient adsorbent. World J Eng 15:3–13. https://doi.org/10.1108/WJE-06-2017-0132

Dhal B, Thatoi HN, Das NN, Pandey BD (2013) Chemical and microbial remediation of hexavalent chromium from contaminated soil and mining/metallurgical solid waste: A review. J Hazard Mater 250–251:272–291. https://doi.org/10.1016/j.jhazmat.2013.01.048

Dharnaik AS, Ghosh PK (2014) Hexavalent chromium [Cr (VI)] removal by the electrochemical ion-exchange process. Environ Technol 35(18):2272–2279. https://doi.org/10.1080/09593330.2014.902108

Di Natale F, Lancia A, Molino A, Musmarra D (2007) Removal of chromium ions form aqueous solutions by adsorption on activated carbon and char. J Hazard Mater 145(3):381–390. https://doi.org/10.1016/j.jhazmat.2006.11.028

Di Z-C, Li Y-H, Luan Z-K, Liang J (2004) Adsorption of chromium (VI) ions from water by carbon nanotubes. Adsorpt Sci Technol 22(6):467–474. https://doi.org/10.1260/0263617042879537

Di Z-C, Ding J, Peng X-J, Li Y-H, Luan Z-K, Liang J (2006) Chromium adsorption by aligned carbon nanotubes supported ceria nanoparticles. Chemosphere 62(5):861–865. https://doi.org/10.1016/j.chemosphere.2004.06.044

Dimos V, Haralambous KJ, Malamis S (2012) A review on the recent studies for chromium species adsorption on raw and modified natural minerals. Crit Rev Environ Sci Technol 42(19):1977–2016. https://doi.org/10.1080/10643389.2011.574102

Dinker MK, Kulkarni PS (2015) Recent advances in silica-based materials for the removal of hexavalent chromium: A review. J Chem Eng Data 60(9):2521–2540. https://doi.org/10.1021/acs.jced.5b00292

Doke SM, Yadav GD (2014) Process efficacy and novelty of titania membrane prepared by polymeric sol–gel method in removal of chromium (VI) by surfactant enhanced microfiltration. Chem Eng J 255:483–491. https://doi.org/10.1016/j.cej.2014.05.098

Dong Z, Zhao J, Du J, Li C, Zhao L (2016) Radiation synthesis of spherical cellulose-based adsorbent for efficient adsorption and detoxification of Cr (VI). Rad Phys Chem 126:68–74. https://doi.org/10.1016/j.radphyschem.2016.05.013

Dragan ES, Humelnicu D, Dinu MV, Olariu RI (2017) Kinetics, equilibrium modeling, and thermodynamics on removal of Cr (VI) ions from aqueous solution using novel composites with strong base anion exchanger microspheres embedded into chitosan/poly (vinyl amine) cryogels. Chem Eng J 330:675–691. https://doi.org/10.1016/j.cej.2017.08.004

Duan W, Chen G, Chen C, Sanghvi R, Iddya A, Walker S et al (2017) Electrochemical removal of hexavalent chromium using electrically conducting carbon nanotube/polymer composite ultrafiltration membranes. J Membr Sci 531:160–171. https://doi.org/10.1016/j.memsci.2017.02.050

Dubey SP, Gopal K (2007) Adsorption of chromium (VI) on low cost adsorbents derived from agricultural waste material: a comparative study. J Hazard Mater 145(3):465–470. https://doi.org/10.1016/j.jhazmat.2006.11.041

Dula T, Siraj K, Kitte SA (2014) Adsorption of hexavalent chromium from aqueous solution using chemically activated carbon prepared from locally available waste of bamboo (Oxytenanthera abyssinica). ISRN Environ Chem 2014. https://doi.org/10.1155/2014/438245

Dutta BK (2007) Principles of mass transfer and seperation processes. PHI Learning Pvt. Ltd., New Delhi

Edition F (2011) Guidelines for drinking-water quality. WHO Chron 38(4):104–108

Egodawatte S, Datt A, Burns EA, Larsen SC (2015) Chemical insight into the adsorption of chromium (III) on iron oxide/mesoporous silica nanocomposites. Langmuir 31(27):7553–7562. https://doi.org/10.1021/acs.langmuir.5b01483

El Nemr A, El-Sikaily A, Khaled A, Abdelwahab O (2015) Removal of toxic chromium from aqueous solution, wastewater and saline water by marine red alga Pterocladia capillacea and its activated carbon. Arab J Chem 8(1):105–117. https://doi.org/10.1016/j.arabjc.2011.01.016

El-Mehalmey WA, Ibrahim AH, Abugable AA, Hassan MH, Haikal RR, Karakalos SG, Alkordi MH (2018) Metal–organic framework@ silica as a stationary phase sorbent for rapid and cost-effective removal of hexavalent chromium. J Mater Chem A 6(6):2742–2751. https://doi.org/10.1039/C7TA08281A

Elmolla ES, Hamdy W, Kassem A, Abdel Hady A (2016) Comparison of different rice straw based adsorbents for chromium removal from aqueous solutions. Desalin Water Treat 57(15):6991–6999. https://doi.org/10.1080/19443994.2015.1015175

El-Taweel YA, Nassef EM, Elkheriany I, Sayed D (2015) Removal of Cr (VI) ions from waste water by electrocoagulation using iron electrode. Egypt J Pet 24(2):183–192. https://doi.org/10.1016/j.ejpe.2015.05.011

Ertani A, Mietto A, Borin M, Nardi S (2017) Chromium in agricultural soils and crops: a review. Water Air Soil Pollut 228(5):190. https://doi.org/10.1007/s11270-017-3356-y

Esmaeili A, Hejazi E, Vasseghian Y (2015) Comparison study of biosorption and coagulation/air flotation methods for chromium removal from wastewater: experiments and neural network modeling. RSC Adv 5(111):91776–91784. https://doi.org/10.1039/C5RA16997F

Espinoza-Quiñones F, Módenes A, Theodoro P, Palácio S, Trigueros D, Borba C et al (2012) Optimization of the iron electro-coagulation process of Cr, Ni, cu, and Zn galvanization by-products by using response surface methodology. Sep Sci Technol 47(5):688–699. https://doi.org/10.1080/01496395.2011.629396

Eyupoglu V, Tutkun O (2011) The extraction of Cr (VI) by a flat sheet supported liquid membrane using alamine 336 as a carrier. Arab J Sci Eng 36(4):529–539. https://doi.org/10.1007/s13369-011-0057-5

Eyupoglu V, Eren B, Dogan E (2010) Prediction of ionic Cr (VI) extraction efficiency in flat sheet supported liquid membrane using artificial neural networks (ANNs). Int J Environ Res 4(3):463–470. https://doi.org/10.22059/IJER.2010.231

Fahim N, Barsoum B, Eid A, Khalil M (2006) Removal of chromium (III) from tannery wastewater using activated carbon from sugar industrial waste. J Hazard Mater 136(2):303–309. https://doi.org/10.1016/j.jhazmat.2005.12.014

Fan Y, Wang X, Wang M (2013) Separation and recovery of chromium and vanadium from vanadium-containing chromate solution by ion exchange. Hydrometallurgy 136:31–35. https://doi.org/10.1016/j.hydromet.2013.03.008

Fang W, Jiang X, Luo H, Geng J (2018) Synthesis of graphene/SiO 2@ polypyrrole nanocomposites and their application for Cr (VI) removal in aqueous solution. Chemosphere. https://doi.org/10.1016/j.chemosphere.2017.12.163

Farajzadeh MA, Monji AB (2004) Treated rice bran for scavenging Cr (III) and hg (II) from acidic solution. J Chin Chem Soc 51(4):751–759. https://doi.org/10.1002/jccs.200400114

Faust S, Ali O (1998) Chemistry of water treatment. Ann Arbor Press, Chelsea

Fellenz N, Martin P, Marchetti S, Bengoa F (2015) Aminopropyl-modified mesoporous silica nanospheres for the adsorption of Cr (VI) from water. J Porous Mater 22(3):729–738. https://doi.org/10.1007/s10934-015-9946-4

Feng N, Guo X, Liang S, Zhu Y, Liu J (2011) Biosorption of heavy metals from aqueous solutions by chemically modified orange peel. J Hazard Mater 185(1):49–54. https://doi.org/10.1016/j.jhazmat.2010.08.114

Feng B, Shen W, Shi L, Qu S (2018) Adsorption of hexavalent chromium by polyacrylonitrile-based porous carbon from aqueous solution. R Soc Open Sci 5(1):171662. https://doi.org/10.1098/rsos.171662

Foroutan R, Mohammadi R, Ramavandi B (2018) Treatment of chromium-laden aqueous solution using CaCl 2-modified Sargassum oligocystum biomass: characteristics, equilibrium, kinetic, and thermodynamic studies. Korean J Chem Eng 35(1):234–245. https://doi.org/10.1007/s11814-017-0239-2

Frenzel I, Holdik H, Barmashenko V, Stamatialis DF, Wessling M (2006) Electrochemical reduction of dilute chromate solutions on carbon felt electrodes. J Appl Electrochem 36(3):323–332. https://doi.org/10.1007/s10800-005-9074-y

Fu F, Wang Q (2011) Removal of heavy metal ions from wastewaters: a review. J Environ Manag 92(3):407–418. https://doi.org/10.1016/j.jenvman.2010.11.011

Fu F, Ma J, Xie L, Tang B, Han W, Lin S (2013) Chromium removal using resin supported nanoscale zero-valent iron. J Environ Manag 128:822–827. https://doi.org/10.1016/j.jenvman.2013.06.044

Gadd GM, White C (1993) Microbial treatment of metal pollution—a working biotechnology. Trends Biotechnol 11(8):353–359. https://doi.org/10.1016/0167-7799(93)90158-6

Gangadharan P, Nambi IM, Senthilnathan J (2015) Liquid crystal polaroid glass electrode from e-waste for synchronized removal/recovery of Cr+ 6 from wastewater by microbial fuel cell. Bio/Technology 195:96–101. https://doi.org/10.1016/j.biortech.2015.06.078

Gardea-Torresdey JL, de la Rosa G, Peralta-Videa JR (2004) Use of phytofiltration technologies in the removal of heavy metals: A review. Pure Appl Chem 76:801. https://doi.org/10.1351/pac200476040801

Garg V, Gupta R, Kumar R, Gupta R (2004) Adsorption of chromium from aqueous solution on treated sawdust. Bioresour Technol 92(1):79–81. https://doi.org/10.1016/j.biortech.2003.07.004

Garg UK, Kaur M, Garg V, Sud D (2007) Removal of hexavalent chromium from aqueous solution by agricultural waste biomass. J Hazard Mater 140(1–2):60–68. https://doi.org/10.1016/j.jhazmat.2006.06.056

Ge S, Dong X, Zhou J, Ge S (2013) Comparative evaluations on bio-treatment of hexavalent chromate by resting cells of Pseudochrobactrum sp. and Proteus sp. in wastewater. J Environ Manag 126:7–12. https://doi.org/10.1016/j.jenvman.2013.04.011

Ghasemi R, Sayahi T, Tourani S, Kavianimehr M (2016) Modified magnetite nanoparticles for hexavalent chromium removal from water. J Dispers Sci Technol 37(9):1303–1314. https://doi.org/10.1080/01932691.2015.1090906

Gholami M, Aghaie M (2015) Thermodynamic study of Cr+ 3 ions removal by "MnO2/MWCNT" nanocomposite. Orient J Chem 31(3):1429–1436. https://doi.org/10.13005/ojc/310321

Ghosh G, Bhattacharya PK (2006) Hexavalent chromium ion removal through micellar enhanced ultrafiltration. Chem Eng J 119(1):45–53. https://doi.org/10.1016/j.cej.2006.02.014

Golder AK, Chanda AK, Samanta AN, Ray S (2011) Removal of hexavalent chromium by electrochemical reduction–precipitation: investigation of process performance and reaction stoichiometry. Sep Purif Technol 76(3):345–350. https://doi.org/10.1016/j.seppur.2010.11.002

Gonçalves AC, Schwantes D, Campagnolo MA, Dragunski DC, Tarley CRT, dos Santos Silva AK (2018) Removal of toxic metals using endocarp of açaí berry as biosorbent. Water Sci Technol 77(6):1547–1557. https://doi.org/10.2166/wst.2018.032

Gong K, Hu Q, Yao L, Li M, Sun D, Shao Q, Guo Z (2018) Ultrasonic pretreated sludge derived stable magnetic active carbon for Cr (VI) removal from wastewater. ACS Sustain Chem Eng 6(6):7283–7291. https://doi.org/10.1021/acssuschemeng.7b04421

Gopinath A, Kadirvelu K (2018) Strategies to design modified activated carbon fibers for the decontamination of water and air. Environ Chem Lett:1–32. https://doi.org/10.1007/s10311-018-0740-9

Gu S, Kang X, Wang L, Lichtfouse E, Wang C (2018) Clay mineral adsorbents for heavy metal removal from wastewater: a review. Environ Chem Lett:1–26. https://doi.org/10.1007/s10311-018-0813-9

Güell R, Antico E, Salvado V, Fontàs C (2008) Efficient hollow fiber supported liquid membrane system for the removal and preconcentration of Cr (VI) at trace levels. Sep Purif Technol 62(2):389–393. https://doi.org/10.1016/j.seppur.2008.02.015

Guerra DL, Oliveira HC, da Costa PCC, Viana RR, Airoldi C (2010) Adsorption of chromium (VI) ions on Brazilian smectite: effect of contact time, pH, concentration, and calorimetric investigation. Catena 82(1):35–44. https://doi.org/10.1016/j.catena.2010.04.008

Gueye M, Richardson Y, Kafack FT, Blin J (2014) High efficiency activated carbons from African biomass residues for the removal of chromium (VI) from wastewater. J Environ Chem Eng 2(1):273–281. https://doi.org/10.1016/j.jece.2013.12.014

Guo L, Zhang J, Zhang D, Liu Y, Deng Y, Chen J (2012) Preparation of poly (vinylidene fluoride-co-tetrafluoroethylene)-based polymer inclusion membrane using bifunctional ionic liquid extractant for Cr (VI) transport. Ind Eng Chem Res 51(6):2714–2722. https://doi.org/10.1021/ie201824s

Guo D-M, An Q-D, Xiao Z-Y, Zhai S-R, Shi Z (2017) Polyethylenimine-functionalized cellulose aerogel beads for efficient dynamic removal of chromium (VI) from aqueous solution. RSC Adv 7(85):54039–54052. https://doi.org/10.1039/C7RA09940A

Gupta VK, Ali I (2004) Removal of lead and chromium from wastewater using bagasse fly ash—a sugar industry waste. J Colloid Interface Sci 271(2):321–328. https://doi.org/10.1016/j.jcis.2003.11.007

Gupta VK, Gupta M, Sharma S (2001) Process development for the removal of lead and chromium from aqueous solutions using red mud—an aluminium industry waste. Water Res 35(5):1125–1134. https://doi.org/10.1016/S0043-1354(00)00389-4

Gupta V, Agarwal S, Saleh TA (2011) Chromium removal by combining the magnetic properties of iron oxide with adsorption properties of carbon nanotubes. Water Res 45(6):2207–2212. https://doi.org/10.1016/j.watres.2011.01.012

Gupta VK, Ali I, Saleh TA, Siddiqui M, Agarwal S (2013a) Chromium removal from water by activated carbon developed from waste rubber tires. Environ Sci Pollut Res 20(3):1261–1268. https://doi.org/10.1007/s11356-012-0950-9

Gupta VK, Pathania D, Sharma S, Agarwal S, Singh P (2013b) Remediation of noxious chromium (VI) utilizing acrylic acid grafted lignocellulosic adsorbent. J Mol Liq 177:343–352. https://doi.org/10.1016/j.molliq.2012.10.017

Hafiane A, Lemordant D, Dhahbi M (2000) Removal of hexavalent chromium by nanofiltration. Desalination 130(3):305–312. https://doi.org/10.1016/S0011-9164(00)00094-1

Hamadi NK, Chen XD, Farid MM, Lu MG (2001) Adsorption kinetics for the removal of chromium (VI) from aqueous solution by adsorbents derived from used tyres and sawdust. Chem Eng J 84(2):95–105. https://doi.org/10.1016/S1385-8947(01)00194-2

Hamdan SS, El-Naas MH (2014a) Characterization of the removal of chromium (VI) from groundwater by electrocoagulation. J Ind Eng Chem 20(5):2775–2781. https://doi.org/10.1016/j.jiec.2013.11.006

Hamdan SS, El-Naas MH (2014b) An electrocoagulation column (ECC) for groundwater purification. J Water Process Eng 4:25–30. https://doi.org/10.1016/j.jwpe.2014.08.004

He M, Li X, Liu H, Miller SJ, Wang G, Rensing C (2011) Characterization and genomic analysis of a highly chromate resistant and reducing bacterial strain Lysinibacillus fusiformis ZC1. J Hazard Mater 185(2–3):682–688. https://doi.org/10.1016/j.jhazmat.2010.09.072

He X, Cheng L, Wang Y, Zhao J, Zhang W, Lu C (2014) Aerogels from quaternary ammonium-functionalized cellulose nanofibers for rapid removal of Cr (VI) from water. Carbohydr Polym 111:683–687. https://doi.org/10.1016/j.carbpol.2014.05.020

Herrmann J-M (1999) Heterogeneous photocatalysis: fundamentals and applications to the removal of various types of aqueous pollutants. Catal Today 53(1):115–129. https://doi.org/10.1016/S0920-5861(99)00107-8

Hezil N, Fellah M, Assala O, Touhami MZ, Guerfi K (2018) Elimination of chromium (VI) by adsorption onto natural and/or modified kaolinite. Paper presented at the Diffusion Foundations. https://doi.org/10.4028/www.scientific.net/DF.18.106

Hlihor RM, Figueiredo H, Tavares T, Gavrilescu M (2017) Biosorption potential of dead and living Arthrobacter viscosus biomass in the removal of Cr (VI): batch and column studies. Process Saf Environ Prot 108:44–56. https://doi.org/10.1016/j.psep.2016.06.016

Hossain MA, Piyatida P, da Silva JAT, Fujita M (2012) Molecular mechanism of heavy metal toxicity and tolerance in plants: central role of glutathione in detoxification of reactive oxygen species and methylglyoxal and in heavy metal chelation. J Bot 2012. https://doi.org/10.1155/2012/872875

Hu J, Chen G, Lo IM (2005) Removal and recovery of Cr (VI) from wastewater by maghemite nanoparticles. Water Res 39(18):4528–4536. https://doi.org/10.1016/j.watres.2005.05.051

Hu J, Chen C, Zhu X, Wang X (2009) Removal of chromium from aqueous solution by using oxidized multiwalled carbon nanotubes. J Hazard Mater 162(2–3):1542–1550. https://doi.org/10.1016/j.jhazmat.2008.06.058

Huang C-P, Wu M-H (1975) Chromium removal by carbon adsorption. J Water Pollut Control Fed 47:2437–2446

Huang L, Zhou S, Jin F, Huang J, Bao N (2014) Characterization and mechanism analysis of activated carbon fiber felt-stabilized nanoscale zero-valent iron for the removal of Cr (VI) from aqueous solution. Colloids Surf, A 447:59–66. https://doi.org/10.1016/j.colsurfa.2014.01.037

Huang R, Ma X, Li X, Guo L, Xie X, Zhang M, Li J (2018) A novel ion-imprinted polymer based on graphene oxide-mesoporous silica nanosheet for fast and efficient removal of chromium

(VI) from aqueous solution. J Colloid Interface Sci 514:544–553. https://doi.org/10.1016/j.jcis.2017.12.065

Ihsanullah K, Al-Khaldi FA, Abu-Sharkh B, Abulkibash AM, Qureshi MI, Laoui T, Atieh MA (2016) Effect of acid modification on adsorption of hexavalent chromium (Cr (VI)) from aqueous solution by activated carbon and carbon nanotubes. Desalin Water Treat 57(16):7232–7244. https://doi.org/10.1080/19443994.2015.1021847

Ismadji S, Soetaredjo FE, Ayucitra A (2015) Clay materials for environmental remediation, vol 25. Springer, Cham

Janik P, Zawisza B, Talik E, Sitko R (2018) Selective adsorption and determination of hexavalent chromium ions using graphene oxide modified with amino silanes. Microchim Acta 185(2):117. https://doi.org/10.1007/s00604-017-2640-2

Janssen L, Koene L (2002) The role of electrochemistry and electrochemical technology in environmental protection. Chem Eng J 85(2):137–146. https://doi.org/10.1016/S1385-8947(01)00218-2

Jin W, Zhang Z, Wu G, Tolba R, Chen A (2014) Integrated lignin-mediated adsorption-release process and electrochemical reduction for the removal of trace Cr (VI). RSC Adv 4(53):27843–27849. https://doi.org/10.1039/C4RA01222D

Jobby R, Jha P, Yadav AK, Desai N (2018) Biosorption and biotransformation of hexavalent chromium [Cr(VI)]: A comprehensive review. Chemosphere 207:255–266. https://doi.org/10.1016/j.chemosphere.2018.05.050

Jung C, Heo J, Han J, Her N, Lee S-J, Oh J et al (2013) Hexavalent chromium removal by various adsorbents: powdered activated carbon, chitosan, and single/multi-walled carbon nanotubes. Sep Purif Technol 106:63–71. https://doi.org/10.1016/j.seppur.2012.12.028

Kabdaşlı I, Tünay O, Orhon D (1999) Wastewater control and management in a leather tanning district. Water Sci Technol 40(1):261–267. https://doi.org/10.1016/S0273-1223(99)00393-5

Kalavathy MH, Karthikeyan T, Rajgopal S, Miranda LR (2005) Kinetic and isotherm studies of cu (II) adsorption onto H3PO4-activated rubber wood sawdust. J Colloid Interface Sci 292(2):354–362. https://doi.org/10.1016/j.jcis.2005.05.087

Karthik R, Meenakshi S (2014) Synthesis, characterization and Cr (VI) uptake studies of polypyrrole functionalized chitin. Synth Met 198:181–187. https://doi.org/10.1016/j.synthmet.2014.10.012

Karthikeyan T, Rajgopal S, Miranda LR (2005) Chromium (VI) adsorption from aqueous solution by Hevea Brasilinesis sawdust activated carbon. J Hazard Mater 124(1–3):192–199. https://doi.org/10.1016/j.jhazmat.2005.05.003

Kathiravan MN, Karthick R, Muthukumar K (2011) Ex situ bioremediation of Cr (VI) contaminated soil by Bacillus sp.: batch and continuous studies. Chem Eng J 169(1–3):107–115. https://doi.org/10.1016/j.cej.2011.02.060

Kaya A, Onac C, Alpoguz HK (2016) A novel electro-driven membrane for removal of chromium ions using polymer inclusion membrane under constant DC electric current. J Hazard Mater 317:1–7. https://doi.org/10.1016/j.jhazmat.2016.05.047

Kazemi M, Jahanshahi M, Peyravi M (2018) Hexavalent chromium removal by multilayer membrane assisted by photocatalytic couple nanoparticle from both permeate and retentate. J Hazard Mater 344:12–22. https://doi.org/10.1016/j.jhazmat.2017.09.059

Khan SA, Khan MA (1995) Adsorption of chromium (III), chromium (VI) and silver (I) on benton-ite. Waste Manag 15(4):271–282. https://doi.org/10.1016/0956-053X(95)00025-U

Khan T, Isa MH, Mustafa MRU, Yeek-Chia H, Baloo L, Manan TSBA, Saeed MO (2016) Cr (VI) adsorption from aqueous solution by an agricultural waste based carbon. RSC Adv 6(61):56365–56374. https://doi.org/10.1039/C6RA05618K

Khosravi R, Fazlzadehdavil M, Barikbin B, Taghizadeh AA (2014) Removal of hexavalent chromium from aqueous solution by granular and powdered Peganum Harmala. Appl Surf Sci 292:670–677. https://doi.org/10.1016/j.apsusc.2013.12.031

Kim J-H, Kang J-K, Lee S-C, Kim S-B (2018) Synthesis of powdered and granular N-(3-trimethoxysilylpropyl) diethylenetriamine-grafted mesoporous silica SBA-15 for Cr (VI)

removal from industrial wastewater. J Taiwan Inst Chem Eng 87:140–149. https://doi.org/10.1016/j.jtice.2018.03.024

Kimbrough DE, Cohen Y, Winer AM, Creelman L, Mabuni C (1999) A critical assessment of chromium in the environment. Crit Rev Environ Sci Technol 29(1):1–46. https://doi.org/10.1080/10643389991259164

Kobya M (2004) Removal of Cr (VI) from aqueous solutions by adsorption onto hazelnut shell activated carbon: kinetic and equilibrium studies. Bioresour Technol 91(3):317–321. https://doi.org/10.1016/j.biortech.2003.07.001

Kobya M, Topcu N, Demircioğlu N (1997) Kinetic analysis of coupled transport of thiocyanate ions through liquid membranes at different temperatures. J Membr Sci 130(1–2):7–15. https://doi.org/10.1016/S0376-7388(96)00348-1

Kocaoba S (2009) Adsorption of cd (II), Cr (III) and Mn (II) on natural sepiolite. Desalination 244(1–3):24–30. https://doi.org/10.1016/j.desal.2008.04.033

Kongsricharoern N, Polprasert C (1995) Electrochemical precipitation of chromium (Cr^{6+}) from an electroplating wastewater. Water Sci Technol 31(9):109–117. https://doi.org/10.1016/0273-1223(95)00412-G

Koujalagi PS, Kulkarni SV, Raviraj M (2018) Asian J Chem 30(5):1083–1087. https://doi.org/10.14233/ajchem.2018.21188

Koushkbaghi S, Zakialamdari A, Pishnamazi M, Ramandi HF, Aliabadi M, Irani M (2018) Aminated-Fe_3O_4 nanoparticles filled chitosan/PVA/PES dual layers nanofibrous membrane for the removal of Cr (VI) and Pb (II) ions from aqueous solutions in adsorption and membrane processes. Chem Eng J 337:169–182. https://doi.org/10.1016/j.cej.2017.12.075

Kozłowski CA (2007) Kinetics of chromium (VI) transport from mineral acids across cellulose triacetate (CTA) plasticized membranes immobilized by tri-n-octylamine. Ind Eng Chem Res 46(16):5420–5428. https://doi.org/10.1021/ie070215i

Kozlowski CA, Walkowiak W (2005) Applicability of liquid membranes in chromium (VI) transport with amines as ion carriers. J Membr Sci 266(1–2):143–150. https://doi.org/10.1016/j.memsci.2005.04.053

Kulshreshtha S (2018) Removal of pollutants using spent mushrooms substrates. Environ Chem Lett:1–15. https://doi.org/10.1007/s10311-018-00840-2

Kumar A, Jena HM (2017) Adsorption of Cr (VI) from aqueous solution by prepared high surface area activated carbon from fox nutshell by chemical activation with H3PO4. J Environ Chem Eng 5(2):2032–2041. https://doi.org/10.1016/j.jece.2017.03.035

Kumar V, Pandey BD (2006) Remediation options for the treatment of electroplating and leather tanning effluent containing chromium—a review. Miner Process Extr Metall Rev 27(2):99–130. https://doi.org/10.1080/08827500600563319

Kumar A, Rao N, Kaul S (2000) Alkali-treated straw and insoluble straw xanthate as low cost adsorbents for heavy metal removal–preparation, characterization and application. Bioresour Technol 71(2):133–142. https://doi.org/10.1016/S0960-8524(99)00064-4

Kumar ASK, Kalidhasan S, Rajesh V, Rajesh N (2011) Application of cellulose-clay composite biosorbent toward the effective adsorption and removal of chromium from industrial wastewater. Ind Eng Chem Res 51(1):58–69. https://doi.org/10.1021/ie201349h

Kumar ASK, Ramachandran R, Kalidhasan S, Rajesh V, Rajesh N (2012) Potential application of dodecylamine modified sodium montmorillonite as an effective adsorbent for hexavalent chromium. Chem Eng J 211:396–405. https://doi.org/10.1016/j.cej.2012.09.029

Kumar ASK, Jiang S-J, Tseng W-L (2015) Effective adsorption of chromium (VI)/Cr (III) from aqueous solution using ionic liquid functionalized multiwalled carbon nanotubes as a super sorbent. J Mater Chem A 3(13):7044–7057. https://doi.org/10.1039/C4TA06948J

Kumar TP, Mandlimath TR, Sangeetha P, Revathi S, Kumar SA (2018) Nanoscale materials as sorbents for nitrate and phosphate removal from water. Environ Chem Lett 16(2):389–400. https://doi.org/10.1007/s10311-017-0682-7

Kumari V, Yadav A, Haq I, Kumar S, Bharagava RN, Singh SK, Raj A (2016) Genotoxicity evaluation of tannery effluent treated with newly isolated hexavalent chromium reducing Bacillus cereus. J Environ Manag 183:204–211. https://doi.org/10.1016/j.jenvman.2016.08.017

Kunungo S, Mohapatra R (1995) Coupled transport of Zn (II) through a supported liquid membrane containing bis (2, 4, 4-trimethylpentyl) phosphinic acid in kerosene. II experimental evaluation of model equations for rate process under different limiting conditions. J Membr Sci 105(3):227–235. https://doi.org/10.1016/0376-7388(95)00062-H

Kuppusamy S, Thavamani P, Megharaj M, Venkateswarlu K, Lee YB, Naidu R (2016) Potential of Melaleuca diosmifolia leaf as a low-cost adsorbent for hexavalent chromium removal from contaminated water bodies. Process Saf Environ Prot 100:173–182. https://doi.org/10.1016/j.psep.2016.01.009

Kurniawan T, Babel S (2003) A research study on Cr (VI) removal from contaminated wastewater using low-cost adsorbents and commercial activated carbon. Paper presented at the Second Int. Conf. on Energy Technology towards a Clean Environment (RCETE)

Kurniawan TA, Chan GY, Lo W-H, Babel S (2006) Physico–chemical treatment techniques for wastewater laden with heavy metals. Chem Eng J 118(1):83–98. https://doi.org/10.1016/j.cej.2006.01.015

Kyzas GZ, Bikiaris DN (2015) Recent modifications of chitosan for adsorption applications: a critical and systematic review. Mar Drugs 13(1):312–337. https://doi.org/10.3390/md13010312

Lara P, Morett E, Juárez K (2017) Acetate biostimulation as an effective treatment for cleaning up alkaline soil highly contaminated with Cr (VI). Environ Sci Pollut Res 24(33):25513–25521. https://doi.org/10.1007/s11356-016-7191-2

Latimer WM (1964) The oxidation states of the elements and their potentials in aqueous solution. Prentice-Hall Internacional, Englewood Cliffs

Lee G, Hering JG (2003) Removal of chromium (VI) from drinking water by redox-assisted coagulation with iron (II). J Water Supp Res Technol AQUA 52(5):319–332. https://doi.org/10.2166/aqua.2003.0030

Lee C-G, Lee S, Park J-A, Park C, Lee SJ, Kim S-B et al (2017) Removal of copper, nickel and chromium mixtures from metal plating wastewater by adsorption with modified carbon foam. Chemosphere 166:203–211. https://doi.org/10.1016/j.chemosphere.2016.09.093

Lee J, Kim J-H, Choi K, Kim H-G, Park J-A, Cho S-H, Lee S (2018) Investigation of the mechanism of chromium removal in (3-aminopropyl) trimethoxysilane functionalized mesoporous silica. Sci Rep 8. https://doi.org/10.1038/s41598-018-29679-x

Legrouri K, Khouya E, Hannache H, El Hartti M, Ezzine M, Naslain R (2017) Activated carbon from molasses efficiency for Cr (VI), Pb (II) and cu (II) adsorption: a mechanistic study. Chem Int 3(3):301–310

Li H, Li Z, Liu T, Xiao X, Peng Z, Deng L (2008) A novel technology for biosorption and recovery hexavalent chromium in wastewater by bio-functional magnetic beads. Bioresour Technol 99(14):6271–6279. https://doi.org/10.1016/j.biortech.2007.12.002

Li Y, Lu A, Ding H, Jin S, Yan Y, Wang C, Wang X (2009) Cr (VI) reduction at rutile-catalyzed cathode in microbial fuel cells. Electrochem Commun 11(7):1496–1499. https://doi.org/10.1016/j.elecom.2009.05.039

Li S, Lu X, Li X, Xue Y, Zhang C, Lei J, Wang C (2012) Preparation of bamboo-like PPy nanotubes and their application for removal of Cr (VI) ions in aqueous solution. J Colloid Interface Sci 378(1):30–35. https://doi.org/10.1016/j.jcis.2012.03.065

Li M, Gong Y, Lyu A, Liu Y, Zhang H (2016) The applications of populus fiber in removal of Cr (VI) from aqueous solution. Appl Surf Sci 383:133–141. https://doi.org/10.1016/j.apsusc.2016.04.167

Li Y, Zhu H, Zhang C, Cheng M, He H (2018) PEI-grafted magnetic cellulose for Cr (VI) removal from aqueous solution. Cellulose 25(8):4757–4769. https://doi.org/10.1007/s10570-018-1868-2

Liu G, Yang H, Wang J, Jin R, Zhou J, Lv H (2010a) Enhanced chromate reduction by resting Escherichia coli cells in the presence of quinone redox mediators. Bioresour Technol 101(21):8127–8131. https://doi.org/10.1016/j.biortech.2010.06.050

Liu Y, Li H, Tan G-Q, Zhu X-h (2010b) Fe^{2+}-modified vermiculite for the removal of chromium (VI) from aqueous solution. Sep Sci Technol 46(2):290–299. https://doi.org/10.1080/014963 95.2010.491493

Liu Y-X, Yuan D-X, Yan J-M, Li Q-L, Ouyang T (2011) Electrochemical removal of chromium from aqueous solutions using electrodes of stainless steel nets coated with single wall carbon nanotubes. J Hazard Mater 186(1):473–480. https://doi.org/10.1016/j.jhazmat.2010.11.025

Liu H, Liang S, Gao J, Ngo HH, Guo W, Guo Z, Li Y (2014) Enhancement of Cr (VI) removal by modifying activated carbon developed from Zizania caduciflora with tartaric acid during phosphoric acid activation. Chem Eng J 246:168 174. https://doi.org/10.1016/j.cej.2014.02.046

Lofrano G, Belgiorno V, Gallo M, Raimo A, Meric S (2006) Toxicity reduction in leather tanning wastewater by improved coagulation flocculation process. Glob Nest J 8(2):151–158

Luo T, Tian X, Yang C, Luo W, Nie Y, Wang Y (2017) Polyethylenimine-functionalized corn bract, an agricultural waste material, for efficient removal and recovery of Cr (VI) from aqueous solution. J Agric Food Chem 65(33):7153–7158. https://doi.org/10.1021/acs.jafc.7b02699

Mahmood S, Khalid A, Mahmood T, Arshad M, Ahmad R (2013) Potential of newly isolated bacterial strains for simultaneous removal of hexavalent chromium and reactive black-5 azo dye from tannery effluent. J Chem Technol Biotechnol 88(8):1506–1513. https://doi.org/10.1002/jctb.3994

Mancuso TF (1997) Chromium as an industrial carcinogen: part I. Am J Ind Med 31(2):129–139. https://doi.org/10.1002/(SICI)1097-0274(199702)31:2<129::AID-AJIM1>3.0.CO;2-V

Mangaiyarkarasi MM, Vincent S, Janarthanan S, Rao TS, Tata B (2011) Bioreduction of Cr (VI) by alkaliphilic Bacillus subtilis and interaction of the membrane groups. Saudi J Biol Sci 18(2):157–167. https://doi.org/10.1016/j.sjbs.2010.12.003

Manju G, Anirudhan T (1997) Use of coconut fibre pith-based pseudo-activated carbon for chromium (VI) removal. Indian J Environ Health 39(4):289–298

Marjanović V, Lazarević S, Janković-Častvan I, Jokić B, Janaćković D, Petrović R (2013) Adsorption of chromium (VI) from aqueous solutions onto amine-functionalized natural and acid-activated sepiolites. Appl Clay Sci 80:202–210. https://doi.org/10.1016/j.clay.2013.04.008

Mehrotra R, Dwivedi N (1988) Removal of chromium (VI) from water using unconventional materials. J Ind Water Works Assoc 20:323–327

Memon S, Roundhill MD, Yilmaz M (2004) Remediation and liquid-liquid phase transfer extraction of chromium(VI). A review. Collect Czechoslov Chem Commun 69(6):1231–1250. https://doi.org/10.1135/cccc20041231

Meng X, Wang C, Ren T, Wang L, Wang X (2018) Electrodriven transport of chromium (VI) using 1-octanol/PVC in polymer inclusion membrane under low voltage. Chem Eng J 346:506–514. https://doi.org/10.1016/j.cej.2018.04.004

Miretzky P, Cirelli AF (2010) Cr (VI) and Cr (III) removal from aqueous solution by raw and modified lignocellulosic materials: a review. J Hazard Mater 180(1–3):1–19. https://doi.org/10.1016/j.jhazmat.2010.04.060

Mohammad M, Yakub I, Yaakob Z, Asim N, Sopian K (2017) Adsorption Isotherm of Chromium (Vi) into Zncl2 Impregnated Activated Carbon Derived by Jatropha Curcas Seed Hull. Paper presented at the IOP Conference Series: Materials Science and Engineering

Mohan D, Pittman CU Jr (2006) Activated carbons and low cost adsorbents for remediation of tri-and hexavalent chromium from water. J Hazard Mater 137(2):762–811. https://doi.org/10.1016/j.jhazmat.2006.06.060

Mohan D, Singh KP, Singh VK (2006) Trivalent chromium removal from wastewater using low cost activated carbon derived from agricultural waste material and activated carbon fabric cloth. J Hazard Mater 135(1):280–295. https://doi.org/10.1016/j.jhazmat.2005.11.075

Mohanty K, Jha M, Meikap B, Biswas M (2005) Removal of chromium (VI) from dilute aqueous solutions by activated carbon developed from Terminalia arjuna nuts activated with zinc chloride. Chem Eng Sci 60(11):3049–3059. https://doi.org/10.1016/j.ces.2004.12.049

Moosa AA, Ridha AM, Abdullha IN (2015) Chromium ions removal from wastewater using carbon nanotubes. Int J Innov Res Sci Eng Technol 4:275–282. https://doi.org/10.15680/IJIRSET.2015.0402057

Morin-Crini N, Fourmentin M, Fourmentin S, Torri G, Crini G (2018) Synthesis of silica materials containing cyclodextrin and their applications in wastewater treatment. Environ Chem Lett:1–14. https://doi.org/10.1007/s10311-018-00818-0

Mubarak NM, Thobashinni M, Abdullah EC, Sahu JN (2016) Comparative kinetic study of removal of Pb^{2+} ions and Cr^{3+} ions from waste water using carbon nanotubes produced using microwave heating. J Carbon Res 2(1):7. https://doi.org/10.3390/c2010007

Mudhoo A, Gautam RK, Ncibi MC, Zhao F, Garg VK, Sillanpää M (2018) Green synthesis, activation and functionalization of adsorbents for dye sequestration. Environ Chem Lett:1–37. https://doi.org/10.1007/s10311-018-0784-x

Muhammad Ekramul Mahmud HN, Huq AKO, Yahya R b (2016) The removal of heavy metal ions from wastewater/aqueous solution using polypyrrole-based adsorbents: a review. RSC Adv 6(18):14778–14791. https://doi.org/10.1039/c5ra24358k

Mukhopadhyay B, Sundquist J, Schmitz RJ (2007) Removal of Cr (VI) from Cr-contaminated groundwater through electrochemical addition of Fe (II). J Environ Manag 82(1):66–76. https://doi.org/10.1016/j.jenvman.2005.12.005

Muliwa AM, Leswifi TY, Onyango MS, Maity A (2016) Magnetic adsorption separation (MAS) process: An alternative method of extracting Cr (VI) from aqueous solution using polypyrrole coated Fe3O4 nanocomposites. Sep Purif Technol 158:250–258. https://doi.org/10.1016/j.seppur.2015.12.021

Muthukrishnan M, Guha B (2008) Effect of pH on rejection of hexavalent chromium by nanofiltration. Desalination 219(1–3):171–178. https://doi.org/10.1016/j.desal.2007.04.054

Muthumareeswaran M, Alhoshan M, Agarwal GP (2017) Ultrafiltration membrane for effective removal of chromium ions from potable water. Sci Rep 7:41423. https://doi.org/10.1038/srep41423

Na C, Wang H (2016) Binder-free carbon nanotube electrode for electrochemical removal of chromium: Google Patents

Nakkeeran E, Saranya N, Giri Nandagopal M, Santhiagu A, Selvaraju N (2016) Hexavalent chromium removal from aqueous solutions by a novel powder prepared from Colocasia esculenta leaves. Int J Phytoremediation 18(8):812–821. https://doi.org/10.1080/15226514.2016.1146229

Nasseh N, Taghavi L, Barikbin B, Harifi-Mood AR (2017a) The removal of Cr (VI) from aqueous solution by almond green hull waste material: kinetic and equilibrium studies. J Water Reuse Desalination 7(4):449–460. https://doi.org/10.2166/wrd.2016.047

Nasseh N, Taghavi L, Barikbin B, Khodadadi M (2017b) Advantage of almond green hull over its resultant ash for chromium (VI) removal from aqueous solutions. Int J Environ Sci Technol 14(2):251–262. https://doi.org/10.1007/s13762-016-1210-

Nawaz R, Ali K, Ali N, Khaliq A (2016) Removal of chromium (VI) from industrial effluents through supported liquid membrane using trioctylphosphine oxide as a carrier. J Braz Chem Soc 27(1):209–220. https://doi.org/10.5935/0103-5053.20150272

Nayak V, Jyothi MS, Balakrishna RG, Padaki M, Ismail AF (2015) Preparation and characterization of chitosan thin films on mixed-matrix membranes for complete removal of chromium. Chem Open 4(3):278–287. https://doi.org/10.1002/open.201402133

Nethaji S, Sivasamy A, Mandal A (2013) Preparation and characterization of corn cob activated carbon coated with nano-sized magnetite particles for the removal of Cr (VI). Bioresour Technol 134:94–100. https://doi.org/10.1016/j.biortech.2013.02.012

Ng TW, Cai Q, Wong C-K, Chow AT, Wong P-K (2010) Simultaneous chromate reduction and azo dye decolourization by Brevibacterium casei: azo dye as electron donor for chromate reduction. J Hazard Mater 182(1–3):792–800. https://doi.org/10.1016/j.jhazmat.2010.06.106

Nguyen T, Ngo H, Guo W, Zhang J, Liang S, Yue Q, Nguyen T (2013) Applicability of agricultural waste and by-products for adsorptive removal of heavy metals from wastewater. Bioresour Technol 148:574–585. https://doi.org/10.1016/j.biortech.2013.08.124

Okoli CP, Diagboya PN, Anigbogu IO, Olu-Owolabi BI, Adebowale KO (2017) Competitive biosorption of Pb (II) and Cd (II) ions from aqueous solutions using chemically modified moss biomass (Barbula lambarenensis). Environ Earth Sci 76(1):33. https://doi.org/10.1007/s12665-016-6368-9

Onac C, Kaya A, Sener I, Alpoguz HK (2018) An Electromembrane extraction with polymeric membrane under constant current for the recovery of Cr (VI) from industrial water. J Electrochem Soc 165(2):E76–E80

Osikoya A, Wankasi D, Vala R, Afolabi A, Dikio E (2014) Synthesis, characterization and adsorption studies of fluorine–doped carbon nanotubes. Dig J Nanomater Bios 9:1187–1197

Owalude SO, Tella AC (2016) Removal of hexavalent chromium from aqueous solutions by adsorption on modified groundnut hull. Beni-suef Univ J Basic Appl Sci 5(4):377–388. https://doi.org/10.1016/j.bjbas.2016.11.005

Owlad M, Aroua MK, Daud WMAW (2010) Hexavalent chromium adsorption on impregnated palm shell activated carbon with polyethyleneimine. Bioresour Technol 101(14):5098–5103. https://doi.org/10.1016/j.biortech.2010.01.135

Oze C, Bird DK, Fendorf S (2007) Genesis of hexavalent chromium from natural sources in soil and groundwater. Proc Natl Acad Sci 104(16):6544–6549. https://doi.org/10.1073/pnas.0701085104

Pakade VE, Maremeni LC, Ntuli TD, Tavengwa NT (2016) Application of quaternized activated carbon derived from Macadamia nutshells for the removal of hexavalent chromium from aqueous solutions. S Afr J Chem 69(1):180–188. https://doi.org/10.17159/0379-4350/2016/v69a22

Pakade VE, Ntuli TD, Ofomaja AE (2017) Biosorption of hexavalent chromium from aqueous solutions by Macadamia nutshell powder. Appl Water Sci 7(6):3015–3030. https://doi.org/10.1007/s13201-016-0412-5

Pan B, Xing B (2008) Adsorption mechanisms of organic chemicals on carbon nanotubes. Environ Sci Technol 42(24):9005–9013. https://doi.org/10.1021/es801777n

Pan X, Liu Z, Chen Z, Cheng Y, Pan D, Shao J, Guan X (2014) Investigation of Cr (VI) reduction and Cr (III) immobilization mechanism by planktonic cells and biofilms of Bacillus subtilis ATCC-6633. Water Res 55:21–29. https://doi.org/10.1016/j.watres.2014.01.066

Pan Y, Cai P, Farmahini-Farahani M, Li Y, Hou X, Xiao H (2016) Amino-functionalized alkaline clay with cationic star-shaped polymer as adsorbents for removal of Cr (VI) in aqueous solution. Appl Surf Sci 385:333–340. https://doi.org/10.1016/j.apsusc.2016.05.112

Papp JF (1999) Chromium (Cr). Metal Prices in the United States Through 2010, 28

Park S-J, Jung W-Y (2001) Removal of chromium by activated carbon fibers plated with copper metal. Carbon Lett 2(1):15–21

Parlayıcı Ş, Pehlivan E (2015) Natural biosorbents (garlic stem and horse chesnut shell) for removal of chromium (VI) from aqueous solutions. Environ Monit Assess 187(12):763. https://doi.org/10.1007/s10661-015-4984-6

Peng Y, Tang D, Wang Q, He T, Chu Y, Li R (2018) Adsorption of chromate from aqueous solution by Polyethylenimine modified multi-walled carbon nanotubes. J Nanosci Nanotechnol 18(6):4006–4013. https://doi.org/10.1166/jnn.2018.15197

Pérez E, Ayele L, Getachew G, Fetter G, Bosch P, Mayoral A, Díaz I (2015) Removal of chromium (VI) using nano-hydrotalcite/SiO2 composite. J Environ Chem Eng 3(3):1555–1561. https://doi.org/10.1016/j.jece.2015.05.009

Pérez-Candela M, Martín-Martínez J, Torregrosa-Maciá R (1995) Chromium (VI) removal with activated carbons. Water Res 29(9):2174–2180. https://doi.org/10.1016/0043-1354(95)00035-J

Periyasamy S, Gopalakannan V, Viswanathan N (2017) Fabrication of magnetic particles imprinted cellulose based biocomposites for chromium (VI) removal. Carbohydr Polym 174:352–359. https://doi.org/10.1016/j.carbpol.2017.06.029

Petala E, Dimos K, Douvalis A, Bakas T, Tucek J, Zbořil R, Karakassides MA (2013) Nanoscale zero-valent iron supported on mesoporous silica: characterization and reactivity for Cr (VI) removal from aqueous solution. J Hazard Mater 261:295–306. https://doi.org/10.1016/j.jhazmat.2013.07.046

Polti MA, Atjián MC, Amoroso MJ, Abate CM (2011) Soil chromium bioremediation: synergic activity of actinobacteria and plants. Int Biodeterior Biodegrad 65(8):1175–1181. https://doi.org/10.1016/j.ibiod.2011.09.008

Pourbaix M (1974) Atlas of electrochemical equilibria in aqueous solution. NACE, Houstan, p 307

Pugazhenthi G, Sachan S, Kishore N, Kumar A (2005) Separation of chromium (VI) using modified ultrafiltration charged carbon membrane and its mathematical modeling. J Membr Sci 254(1):229–239. https://doi.org/10.1016/j.memsci.2005.01.011

Pyrzynska K (2008) Carbon nanotubes as a new solid-phase extraction material for removal and enrichment of organic pollutants in water. Sep Purif Rev 37(4):372–389. https://doi.org/10.1080/15422110802178843

Qin G, McGuire MJ, Blute NK, Seidel C, Fong L (2005) Hexavalent chromium removal by reduction with ferrous sulfate, coagulation, and filtration: A pilot-scale study. Environ Sci Technol 39(16):6321–6327. https://doi.org/10.1021/es050486p

Qiu J, Wang Z, Li H, Xu L, Peng J, Zhai M, Wei G (2009) Adsorption of Cr (VI) using silica-based adsorbent prepared by radiation-induced grafting. J Hazard Mater 166(1):270–276. https://doi.org/10.1016/j.jhazmat.2008.11.053

Qiu B, Guo J, Zhang X, Sun D, Gu H, Wang Q, Weeks BL (2014) Polyethylenimine facilitated ethyl cellulose for hexavalent chromium removal with a wide pH range. ACS Appl Mater Interfaces 6(22):19816–19824. https://doi.org/10.1021/am505170j

Rahman ML, Mandal BH, Sarkar SM, Wahab NAA, Yusoff MM, Arshad SE, Musta B (2016) Synthesis of poly (hydroxamic acid) ligand from polymer grafted khaya cellulose for transition metals extraction. Fibers Polym 17(4):521–532. https://doi.org/10.1007/s12221-016-6001-

Rai D, Sass BM, Moore DA (1987) Chromium (III) hydrolysis constants and solubility of chromium (III) hydroxide. Inorg Chem 26(3):345–349. https://doi.org/10.1021/ic00250a002

Rai M, Shahi G, Meena V, Meena R, Chakraborty S, Singh R, Rai B (2016) Removal of hexavalent chromium Cr (VI) using activated carbon prepared from mango kernel activated with H₃PO₄. Resour-Effic Technol 2:S63–S70. https://doi.org/10.1016/j.reffit.2016.11.011

Raji C, Anirudhan T (1998) Batch Cr (VI) removal by polyacrylamide-grafted sawdust: kinetics and thermodynamics. Water Res 32(12):3772–3780. https://doi.org/10.1016/S0043-1354(98)00150-X

Ramírez-Estrada A, Mena-Cervantes V, Fuentes-García J, Vazquez-Arenas J, Palma-Goyes R, Flores-Vela A, Altamirano RH (2018) Cr (III) removal from synthetic and real tanning effluents using an electro-precipitation method. J Environ Chem Eng 6(1):1219–1225. https://doi.org/10.1016/j.jece.2018.01.038

Ramrakhiani L, Majumder R, Khowala S (2011) Removal of hexavalent chromium by heat inactivated fungal biomass of Termitomyces clypeatus: surface characterization and mechanism of biosorption. Chem Eng J 171(3):1060–1068. https://doi.org/10.1016/j.cej.2011.05.002

Rana P, Mohan N, Rajagopal C (2004) Electrochemical removal of chromium from wastewater by using carbon aerogel electrodes. Water Res 38(12):2811–2820. https://doi.org/10.1016/j.watres.2004.02.029

Rana-Madaria P, Nagarajan M, Rajagopal C, Garg BS (2005) Removal of chromium from aqueous solutions by treatment with carbon aerogel electrodes using response surface methodology. Ind Eng Chem Res 44(17):6549–6559. https://doi.org/10.1021/ie050321p

Ranjbar N, Hashemi S, Ramavandi B, Ravanipour M (2018) Chromium(VI) removal by bone char–ZnO composite: parameters optimization by response surface methodology and modeling. Environ Prog Sustain Energy 37(5):1684–1695. https://doi.org/10.1002/ep.12854

Rao M, Parwate A, Bhole A (2002) Removal of Cr^{6+} and Ni^{2+} from aqueous solution using bagasse and fly ash. Waste Manag 22(7):821–830. https://doi.org/10.1016/S0956-053X(02)00011-9

Rao GP, Lu C, Su F (2007) Sorption of divalent metal ions from aqueous solution by carbon nanotubes: a review. Sep Purif Technol 58(1):224–231. https://doi.org/10.1016/j.seppur.2006.12.006

Rapti S, Pournara A, Sarma D, Papadas IT, Armatas GS, Tsipis AC, Manos MJ (2016) Selective capture of hexavalent chromium from an anion-exchange column of metal organic resin–alginic acid composite. Chem Sci 7(3):2427–2436. https://doi.org/10.1039/C5SC03732H

Ravikumar K, Argulwar S, Sudakaran SV, Pulimi M, Chandrasekaran N, Mukherjee A (2018) Nano-bio sequential removal of hexavalent chromium using polymer-nZVI composite film and sulfate reducing bacteria under anaerobic condition. Environ Technol Innov 9:122–133. https://doi.org/10.1016/j.eti.2017.11.006

Ren X, Chen C, Nagatsu M, Wang X (2011) Carbon nanotubes as adsorbents in environmental pollution management: a review. Chem Eng J 170(2):395–410. https://doi.org/10.1016/j.cej.2010.08.045

Rivas BL, Morales DV, Kabay N, Bryjak M (2018) Cr (III) removal from aqueous solution byion exchange resins containing carboxylic acid and sulphonic acid groups. J Chil Chem Soc 63(2):4012–4018. https://doi.org/10.4067/s0717-97072018000204012

Rodriguez-Valadez F, Ortiz-Éxiga C, Ibanez JG, Alatorre-Ordaz A, Gutierrez-Granados S (2005) Electroreduction of Cr (VI) to Cr (III) on reticulated vitreous carbon electrodes in a parallel-plate reactor with recirculation. Environ Sci Technol 39(6):1875–1879. https://doi.org/10.1021/es049091g

Rosales E, Meijide J, Tavares T, Pazos M, Sanromán M (2016) Grapefruit peelings as a promising biosorbent for the removal of leather dyes and hexavalent chromium. Pro Safety Environ Prot 101:61–71. https://doi.org/10.1016/j.psep.2016.03.006

Saha R, Mukherjee K, Saha I, Ghosh A, Ghosh SK, Saha B (2013) Removal of hexavalent chromium from water by adsorption on mosambi (Citrus limetta) peel. Res Chem Intermed 39(5):2245–2257. https://doi.org/10.1007/s11164-012-0754-z

Sahmoune MN (2018) Evaluation of thermodynamic parameters for adsorption of heavy metals by green adsorbents. Environ Chem Lett:1–8. https://doi.org/10.1007/s10311-018-00819-z

Sánchez J, Rivas BL (2011) Cationic hydrophilic polymers coupled to ultrafiltration membranes to remove chromium (VI) from aqueous solution. Desalination 279(1–3):338–343. https://doi.org/10.1016/j.desal.2011.06.029

Sánchez J, Espinosa C, Pooch F, Tenhu H, Pizarro G d C, Oyarzún DP (2018) Poly (N, N-dimethylaminoethyl methacrylate) for removing chromium (VI) through polymer-enhanced ultrafiltration technique. React Funct Polym 127:67–73. https://doi.org/10.1016/j.reactfunctpolym.2018.04.002

Sanjay K, Arora A, Shekhar R, Das R (2003) Electroremediation of Cr (VI) contaminated soils: kinetics and energy efficiency. Colloids Surf, A 222(1–3):253–259. https://doi.org/10.1016/S0927-7757(03)00229-2

Saranraj P, Sujitha D (2013) Microbial bioremediation of chromium in tannery effluent: a review. Int J Microbiol Res 4(3):305–320. https://doi.org/10.5829/idosi.ijmr.2013.4.3.81228

Sari A, Tuzen M (2008) Removal of Cr (VI) from aqueous solution by Turkish vermiculite: equilibrium, thermodynamic and kinetic studies. Sep Sci Technol 43(13):3563–3581. https://doi.org/10.1080/01496390802222657

Sarin V, Pant KK (2006) Removal of chromium from industrial waste by using eucalyptus bark. Bioresour Technol 97(1):15–20. https://doi.org/10.1016/j.biortech.2005.02.010

Sathvika T, Soni A, Sharma K, Praneeth M, Mudaliyar M, Rajesh V, Rajesh N (2018) Potential application of Saccharomyces cerevisiae and rhizobium immobilized in multi walled carbon nanotubes to adsorb hexavalent chromium. Sci Rep 8(1):9862. https://doi.org/10.1038/s41598-018-28067-9

Schwantes D, Gonçalves AC, Coelho GF, Campagnolo MA, Dragunski DC, Tarley CRT, Leismann EAV (2016) Chemical modifications of cassava peel as adsorbent material for metals ions from wastewater. J Chem 2016. https://doi.org/10.1155/2016/3694174

Selomulya C, Meeyoo V, Amal R (1999) Mechanisms of Cr (VI) removal from water by various types of activated carbons. J Chem Technol Biotechnol 74(2):111–122. https://doi.org/10.1002/(SICI)1097-4660(199902)74:2<111::AID-JCTB990>3.0.CO;2-D

Selvaraj R, Santhanam M, Selvamani V, Sundaramoorthy S, Sundaram M (2018) A membrane electroflotation process for recovery of recyclable chromium (III) from tannery spent liquor effluent. J Hazard Mater 346:133–139. https://doi.org/10.1016/j.jhazmat.2017.11.052

Setshedi KZ, Bhaumik M, Songwane S, Onyango MS, Maity A (2013) Exfoliated polypyrrole-organically modified montmorillonite clay nanocomposite as a potential adsorbent for Cr (VI) removal. Chem Eng J 222:186–197. https://doi.org/10.1016/j.cej.2013.02.061

Setshedi KZ, Bhaumik M, Onyango MS, Maity A (2015) High-performance towards Cr (VI) removal using multi-active sites of polypyrrole–graphene oxide nanocomposites: batch and column studies. Chem Eng J 262:921–931. https://doi.org/10.1016/j.cej.2014.10.034

Shan C (2018) Preparation method of NZVI-PVDF hybrid films with cation-exchange function for reductive transformation of Cr (VI). J Environ Eng Landsc Manag 26(1):19–27. https://doi.org/10.3846/16486897.2017.1339048

Shanker AK, Cervantes C, Loza-Tavera H, Avudainayagam S (2005) Chromium toxicity in plants. Environ Int 31(5):739–753. https://doi.org/10.1016/j.envint.2005.02.003

Shariati S, Khabazipour M, Safa F (2017) Synthesis and application of amine functionalized silica mesoporous magnetite nanoparticles for removal of chromium (VI) from aqueous solutions. J Porous Mater 24(1):129–139. https://doi.org/10.1007/s10934-016-0245-5

Sharma C (1996) Removal of hexavalent chromium from aqueous solutions by granular activated carbon. Water SA 22(2):153–160. https://journals.co.za/content/waters/22/2/AJA03784738_1880

Sharma RK, Chauhan GS (2009) Synthesis and characterization of graft copolymers of 2-hydroxyethyl methacrylate and some comonomers onto extracted cellulose for use in separation technologies. Bioresources 4(3):986–1005

Sharma A, Thakur KK, Mehta P, Pathania D (2018) Efficient adsorption of chlorpheniramine and hexavalent chromium (Cr (VI)) from water system using agronomic waste material. Sustain Chem Pharm 9:1–11. https://doi.org/10.1016/j.scp.2018.04.002

Shi L-n, Zhang X, Chen Z-l (2011) Removal of chromium (VI) from wastewater using bentonite-supported nanoscale zero-valent iron. Water Res 45(2):886–892. https://doi.org/10.1016/j.watres.2010.09.025

Shi Y, Chai L, Yang Z, Jing Q, Chen R, Chen Y (2012) Identification and hexavalent chromium reduction characteristics of Pannonibacter phragmitetus. Bioprocess Biosyst Eng 35(5):843–850. https://doi.org/10.1007/s00449-011-0668-

Shin K-Y, Hong J-Y, Jang J (2011) Heavy metal ion adsorption behavior in nitrogen-doped magnetic carbon nanoparticles: isotherms and kinetic study. J Hazard Mater 190(1–3):36–44. https://doi.org/10.1016/j.jhazmat.2010.12.102

Singh R, Kumar A, Kirrolia A, Kumar R, Yadav N, Bishnoi NR, Lohchab RK (2011) Removal of sulphate, COD and Cr (VI) in simulated and real wastewater by sulphate reducing bacteria enrichment in small bioreactor and FTIR study. Bioresour Technol 102(2):677–682. https://doi.org/10.1016/j.biortech.2010.08.041

Sitko R, Zawisza B, Malicka E (2012) Modification of carbon nanotubes for preconcentration, separation and determination of trace-metal ions. TrAC, Trends Anal Chem 37:22–31. https://doi.org/10.1016/j.trac.2012.03.016

Smutok O, Broda D, Smutok H, Dmytruk K, Gonchar M (2011) Chromate-reducing activity of Hansenula polymorpha recombinant cells over-producing flavocytochrome b2. Chemosphere 83(4):449–454. https://doi.org/10.1016/j.chemosphere.2010.12.078

Soko L, Cukrowska E, Chimuka L (2002) Extraction and preconcentration of Cr (VI) from urine using supported liquid membrane. Anal Chim Acta 474(1–2):59–68. https://doi.org/10.1016/S0003-2670(02)01003-6

Somasundaram V, Philip L, Bhallamudi SM (2011) Laboratory scale column studies on transport and biotransformation of Cr (VI) through porous media in presence of CRB, SRB and IRB. Chem Eng J 171(2):572–581. https://doi.org/10.1016/j.cej.2011.04.032

Sreenivas K, Inarkar M, Gokhale S, Lele S (2014) Re-utilization of ash gourd (Benincasa hispida) peel waste for chromium (VI) biosorption: equilibrium and column studies. J Environ Chem Eng 2(1):455–462. https://doi.org/10.1016/j.jece.2014.01.017

Stanković N, Logar M, Luković J, Pantić J, Miljević M, Babić B, Radosavljević-Mihajlović A (2011) Characterization of bentonite clay from'Greda'deposit. Proc Appl Ceram 5(2):97–101

Stasinakis AS, Mamais D, Thomaidis NS, Lekkas TD (2002) Effect of chromium (VI) on bacterial kinetics of heterotrophic biomass of activated sludge. Water Res 36(13):3341–3349. https://doi.org/10.1016/S0043-1354(02)00018-0

Su M, Fang Y, Li B, Yin W, Gu J, Liang H, Wu J (2019) Enhanced hexavalent chromium removal by activated carbon modified with micro-sized goethite using a facile impregnation method. Sci Total Environ 647:47–56. https://doi.org/10.1016/j.scitotenv.2018.07.372

Sud D, Mahajan G, Kaur M (2008) Agricultural waste material as potential adsorbent for sequestering heavy metal ions from aqueous solutions–A review. Bioresour Technol 99(14):6017–6027. https://doi.org/10.1016/j.biortech.2007.11.064

Sugashini S, Begum KMMS (2015) Preparation of activated carbon from carbonized rice husk by ozone activation for Cr (VI) removal. New Carbon Mater 30(3):252–261. https://doi.org/10.1016/S1872-5805(15)60190-1

Sun X, Yang L, Li Q, Zhao J, Li X, Wang X, Liu H (2014a) Amino-functionalized magnetic cellulose nanocomposite as adsorbent for removal of Cr (VI): synthesis and adsorption studies. Chem Eng J 241:175–183. https://doi.org/10.1016/j.cej.2013.12.051

Sun Y, Yue Q, Mao Y, Gao B, Gao Y, Huang L (2014b) Enhanced adsorption of chromium onto activated carbon by microwave-assisted H3PO4 mixed with Fe/Al/Mn activation. J Hazard Mater 265:191–200. https://doi.org/10.1016/j.jhazmat.2013.11.057

Sun Y, Liu C, Zan Y, Miao G, Wang H, Kong L (2018) Hydrothermal carbonization of microalgae (Chlorococcum sp.) for porous carbons with high Cr (VI) adsorption performance. Appl Biochem Biotechnol:1–11. https://doi.org/10.1007/s12010-018-2752-0

Sundaramoorthy P, Chidambaram A, Ganesh KS, Unnikannan P, Baskaran L (2010) Chromium stress in paddy: (i) nutrient status of paddy under chromium stress;(II) phytoremediation of chromium by aquatic and terrestrial weeds. C R Biol 333(8):597–607. https://doi.org/10.1016/j.crvi.2010.03.002

Sundarapandiyan S, Chandrasekar R, Ramanaiah B, Krishnan S, Saravanan P (2010) Electrochemical oxidation and reuse of tannery saline wastewater. J Hazard Mater 180(1–3):197–203. https://doi.org/10.1016/j.jhazmat.2010.04.013

Sutton R (2010) Chromium-6 in US tap water. Environmental Working Group, Washington, DC

Taa N, Benyahya M, Chaouch M (2016) Using a bio-flocculent in the process of coagulation flocculation for optimizing the chromium removal from the polluted water. J Mater Environ Sci 7(5):1581–1588

Taha AA, Wu Y-n, Wang H, Li F (2012) Preparation and application of functionalized cellulose acetate/silica composite nanofibrous membrane via electrospinning for Cr (VI) ion removal from aqueous solution. J Environ Manag 112:10–16. https://doi.org/10.1016/j.jenvman.2012.05.031

Tahar LB, Oueslati MH, Abualreish MJA (2018) Synthesis of magnetite derivatives nanoparticles and their application for the removal of chromium (VI) from aqueous solutions. J Colloid Interface Sci 512:115–126. https://doi.org/10.1016/j.jcis.2017.10.044

Tang S, Chen Y, Xie R, Jiang W, Jiang Y (2016) Preparation of activated carbon from corn cob and its adsorption behavior on Cr (VI) removal. Water Sci Technol 73(11):2654–2661. https://doi.org/10.2166/wst.2016.120

Tao SS-H, Michael Bolger P (1999) Dietary arsenic intakes in the United States: FDA total diet study, September 1991-December 1996. Food Addit Contam 16(11):465–472. https://doi.org/10.1080/026520399283759

Tavani E, Volzone C (1997) Adsorption of chromium(III) from a tanning wastewater on kaolinite. J Soc Leather Technol Chem 81(4):143–148

Thakur SS, Chauhan GS (2014) Gelatin–silica-based hybrid materials as efficient candidates for removal of chromium (VI) from aqueous solutions. Ind Eng Chem Res 53(12):4838–4849. https://doi.org/10.1021/ie401997g

Tofan L, Paduraru C, Teodosiu C, Toma O (2015) Fixed bed column study on the removal of chromium (III) ions from aqueous solutions by using hemp fibers with improved sorption performance. Cellul Chem Technol 49(2):219–229

Un UT, Onpeker SE, Ozel E (2017) The treatment of chromium containing wastewater using electrocoagulation and the production of ceramic pigments from the resulting sludge. J Environ Manag 200:196–203. https://doi.org/10.1016/j.jenvman.2017.05.075

Van der Bruggen B, Lejon L, Vandecasteele C (2003) Reuse, treatment, and discharge of the concentrate of pressure-driven membrane processes. Environ Sci Technol 37(17):3733–3738. https://doi.org/10.1021/es0201754

Varghese AG, Paul SA, Latha M (2018) Remediation of heavy metals and dyes from wastewater using cellulose-based adsorbents. Environ Chem Lett:1–11. https://doi.org/10.1007/s10311-018-00843-z

Vasanth D, Pugazhenthi G, Uppaluri R (2012) Biomass assisted microfiltration of chromium (VI) using Baker's yeast by ceramic membrane prepared from low cost raw materials. Desalination 285:239–244. https://doi.org/10.1016/j.desal.2011.09.055

Velempini T, Pillay K, Mbianda XY, Arotiba OA (2017) Epichlorohydrin crosslinked carboxymethyl cellulose-ethylenediamine imprinted polymer for the selective uptake of Cr (VI). Int J Biol Macromol 101:837–844. https://doi.org/10.1016/j.ijbiomac.2017.03.048

Venkatesan S, Meera Sheriffa Begum K (2009) Removal of trivalent chromium from dilute aqueous solutions and industrial effluents using emulsion liquid membrane technique. Int J Environ Eng 2(1–3):250–268. https://doi.org/10.1504/IJEE.2010.029832

Venkateswarlu P, Ratnam MV, Rao DS, Rao MV (2007) Removal of chromium from an aqueous solution using Azadirachta indica (nccm) leaf powder as an adsorbent. Int J Phys Sci 2(8):188–195

Verma SK, Khandegar V, Saroha AK (2013) Removal of chromium from electroplating industry effluent using electrocoagulation. J Hazard Toxic Radioact Waste 17(2):146–152. https://doi.org/10.1061/(ASCE)HZ.2153-5515.0000170

Wanees SA, Ahmed AMM, Adam MS, Mohamed MA (2012) Adsorption studies on the removal of hexavalent chromium-contaminated wastewater using activated carbon and bentonite. Chem J 2(3):95–105

Wang W (2018) Chromium (VI) removal from aqueous solutions through powdered activated carbon countercurrent two-stage adsorption. Chemosphere 190:97–102. https://doi.org/10.1016/j.chemosphere.2017.09.141

Wang H, Na C (2014) Binder-free carbon nanotube electrode for electrochemical removal of chromium. ACS Appl Mater Interfaces 6(22):20309–20316. https://doi.org/10.1021/am505838r

Wang Y, Zou B, Gao T, Wu X, Lou S, Zhou S (2012) Synthesis of orange-like Fe$_3$O$_4$/PPy composite microspheres and their excellent Cr (VI) ion removal properties. J Mater Chem 22(18):9034–9040. https://doi.org/10.1039/C2JM30440F

Wang J, Pan K, He Q, Cao B (2013a) Polyacrylonitrile/polypyrrole core/shell nanofiber mat for the removal of hexavalent chromium from aqueous solution. J Hazard Mater 244:121–129. https://doi.org/10.1016/j.jhazmat.2012.11.020

Wang Z, Ye C, Wang X, Li J (2013b) Adsorption and desorption characteristics of imidazole-modified silica for chromium (VI). App Surface Sci 287:232–241. https://doi.org/10.1016/j.apsusc.2013.09.133

Wang H, Yuan X, Wu Y, Chen X, Leng L, Wang H, Zeng G (2015) Facile synthesis of polypyrrole decorated reduced graphene oxide–Fe$_3$O$_4$ magnetic composites and its application for the Cr (VI) removal. Chem Eng J 262:597–606. https://doi.org/10.1016/j.cej.2014.10.020

Wang J, Ji B, Shu Y, Chen W, Zhu L, Chen F (2018) Cr (VI) removal from aqueous solution using starch and sodium Carboxymethyl cellulose-coated Fe and Fe/Ni nanoparticles. Pol J Environ Stud 27(6). https://doi.org/10.15244/pjoes/83729

Wen T, Wang J, Yu S, Chen Z, Hayat T, Wang X (2017) Magnetic porous carbonaceous material produced from tea waste for efficient removal of as (V), Cr (VI), humic acid, and dyes. ACS Sustain Chem Eng 5(5):4371–4380. https://doi.org/10.1021/acssuschemeng.7b00418

Wójcik G, Neagu V, Bunia I (2011) Sorption studies of chromium (VI) onto new ion exchanger with tertiary amine, quaternary ammonium and ketone groups. J Hazard Mater 190(1–3):544–552. https://doi.org/10.1016/j.jhazmat.2011.03.080

Wu L, Liao L, Lv G, Qin F, He Y, Wang X (2013) Micro-electrolysis of Cr (VI) in the nanoscale zero-valent iron loaded activated carbon. J Hazard Mater 254:277–283. https://doi.org/10.1016/j.jhazmat.2013.03.009

Xiao K, Xu F, Jiang L, Duan N, Zheng S (2016) Resin oxidization phenomenon and its influence factor during chromium (VI) removal from wastewater using gel-type anion exchangers. Chem Eng J 283:1349–1356. https://doi.org/10.1016/j.cej.2015.08.084

Xing J, Zhu C, Chowdhury I, Tian Y, Du D, Lin Y (2018) Electrically switched ion exchange based on Polypyrrole and carbon nanotube nanocomposite for the removal of chromium (VI) from aqueous solution. Ind Eng Chem Res 57(2):768–774. https://doi.org/10.1021/acs.iecr.7b03520

Xu W-H, Liu Y-G, Zeng G-M, Xin L, Song H-X, Peng Q-Q (2009) Characterization of Cr (VI) resistance and reduction by Pseudomonas aeruginosa. Trans Nonferrous Metals Soc China 19(5):1336–1341. https://doi.org/10.1016/S1003-6326(08)60446-X

Xu L, Luo M, Li W, Wei X, Xie K, Liu L, Liu H (2011a) Reduction of hexavalent chromium by Pannonibacter phragmitetus LSSE-09 stimulated with external electron donors under alkaline conditions. J Hazard Mater 185(2–3):1169–1176. https://doi.org/10.1016/j.jhazmat.2010.10.028

Xu Y-j, Rosa A, Xi L, Su D-s (2011b) Characterization and use of functionalized carbon nanotubes for the adsorption of heavy metal anions. New Carbon Mater 26(1):57–62. https://doi.org/10.1016/S1872-5805(11)60066-8

Yagub MT, Sen TK, Afroze S, Ang HM (2014) Dye and its removal from aqueous solution by adsorption: a review. Adv Colloid Interface Sci 209:172–184. https://doi.org/10.1016/j.cis.2014.04.002

Yang L, Chen JP (2008) Biosorption of hexavalent chromium onto raw and chemically modified Sargassum sp. Bioresour Technol 99(2):297–307. https://doi.org/10.1016/j.biortech.2006.12.021

Yang R, Aubrecht KB, Ma H, Wang R, Grubbs RB, Hsiao BS, Chu B (2014a) Thiol-modified cellulose nanofibrous composite membranes for chromium (VI) and lead (II) adsorption. Polymer 55(5):1167–1176. https://doi.org/10.1016/j.polymer.2014.01.043

Yang Y, Diao M h, Gao M m, Sun X f, Liu XW, Zhang G h et al (2014b) Facile preparation of graphene/polyaniline composite and its application for electrocatalysis hexavalent chromium reduction. Electrochim Acta 132:496–503. https://doi.org/10.1016/j.electacta.2014.03.152

Yang J, Yu M, Chen W (2015) Adsorption of hexavalent chromium from aqueous solution by activated carbon prepared from longan seed: kinetics, equilibrium and thermodynamics. J Ind Eng Chem 21:414–422. https://doi.org/10.1016/j.jiec.2014.02.054

Yang Z, Ren L, Jin L, Huang L, He Y, Tang J, Wang H (2018) In-situ functionalization of poly (m-phenylenediamine) nanoparticles on bacterial cellulose for chromium removal. Chem Eng J 344:441–452. https://doi.org/10.1016/j.cej.2018.03.086

Yao T, Cui T, Wu J, Chen Q, Lu S, Sun K (2011) Preparation of hierarchical porous polypyrrole nanoclusters and their application for removal of Cr (VI) ions in aqueous solution. Polym Chem 2(12):2893–2899. https://doi.org/10.1039/C1PY00311A

Yao W, Ni T, Chen S, Li H, Lu Y (2014) Graphene/Fe_3O_4@ polypyrrole nanocomposites as a synergistic adsorbent for Cr (VI) ion removal. Compos Sci Technol 99:15–22. https://doi.org/10.1016/j.compscitech.2014.05.007

Yao X, Deng S, Wu R, Hong S, Wang B, Huang J, Yu G (2016) Highly efficient removal of hexavalent chromium from electroplating wastewater using aminated wheat straw. RSC Adv 6(11):8797–8805. https://doi.org/10.1039/C5RA24508G

Yu X-Y, Luo T, Jia Y, Xu R-X, Gao C, Zhang Y-X, Huang X-J (2012) Three-dimensional hierarchical flower-like mg–Al-layered double hydroxides: highly efficient adsorbents for as (V) and Cr (VI) removal. Nanoscale 4(11):3466–3474. https://doi.org/10.1039/C2NR30457K

Zahoor A, Rehman A (2009) Isolation of Cr (VI) reducing bacteria from industrial effluents and their potential use in bioremediation of chromium containing wastewater. J Environ Sci 21(6):814–820. https://doi.org/10.1016/S1001-0742(08)62346-3

Zang Y, Yue Q, Kan Y, Zhang L, Gao B (2018) Research on adsorption of Cr (VI) by poly-epichlorohydrin-dimethylamine (EPIDMA) modified weakly basic anion exchange resin D301. Ecotoxicol Environ Saf 161:467–473. https://doi.org/10.1016/j.ecoenv.2018.06.020

Zarraa MA (1995) A study on the removal of chromium (VI) from waste solutions by adsorption on to sawdust in stirred vessels. Adsorpt Sci Technol 12(2):129–138. https://doi.org/10.1177/026361749501200205

Zeng J, Guo Q, Ou-Yang Z, Zhou H, Chen H (2014) Chromium (VI) removal from aqueous solutions by polyelectrolyte-enhanced ultrafiltration with polyquaternium. Asia Pac J Chem Eng 9(2):248–255. https://doi.org/10.1002/apj.1764

Zewail T, Yousef N (2014) Chromium ions (Cr^{6+} & Cr^{3+}) removal from synthetic wastewater by electrocoagulation using vertical expanded Fe anode. J Electroanal Chem 735:123–128. https://doi.org/10.1016/j.jelechem.2014.09.002

Zhang H, Tang Y, Cai D, Liu X, Wang X, Huang Q, Yu Z (2010) Hexavalent chromium removal from aqueous solution by algal bloom residue derived activated carbon: equilibrium and kinetic studies. J Hazard Mater 181(1–3):801–808. https://doi.org/10.1016/j.jhazmat.2010.05.084

Zhang J, Shang T, Jin X, Gao J, Zhao Q (2015) Study of chromium (VI) removal from aqueous solution using nitrogen-enriched activated carbon based bamboo processing residues. RSC Adv 5(1):784–790. https://doi.org/10.1039/C4RA11016A

Zhang J-K, Wang Z-H, Ye Y (2016) Heavy metal resistances and chromium removal of a novel Cr (VI)-reducing pseudomonad strain isolated from circulating cooling water of Iron and steel plant. Appl Biochem Biotechnol 180(7):1328–1344. https://doi.org/10.1007/s12010-016-2170-0

Zhang X, Lv L, Qin Y, Xu M, Jia X, Chen Z (2018) Removal of aqueous Cr (VI) by a magnetic biochar derived from Melia azedarach wood. Bioresour Technol 256:1–10. https://doi.org/10.1016/j.biortech.2018.01.145

Zhao J, Zhang X, He X, Xiao M, Zhang W, Lu C (2015) A super biosorbent from dendrimer poly (amidoamine)-grafted cellulose nanofibril aerogels for effective removal of Cr (VI). J Mater Chem A 3(28):14703–14711. https://doi.org/10.1039/C5TA03089G

Zhou K, Wang X, Ma Z, Lu X, Wang Z, Wang L (2018) Preparation and characterization of modified Polyvinylidene fluoride/2-Amino-4-thiazoleacetic acid ultrafiltration membrane for purification of Cr (VI) in water. J Chem Eng Japan 51(6):501–506. https://doi.org/10.1252/jcej.17we286

Zouhri A, Ernst B, Burgard M (1999) Bulk liquid membrane for the recovery of chromium (VI) from a hydrochloric acid medium using dicyclohexano-18-crown-6 as extractant-carrier. Sep Sci Technol 34(9):1891–1905. https://doi.org/10.1081/SS-100100745

Zuo X, Balasubramanian R (2013) Evaluation of a novel chitosan polymer-based adsorbent for the removal of chromium (III) in aqueous solutions. Carbohydr Polym 92(2):2181–2186. https://doi.org/10.1016/j.carbpol.2012.12.009

Chapter 5
Water Quality Assessment Techniques

Priti Saha and Biswajit Paul

Abstract Water is a major resource for the sustenance of the living organisms. However, water resources are exploited to satisfy the demand of the growing population for drinking, domestic activities as well as economic expansion of the nation through industries, irrigation and other activities. This led to scarcity of fresh water and impacts on economy, environment and human health. This review details the standard limit of major cations, anions and heavy metals recommended by various organizations, as well as sources and health impacts. It also describes water quality assessment steps. Therefore, it draws an attention to develop interdisciplinary techniques to minimize the large data volume into an understandable format. The suitability of water for drinking, irrigation and industries through indices, health risk assessment through mathematical model, and source identification through statistical procedures as well as application of geostatistics are also reviewed. It reveals that water quality index is very effective technique to determine the suitability of water but it is limited to drinking water quality only. Moreover, the conventional indices evaluating the suitability of water for irrigation or industries does not clearly specify the status as these indices does not aggregate all required compounds. Even, the existing assessment techniques do not estimate the required percentage of pollutants to be removed from the water bodies.

Keywords Water pollution · Pollutants · Irrigation · Health impact · Data management · Geographical information system (GIS)

Abbreviations

%Na	Percent sodium
AD	Average Dose
AHP	Analytical hierarchy process

P. Saha · B. Paul (✉)
Department of Environmental Science and Engineering, Indian Institute of Technology (Indian School of Mines), Dhanbad, Jharkhand, India

© Springer Nature Switzerland AG 2020
E. Lichtfouse (ed.), *Sustainable Agriculture Reviews 40*, Sustainable Agriculture
Reviews 40, https://doi.org/10.1007/978-3-030-33281-5_5

BIS   Bureau of Indian Standard
CA   Cluster Analysis
CCMEWQI   Council of Minister of the Environmental Water Quality Index
CDI   Chronic daily intake
CPCB   Central Pollution Control Board
CR   Corrositivity ratio
EC   Electrical conductivity
EPA   Environmental Protection Agency
EU   Council of the European Union
FA   Factor analysis
GIS   Geographical information system
HI   Hazard index
HQ   Hazard quotient
ISO   International Organization for Standardization
LOAEL   Lowest observed adverse effect level
LSI   Langelier saturation index
MH   Magnesium hazard
NOAEL   No observable toxic effect
NGL   No guideline limit
NM   Not mentioned
NSFWQI   National Sanitation Foundation Water Quality Index
OWQI   Oregon water quality index
PC   Principal component
PCA   Principal component analysis
PI   Permeability index
PSI   Practical scale index
RfD   Reference dose
RSC   Residual sodium carbonate
RSI   Ryznar stability index
SAR   Sodium adsorption ratio
SF   Slope factor
TDS   Total dissolved solid
USEPA   United States Environment Protection Agency
WAWQI   Weight Arithmetic Water Quality Index
WHO   World Health Organization
WQI   Water Quality Index

5.1   Introduction

Water, the driver of nature as well as hub of life is the most important asset required for the sustenance of environment and life in this planet. It serves for the basic metabolic reaction of all living organism; ensure food security through irrigation, industrial production; conserve biodiversity and environment. More than 96% of 1.386

Table 5.1 Water distribution of the planet Earth, showing a very small percentage of usable-fresh water present in underground aquifers, lakes, swamp and rivers

Water Source	Water volume (Km³)	Percent of freshwater	Percent of total water
Oceans, seas, & bays	1,338,000,000	–	96.5
Ice caps, glaciers, and Permanent snow	24,064,000	68.7	1.74
Ground water	23,400,000	–	1.69
Fresh	10,530,000	30.1	0.76
Saline	12,870,000	–	0.93
Soil moisture	16,500	0.05	0.001
Ground ice and Permafrost	300,000	0.86	0.022
Lakes	176,400	–	0.013
Fresh	91,000	0.26	0.007
Saline	85,400	–	0.006
Atmosphere	12,900	0.04	0.001
Swamp water	11,470	0.03	0.0008
Rivers	2,120	0.006	0.0002
Biological water	1,120	0.003	0.0001

Source: Gleick (1993)

million km³ of water is saline, whereas the reaming percentage 4% is fresh water. However, 68.7% of the fresh water is locked as ice cape, glaciers and permanent snow, which cannot be used for any productive purpose (USGS 2016). The remaining 30.1% and 1.2% of fresh water is groundwater and surface water respectively, which is also not readily available for human use. The water distribution shows that less 0.3% of total fresh water resource is useable (Table 5.1).

The development of global economic system due to industrialization and modernization has deteriorated water quality, which proportionally diminish the total useable amount. Therefore, apart from the investigation on the quantity of this resource, research on quality is equally important for sustainable development. Hence, water quality assessment as a component of water quality management has become a popular topic among researchers and scientists for the protection of the water environment. This review emphasis on the importance of water quality assessment, explore the background of water pollutants, which includes its source and health impacts and conclusively quality assessment techniques.

5.2 Inorganic Water Pollutants

The fresh water that is readily accessible for public use contains many dissolved constituents. These constituents are attributed to the medium of contact, which may pose threat to human health and environment. Therefore, suitability assessment of this water for drinking, domestic use, irrigation, industrial activities, fishery, and other purposes is vital, which is affirmed by quality analysis. The quality of water in a particular area is assessed through evaluating its biological, organic and inorganic constituents. The inorganic component comprises the cations such as (Na^+), potassium (K^+), calcium (Ca^{2+}), magnesium (Mg^{2+}), anions such as carbonate/bicarbonate ($CO_3{}^{2-}$ / $HCO_3{}^-$), sulfate ($SO_4{}^{2-}$), chloride (Cl^-), phosphate ($PO_4{}^{3-}$), nitrate ($NO_3{}^-$) and fluoride (F^-) and heavy metals. The metallic elements with relative density greater than 4 g/cm^3, i.e. 5 times or more than water and is toxic or poisonous even at low concentration are termed as heavy metals (Duruibe et al. 2007; Nagajyoti ct al. 2010; Sujitha et al. 2014). These constituents when go beyond the concentrations endorsed by the World Health Organisation (WHO) and different organisations, as well as have undesirable effects are termed as "pollutant".

Table 5.2 illustrates recommended standards given by various organisations across the world. The "Water pollution" is defined by various researchers and scientists as, contamination of the water bodies by the content of substances beyond the desirable limits set by different organisations that are responsible to cause hazards to living organisms or interferes with legitimate use of the water resources (Holdgate 1980; Chapman 2007; Dung et al. 2013).

5.3 Sources and Health Impacts

The pollutants enter into the water bodies through natural and anthropogenic sources (Jarvie et al. 1998; Ravichandran 2003; Wong et al. 2003). Atmospheric deposition of natural salts, water-soil and water-rock interaction are the major categories of natural source of water pollution (Liu et al. 2003), whereas, the rapid urbanization and haphazard industrialization are the most important anthropogenic sources (Olajire and Imeokparia 2001; Akoto et al. 2008; Karbassi et al. 2008; Murray et al. 2010).

5.3.1 Natural Sources

Geology and geochemical characteristics of the aquifer regulate the occurrence of pollutants in water environment (Gibert et al. 1994; Edmunds et al. 2003; Wang and Mulligan 2006; Warner et al. 2012). Table 5.3 shows natural sources of the major inorganic components that pollute both surface and groundwater. Weathering of minerals and rocks are considered as the major natural activities, which releases the toxic pollutants into water bodies. The basic sources of pollutants are sedimentary rocks such as limestone, dolomite, shale and sandstone. Interaction of water with

Table 5.2 Acceptable limits of major cations, anions and heavy metals recommended by different organizations for human consumption through drinking or cooking

Compounds	USEPA (2009)	WHO (2011)	EU (1998)	BIS (2012)
Na (mg/L)	NGL	NGL	200	NM
K (mg/L)	NGL	NGL	NM	NM
Ca (mg/L)	NM	NM	NM	75
Mg (mg/L)	NM	NM	NM	30
CO_3/HCO_3	NM	NM	NM	200
SO_4 (mg/L)	250	NGL	250	200
Cl (mg/L)	250	NGL	250	250
PO_4 (mg/L)	NM	NGL	NM	NM
NO_3 (mg/L)	10	50	50	45
F (mg/L)	4	1.5	1.5	1
Fe (µg/L)	300	NGL	200	300
Pb (µg/L)	15	10	10	10
Zn (µg/L)	5000	NGL	NM	5000
Cd (µg/L)	5	3	5	3
Cu (µg/L)	1300	2000	2000	50
Hg (µg/L)	2	6	1	1
Cr (µg/L)	100	50	50	50
As (µg/L)	10	10	10	10
Ni (µg/L)	NM	70	20	20
Mn (µg/L)	50	400	50	100
Co (µg/L)	NM	NM	NM	NM

[a]*USEPA* United States Environment Protection Agency (2009), *WHO* World Health Organisation (2011), *EU* Council of the European Union (1998), *BIS* Bureau of Indian Standard, IS 10500 (2012), *NM* Not Mentioned, *NGL* Guideline Limit not given

igneous rocks such as granite, gabbro, nepheline syenite, basalt, andesite and ultra-mafic also contributes some major cations, anions as well as heavy metals. More specifically weathering of andesite rocks consisting of plagioclase feldspar minerals, i.e, albite, anorthite, microcline and orthoclase also increase the levels of Na, K, Ca, Mg, Fe and Mn in aquafier. The major minerals that dissolve the ions and metals are carbonate, silicate and clay. The specific minerals or ores that increases the level of contaminants are: muscovite (K); biotite, augite, hornblen (Mg); hydroxyapatite, fluorapatite (PO_4); fluorspar, cryolite, apatite, mica, hornblende (F); magnetite, hematite, goethite, siderite (Fe); calcite, cuprite, malachite, azuite (Cu) ; chromite (Cr); kaolinite, montmorillonite, arsenic trioxide, orpiment, arsenopyrite (As); calamine, smithsonite (Zn); pyrolusite, rhodochriste (Mn); linnact, smaltyn, karrolit (Co). Moreover, heavy metal such as arsenic (As) concentrated in sulfide-bearing mineral deposits, especially associated with gold mineralization also leads to water pollution (Nordstrom 2002). Moreover, few minor elements such as Cd, Co, Mn

Table 5.3 Sources of major cations, anions and heavy metals in water bodies and its health impacts when the concentration of the compound is beyond the drinking water quality standard recommended by different organizations

Compound	Natural source	Anthropogenic source	Health impact	References
Na	Weathering of sedimentary rock, igneous rock containing mineral plagioclase feldspar, clay	Agricultural runoff, fertilizer, pesticides, domestic effluent industrial waste	Hypertension, build-up of fluid with congestive heart failure, stroke, cirrhosis of liver, kidney, increase in blood lipids and catecholamine levels, mortality	Hofman et al. (1980), Mondal et al. (2005), Naik et al. (2007), Singh et al. (2008), Yammani et al. (2008), Shivasharanappa and Srinivas (2013), Bhat et al. (2014), Bhat and Pandit (2014) and The New York Times (2016)
K	Weathering of parent rock containing plagioclase feldspar, carbonate, silicate, muscovite minerals	Potassium fertilizer, domestic sewage and effluents, agricultural practices	Stroke, Low blood pressure	Ascherio et al. (1998), Singh et al (2008), Yammani et al. (2008), WHO (2012), Shivasharanappa and Srinivas (2013) and Bhat and Pandit (2014)
Ca	Weathering of lime rich rock, basalt comprise of plagioclase feldspar, rock composed of carbonate and silicates minerals	Agricultural runoff, dissolution of lime in streets and sidewalls	Hypercalcaemia, metabolic alkalosis, renal insufficiency, kidney stones, reduce normal absorption of iron, zinc, magnesium and phosphorus within the intestine, cardiovascular and cerebrovascular disease	Yang (1998), Naik et al. (2007), Yammani et al. (2008) WHO (2009), Daly and Ebeling (2010), Cotruvo (2011), Bhat et al. (2014) and Bhat and Pandit (2014)
Mg	Weathering of lime rich rock, basalt comprise of plagioclase feldspar, ferro-magnesium mineral rocks composed of carbonate and silicates minerals	Wastewater, sewage, mining and mineral processing (particularly nickel), atmospheric deposition from the emissions of alloy, steel, iron production, fossil fuels, combustion of fuel additive	Diarrhoea, laxative effect in combination with sulphate, cardiovascular, cerebrovascular disease	Yang (1998), Howe et al. (2004), Naik et al. (2007), Yammani et al. (2008), WHO (2009), Cotruvo (2011), Bhat et al. (2014) and Bhat and Pandit (2014)

Parameter	Geogenic source	Anthropogenic source	Health effect	References
CO_3/ HCO_3	Weathering of carbonate, silicates minerals, basalt containing plagioclase feldspar, CO_2 from root transpiration in soil zone	Decay organic matter in soil zone	Diseases related to calcium and magnesium	Mondal et al. (2005), Naik et al. (2007), Singh et al. (2008) and Yammani et al. (2008)
SO_4	Evaporation process, volcanism process, dissolution of sulphide mineral present in granite, limestone, marls rock	Domestic wastewater, untreated industrial waste, Landfill leaching, fertilizer, atmospheric deposition from combustion of fossil	Gastro intestinal irritation.	Naik et al. (2007), Singh et al. (2008), Srivastava and Ramanathan (2008), Shankar et al. (2008), Shivasharanappa and Srinivas (2013) and Bhat and Pandit (2014)
Cl	Atmospheric deposition of chloride from the ocean, volcanic eruption, weathering of limestone, sandstone, crystalline rocks bearing carbonate and silicates, chloride mineral	Domestic sewage water, atmospheric precipitation of manmade generated chloride salt, landfill leaching of saline residues, tannery industrial activity	Increase blood pressure, build-up of fluid with congestive heart failure, cirrhosis, kidney disorder	Sawka (2005), Mondal et al. (2005), Srivastava and Ramanathan (2008), Shankar et al. (2008), Mullaney et al. (2009), Shivasharanappa and Srinivas (2013), Bhat and Pandit (2014) and NIH (2016)
PO_4	Weathering of non-detrital sedimentary rock containing phosphate mineral	Domestic wastewater, agricultural runoff, detergent, landfill leaching	Renal failure, cardiovascular morbidity	Block et al. (2004), Chi et al. (2006), Srivastava and Ramanathan (2008), Ritz et al. (2012) and Bhat and Pandit (2014)
NO_3	Nitrogen cycle, atmospheric deposition of nitrate salts	Livestock waste, fertilizer, agricultural runoff, industrial sewage, leaching from dumps	Blue baby syndrome/ methaemoglobinemia, cancer	Basappa Reddy (2003); Singh et al. (2008), Srivastava and Ramanathan (2008), Shankar et al. (2008) and Bhat and Pandit (2014)

(continued)

Table 5.3 (continued)

Compound	Natural source	Anthropogenic source	Health impact	References
F	Weathering of sedimentary and igneous rock bearing fluorspar, cryolite, apatite, mica, hornblende, volcanic eruption	Pesticides, atmospheric deposition from coal combustion, manufacture of bricks, iron, glass, operation of coal fired power station, aluminium smelter	Nausea, diarrohea,abdominal pain, dysphagia,hypocalcemia, dental fluorosis, neurological effect, seizuers, muscle weakness, headache, cardiovascular effect, hyperkalemia, hypoglycaemia	Notcutt and Davies (2001), Farooqi et al. (2009), Chouhan and Flora (2010) and Fawell et al. (2013)
Fe	Weathering of serpentinized rock type, basalt comprise of plagioclase feldspar, iron minerals	Landfill leaching, disposal of iron scarp in open area, iron ore mining, industrial effluent, acid-mine drainage, sewage, landfill leachate	Accumulation in muscle, liver, brain and central nervous system damage, sideropenic anaemia, hyperthyroidism, cognitive dysfunction, alzheimer type II astrocytosis, parkinsonism	Basappa Reddy (2003), Liu et al. (2003), Razo et al. (2004), Naik et al. (2007), Shankar et al. (2008), Butterworth (2010) and Fernandez-Luqueno et al. (2013)
Pb	Atmospheric deposition of fluvial inputs and continental dust	Landfill leaching, smelting, corrosion of lead pipe, ceramic, cement industry, paints and petrol additives atmospheric deposition of motor vehicle exhaust fumes, high temperature industrial process	Enzyme inhibitor, damage nervous connections, haematological damage, hypertension, tiredness, irritability, anemia, behavioural change, chronic renal failure, blood, kidney and brain disorders	USGAO (2000), Srivastava and Ramanathan (2008), Gowd and Govil (2008), Jones and Miller (2008), Bawaskar et al. (2010), Hsu et al. (2010), Kang et al. (2011), Mebrahtu and Zerabruk (2011), Mohod and Dhote (2013); and Li et al. (2014)
Zn	Weathering of rock, minerals	Landfill leaching, fertilizer, effluents from brass, bronze, die-casting metal, other alloy, rubber and paint industries, spillage from gasoline and auto lubricant, atmospheric deposition from vulcanization agent released at higher wearing rate and high temperature from tyres, iron ore mining, zinc-rich soils, acid mine drainage	Cancer, birth defects, organ damage, nervous system disorder, damage to the immune system, hematological disorder, detoriates human metabolism	Wedepohl (1995), Parker and Fleischer (1967), Mirenda (1986), Salem et al. (2000), Liu et al. (2003), Ipeaiyeda and Dawodu (2008), Viers et al. (2007), Borrok et al. (2008), Srivastava and Ramanathan (2008), Huffmeyer et al. (2009), Reza and Singh (2010b), Visnjic-Jeftic et al. (2010) and Mohod and Dhote (2013)

Element	Natural sources	Anthropogenic sources	Health effects	References
Cd	Water-rock interaction	Landfill leaching, phosphate fertilizers, metal industry, galvanized pipe breakdown. Atmospheric deposition from coal combustion and waste incineration, acid mine drainage	Accumulate in kidney, interfere with the metallothionein's, nausea, vomiting, respiratory difficulties, cramps, loss of consciousness, anemia, anosmia, cardiovascular disease, hypertension, renal failure, risks of stillbirth, damage membranes and Deoxy ribo-nuclicacid (DNA)	Webb (1979), Salem et al. (2000), Terry and Stone (2002), Von Ehrenstein et al. (2006), Srivastava and Ramanathan (2008), Caruso and Bishop (2009), Bawaskar et al. (2010) and Reza and Singh (2010b)
Cu	Weathering of minerals, volcano	Domestic sewage, agricultural runoff, mining, smelting, copper pipes, effluents from plastic, steel, agrochemical industry, blast furnace, acid mine drainage	Vomiting, abdominal pain, nausea, anemia, diarrhea, neurological complication, hypertension, liver and kidney dysfunctions.	Madsen et al. (1990), Bent and Bohm (1995), Salem et al. (2000), Mercer (2001), Liu et al. (2003), Gowd and Govil (2008), Butterworth (2010), Reza and Singh (2010b), Mohod and Dhote (2013) and Giri and Singh (2014)
Hg	Atmospheric deposition of wind-blown dust, forest fires and vegetation, weathering of rocks and soils, volcanic emissions	Agricultural activity (tillage and logging), domestic sewage discharge, leaching of solid waste, atmospheric deposition of coal and oil combustion, effluents from pyrometallurgical processes, chlor-alkali production, batteries, fluorescent lamps, thermometers, and electronic switches, chemical industry, mining, erosion of mercury contaminated soil	Accumulation in muscle, liver, damage to DNA	Salem et al. (2000), Wang et al. (2004), Kowalski et al. (2007), Li et al. (2009), Visnjic-Jeftic et al. (2010) and Fernández-Luqueño et al. (2013)
Cr	Weathering of basaltic and ultramafic rocks and chromite mineral	Effluent from metallurgical refractories and chemical industries, production of dye, wood preservation, leather tanning, chrome plating, manufacturing of alloy	Ulceration of the skin, kidney, liver, circulatory and nerve tissue damage, cancer	Alloway (1990), ATSDR (2000), Alimonti et al. (2000), Robles-Camacho and Armienta (2000), Kotaś and Stasicka (2000), Ball and Izbicki (2004), Gowd and Govil (2008), Owlad et al. (2009), Smith and Steinmaus (2009) and Giri and Singh (2014)

(continued)

Table 5.3 (continued)

Compound	Natural source	Anthropogenic source	Health impact	References
As	Weathering of clay mineral, sulfide-bearing mineral deposits	Nonferrous mining, mineral extraction, processing, wood preservatives, effluent from poultry and swine feed additives, pesticides, atmosopheric precipitation from incineration of municipal, industrial wastes, roasting of arsenious gold ores, combustion of fossil fuel and wastes, acid mine drainage	Lesions on skin, heart and liver, hyperpigmentation, respiratory complications, change in the hormonal and mucosal immune responses, melanosis, kidney disorder, cerebrovascular diseases, diabetes mellitus, leucomelanosis, keratosis, neurological disorder, damage to DNA, cancer	Vine and Tourtelot (1970), Popovic et al. (2001), Bhattacharya et al. (2002), Nordstrom (2002), Smedley and Kinniburgh (2002), Liu et al. (2003), Mukherjee et al. (2006), Wang and Mulligan (2006), Bawaskar et al. (2010), Fernández-Luqueño et al. (2013) and Sarkar and Paul (2016)
Ni	Atmospheric precipitation of wind-blown dust, forest fires, vegetation, volcanic emissions, weathering of rocks and soils	Mining, smelting, municipal sewage, waste water from sewage treatment plant, landfill leaching, corroded metal pipes and containers, effluent from ceramic, steel-alloy-electroplating, refractory industries	Damage to DNA, systemic toxicity, allergy, hair loss, anemia	Lantzy and Mackenzie (1979), Nriagu (1989), Salem et al. (2000), Cempel and Nikel (2006), Visnjic-Jeftic et al. (2010), Fernández-Luqueño et al. (2013) and Giri and Singh (2014)
Mn	Weathering of ignious, metamorphic, manganese silicate rock, manganese bearing minerals, iron ores	Effluent from iron and steel related industry, power plant, coke oven plant, paper and newsprint mill, electric power generation, dry cell battery, welding industry, acid-mine drainage, sewage discharge, landfill leachate, atmospheric deposition of fossil fuel combustion	Accumulation in muscle, liver and affects brain and central nervous system, damage to DNA	WHO (1986), Roy (1970), Post (1999), Salem et al. (2000), USEPA (2003), Butterworth (2010), Bouchard et al. (2010), Fernández-Luqueño et al. (2013) and Giri and Singh (2014)
Co	Weathering of oxides of accessory minerals, cobalt rich minerals	Metal alloy industry, mining, smelting	Accumulation in muscle, liver	Barakiewicz and Siepak (1999), Barceloux (1999), Visnjic-Jeftic et al. (2010) and Giri and Singh (2014)

occur in earth crust along with other minerals also elevate the level of heavy metals in water environment (Jarup 2003). Apart from the water-rock or water-soil interaction, dry fall out or atmospheric precipitation of the natural salts entrapped in the form of aerosol also increase the level of pollutants in both surface and groundwater (Hsu et al. 2010; Huang et al. 2014; Li et al. 2014). Aerosols formed by the wind-blown dusts, forest fires and volcanic emissions are also included in the natural sources.

5.3.2 Anthropogenic Sources

The carrying capacities of water bodies are decreasing rapidly due to anthropogenic activities. Domestic and industrial wastewater is one of the major anthropogenic sources of water pollution. Some of the industrial units discharge their untreated effluents directly into nearby pits, ponds, tanks and streams (Carpenter et al. 1998; Subrahmanyam et al. 2001; Rim-Rukeh et al. 2006; Karbassi et al. 2007; Gowd and Govil 2008; Sekabira et al. 2010; Amadi et al. 2012; Giri and Singh 2014), while some discharge their effluents indirectly to water bodies through unlined channels (Purandara and Varadarajan 2003; Shankar et al. 2008; Lohani et al. 2008). Moreover, the groundwater also get polluted by leaching from dump of slag, dust, sludge, metal shell, flyash and other industrial waste or bi-products (Nalawade et al. 2012; Peter and Sreedevi 2012). Even runoff from these solid wastes pollutes the surface water bodies. The percolation of rainwater and runoff from the municipal landfill also pollutes the aquifer (Zanoni 1972; Slack et al. 2005; Raman and Narayanan 2008). Moreover, atmospheric deposition of the aerosols due to incineration of municipal solid wastes, emission from industrial activities and vehicular movements also make a count in anthropogenic sources (Subrahmanyam and Yadaiah 2001). Mining activities and agricultural practices are also counted as major anthropogenic sources, which deteriorate the quality of water with the time-line (Singh et al. 2008; Krishna et al. 2009; Giri et al. 2010; Reza and Singh 2010b; Ameh and Akpah 2011; Giri and Singh 2014; Bhat et al. 2014). Acid mine drainage (AMD), is one of the water pollutant in few coal mining areas, which result in acidic pH as well as increase the concentration of dissolved metals such as As, Cd, Cu, Zn and anions such as CO_3, SO_4 in water (Olias et al. 2004; Razo et al. 2004; Johnson and Hallberg 2005; Akcil and Koldas 2006). Table 5.3 lists the major anthropogenic source of the cations, anions and heavy metals in water system.

5.3.3 Health Impacts

The main exposure route of ions and heavy metals in the human body is through oral ingestion or dermal contact from wells, ponds, reservoirs, lakes and other water sources. Although the presence of these elements is vital for the regulation of human biological function, but it act as toxin once it exceeds its threshold limit agreed by different regulatory agencies across the world. These pollutants result in physical, muscular, neurological degenerative processes that can cause acute and chronic diseases, such as muscle weakness, skin lesions, diarrhoea, gastric problems to an extent alzheimer's, parkinson's, multiple sclerosis, methaemoglobinemia, cancer, and other catastrophic diseases (Mishra et al. 2010; Mebrahtu and Zerabruk 2011). Near about 58% of total mortality rate, or 842000 deaths per year, is attributable to unsafe water use (WHO 2016). Table 5.3 details the effect of the cations, anions and heavy metals on human health. Since, the pollutants has a huge impact on human health, therefore water quality assessment is paramount importance.

5.4   Water Quality Assessment Techniques

The researchers and pollution control boards of various countries, emphasis on water quality management after realizing the importance of water quality (Simeonov et al. 2003; Qadir et al. 2008). Water quality assessment or monitoring is one of the important and crucial step of water quality management for sustainable development. The International Organization for Standardization (ISO) defines monitoring as: "the programme of sampling, measurement and subsequent recording or signalling, or both, of various water characteristics, often with the aim of assessing conformity to specified objectives" (CPCB 2007). It is the intensive programme to measure and observe the quality of the aquatic environment for a specific purpose. Water Quality monitoring involves mainly eight steps starting from setting the objective to quality assurance (Fig. 5.1) (CPCB 2007).

The seasonal monitoring and physico-chemical analysis of the samples generate huge data sets. The elemental data sets are not understandable by the general public or regulatory authorities as it does not represent the accurate status of the water body. Therefore, it draws an attention to manage the huge data set into an understandable format that can represent the water quality of an area. Various interdisciplinary techniques and tools are investigated that can effectively minimize the large data volume. These tools help in assessment of water quality, identification of pollution sources and documentation of water quality management report.

The most popular and most convenient interdisciplinary branches that help in assessment of water quality are mathematics, statistics and geomatics. Integration of geographical information system (GIS) with the data sets is also one of the most advanced and popular technique among hydrologist, as it helps in better visualization of water quality in the earth surface. The step of water quality assessment involves the suitability assessment of the water bodies for various purposes such as drinking, irrigation, industries, fisheries and its associated health risk followed by pollution source apportionment.

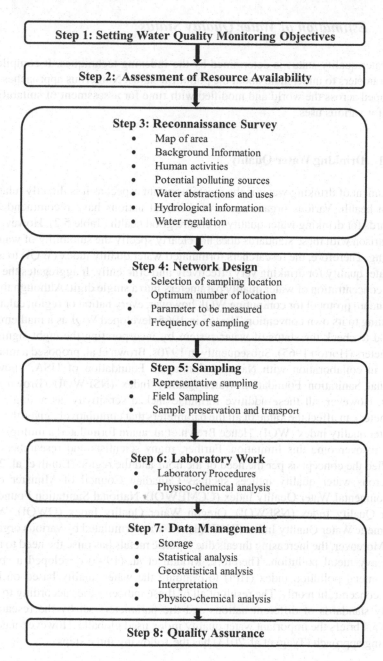

Fig. 5.1 Water quality monitoring steps, starting from setting objectives to quality assurance. (Source: CPCB 2007)

5.4.1   Estimation of Water Quality Status

The water quality status is determined by the indexing techniques. It compiles all the parameters to understand about the pollution status. Numerous approaches were developed across the world and modified with time for assessment of suitability of water for various uses.

5.4.1.1   Drinking Water Quality

Assessment of drinking water is the most important aspect as it is directly related to human health. Various organisations of different nations have recommended the standards for drinking water quality to ensure good health (Table 5.2). However, the comparison with these standards does not clearly specify the suitability of water for drinking. Therefore, the researchers formulated water quality index (WQI) to assess the water quality for drinking very effectively and efficiently. It aggregates the measured concentration of water quality parameters into a single digit. Although there is no standard protocol for computing WQI, however, every nation or region calculates according to its own convention. Horton in 1965 developed WQI as a mathematical method to check the status of water system by incorporating the eight significant parameters (Horton 1965). Subsequently in 1970s, Brown et al. proposed a modified index in collaboration with National Sanitation Foundation of USA, known as National Sanitation Foundation Water Quality Index (NSFWQI) (Brown et al. 1972). However, all these additive formulae lacked sensitivity, as a single toxic parameter can affect the value of all the parameters that cumulatively gives the result of water quality index (WQI). Hence Brown et al. again formulated a multiplicative index to overcome this limitation. Further, many scientists and researchers have modified the concept as per the need of the hour and the region (Lumb et al. 2011). Numerous water quality indices such as Canadian Council of Minister of the Environmental Water Quality Index (CCMEWQI), National Sanitation Foundation Water Quality Index (NSFWQI), Oregon Water Quality Index (OWQI), Weight Arithmetic Water Quality Index (WAWQI) etc. was formulated by various organization. Moreover, the increasing threats due to toxic metals has raise the need to assess the heavy metal pollution. Therefore, Mohan et al. (1996) developed a specific heavy metal pollution index (HPI) to evaluate the water quality based on heavy metal concentration only. The result of all of these indices varies according to water quality standards of different nations and the parameters set by the researchers. Table 5.4 briefs the important water quality index used globally. However, a general indexing approach (Tyagi et al. 2013) has the following three steps:

1. Selection of parameters based on the usage of water, i.e. drinking, ecological impact, swimming etc.
2. Determination of sub-indices, which transform the different scale of data to a non-dimensional scale values.
3. Integration of sub-indices with mathematical expression i.e., through arithmetic or geometric averages.

Table 5.4 Water quality indices and its categorization to evaluate the drinking water suitability

Index	Mathematical expression	WQI rating		References
NSFWQI	$$WQI = \sum_{i=1}^{n} Q_i W_i$$ W_i = Weighting factor; Q_i = Sub-index for ith water quality parameter	91–100	Excellent Water Quality	Katyal (2011), Tyagi et al. (2013), Vaheedunnisha and Shukla (2013) and Darvishi et al. (2016)
		71–90	Good water quality	
		51–70	Medium water quality	
		26–50	Bad water quality	
		0–25	Very bad water quality	
CCME WQI	$$WQI = 100 - \frac{\sqrt{F_1^2 + F_2^2 + F_3^2}}{1.732}$$ F_1 = Number of variables, whose objectives are not met; F_2 = Number of times by which the objectives are not met; F_3 = Amount by which objectives are not met.	95–100	Excellent water quality	Lumb et al. (2006), Katyal (2011) and Tyagi et al. (2013)
		80–94	Good water quality	
		60–79	Fair water quality	
		45–59	Marginal water quality	
		0–44	Poor water quality	
OWQI	$$WQI = \sqrt{\frac{n}{\sum_{i=1}^{n} \frac{1}{SI_i^2}}}$$ n = Number of sub-index; SI= Sub index of ith parameter.	90–100	Excellent water quality	Cude (2001), Katyal (2011), Tyagi et al. (2013) and Darvishi et al. (2016)
		85–89	Good water quality	
		80–84	Fair water quality	
		60–79	Poor water quality	
		0–59	Very poor water quality	

(continued)

Table 5.4 (continued)

Index	Mathematical expression	WQI rating		References
WQI	$$WQI = \sum_{i=1}^{n} W_i q_i$$ $$W_i = \frac{w_i}{\sum_{i=1}^{n} w_i}$$ $$q_i = 100\left[\frac{C_i}{S_i}\right]$$ Wi = Relative weight; wi = Weight of each parameter; qi = Quality rating; Ci = Concentration of each parameter; Si = Drinking water standard.	<50	Excellent water quality	Ramakrishnaiah et al. (2009), Ravikumar et al. (2013) and Tiwari et al. (2014)
		50–100	Good water quality	
		100–200	Poor water quality	
		200–300	Very poor water quality	
		>300	Unsuitable for drinking	
WQI	$$WQI = \frac{\sum_{i=1}^{n} w_i q_i}{\sum_{i=1}^{n} w_i}$$ $$w_i = \frac{k}{S_i}$$ $$q_i = 100\left[\frac{V_i}{S_i}\right]$$ wi = Unit weightage; Si = Recommended standard for ith parameter; Qi = Sub index of ith parameter; Vi =Monitored value of the ith parameter; Si = Standard limit for the ith parameter.	0–25	Very good water quality	Reza and Singh (2010a), Sirajudeen et al. (2013) and Singh and Kamal (2015)
		26–50	Good water quality	
		51–75	Poor water quality	
		>75	Very poor water quality	

5.4.1.2 Irrigation Water Quality

The soil quality and crop yield depends on the quality of water. The dissolved salts in the irrigation water mainly influence the crop production as these remain in the soil and subsequently taken up by the crops. The poor irrigation water quality cause many hazards to soil, which are broadly categorised by the soil scientists as: (1) salinity hazards, (2) infiltration and permeability problems, (3) specific ion toxicity, and (4) miscellaneous problems (Simsek and Gunduz 2007). The long-term use of irrigation water with varying concentrations of total salts such as sodium, calcium, magnesium, carbonate, and bicarbonates alter the physical properties of soil. Irrigation water with high salinity diminishes the plant growth as the salt accumulates in the root zone. It causes a "physiological" drought condition where the roots are unable to absorb the water, even though the field has plenty of moisture. The salinity hazard results in wilting, darker to bluish-green colour as well as thicker and waxier leaves (FAO 2018). The salinity is measured by total dissolved solids (TDS) or electrical conductivity (EC). Moreover, sodium content also impacts the plant growth by influencing the normal infiltration rate. The high level of sodium with respect to calcium and magnesium content in irrigation water results in adsorption of sodium in soil particles. Fine textured soils, especially clay particles are dispersed by this action, which clog soil pores and also regulates the cation exchange capacity. Low calcium or magnesium in the water also results the same problem. Even, excessive bicarbonate in water precipitate calcium as calcite and magnesium as magnesite in soil solution and these allow sodium to be dominant in the soil. It also results in sodicity and its related effects such as reduction in soil permeability (Hwang et al. 2016; Kaur et al. 2017). The infiltration or less permeability of water into the soil causes crusting of seedbeds, excessive weeds, nutritional disorders, decaying of seeds and poor yield (FAO 2018). This hazard is usually investigated through sodium adsorption ratio (SAR). It is the ratio of sodium (Na) to calcium (Ca) and magnesium (Mg) content in water. Apart from these major ions, chloride, boron and other toxic elements often hinders the plant growth. The US Salinity Laboratory with the help of water agencies project planners, agriculturalists, scientists and trained field people has developed a water quality assessment guideline for finding its suitability for irrigation (Table 5.5). This has been adopted by many countries. However, few nations where the soil quality and climatic condition differs have developed their own guideline values. Apart from this, various indices such as percent sodium (%Na), residual sodium carbonate (RSC), permeability index (PI) and magnesium hazard (MH) were developed to assess the irrigation water quality (Eaton 1950; Spandana et al. 2013; Rakotondrabe et al., 2018). Moreover, Doneen plot based on the permeability index (PI) and interaction of calcium, magnesium, sodium and bicarbonate also evaluate the water quality (Doneen 1964). Wilcox (1955) also classified water for irrigation purposes based on percent sodium electrical conductivity (EC).

Table 5.5 Guidelines for evaluating the suitability of water for irrigation based on water quality parameters recommended by the US salinity laboratory

Parameter			Units	Degree of Restriction on Use		
				None	Slight to Moderate	Severe
EC$_w$			dS/m	<0.7	0.7–3.0	>3.0
TDS			mg/L	<450	450–2000	>2000
SAR	=0–3	EC$_w$	=	>0.7	0.7–0.2	<0.2
	=3–6		=	>1.2	1.2–0.3	<0.3
	=6–12		=	>1.9	1.9–0.5	<0.5
	=12–20		=	>2.9	2.9–1.3	<1.3
	=20–40		=	>5.0	5.0–2.9	<2.9
Na						
Surface irrigation			meq/L	<3	3–9	>9
Sprinkler irrigation			meq/L	<3	>3	
Cl						
Surface irrigation			meq/L	<4	4–10	>10
Sprinkler irrigation			meq/L	<3	>3	
Boron			mgq/L	<0.7	0.7–3.0	>3.0
Nitrogen NO$_3$			mg/L	<5	5–30	>30
HCO$_3$						
Overhead sprinkling only			meq/L	<1.5	1.5–8.5	>8.5
Ph				6.5–8.4		

Source: Richards (1954)

5.4.1.3 Industrial Water Quality

The industries uses water for production, processing, washing, cooling or transportation. However, the quality of water is responsible for safeguard of industrial equipment and manufacturing process. It has potential for corrosion, abrasion or precipitation, which hampers production output. The researchers or the authorities consider water to be safe for industrial use if it has neither scale-forming nor corrosive in nature (Hoseinzadeh et al. 2013). The scale formation and corrosion of cooling water is evaluated by certain indices such as Langelier saturation index (LSI), Ryznar stability index (RSI), Puckorius scale index (PSI), corrositivity ratio (CR) (Table 5.6). Although these indices predict the scaling of water, but is limited to calcium carbonate scaling only. The Langelier saturation index depends mainly on five factors which are pH, total alkalinity, calcium hardness, temperature and total dissolved solids. Positive value of Langelier saturation index indicates the scale formation due to super saturation of calcium carbonate in water, whereas negative value indicates under saturated water, which removes calcium carbonate protective coatings. However, the neutral value indicates the suitability of water for industrial use. Further, Langelier saturation index was modified by Ryznar in 1994 as an alternative method for computation of calcium carbonate scale. It is calculated on basis of the pH of calcium carbonate saturation and equilibrium pH

Table 5.6 List of indices to evaluate the suitability of water for industries based on corrosion and scale formation characteristics

Parameter	Classification	References
$LSI = pH_w - pH_s$ $pH_s = (9.3 + A + B) - (C + D)$ $A = (\log_{10}TDS - 1)/10$ $B = -13.12 \times \log_{10}(°C + 273) + 34.55$ $C = \log_{10}(Ca^{2+} \text{ as } CaCO_3) - 0.4$ $D = \log_{10}(Alkalinity \text{ as } CaCO_3)$ $pH_w = $ measured pH	Negative=corrosive Positive=scale forming Neutral= Suitable	Hoseinzadeh et al. (2013), Hwang et al. (2016) and Patel et al. (2016)
$RSI = 2(pH_s) - pH_w$	<5.5 = Heavy scale 5.5–6.2 = Moderate Scale 6.2–6.8 = Suitable for use 6.8–8.5 = Aggressive water and corrosion is likely >8.5 Very aggressive water and corrosion is possible	Hoseinzadeh et al. (2013) and Patel et al. (2016)
$PSI = 2(pH_s) - pH_e$ $pH_e = 1.465 \times \log_{10}(alkalinity) + 4.54$	<6 = scale forming >6 = corrosive tendency	Hoseinzadeh et al. (2013) and Kumar et al. (2012)
$CR = \left[\left(\dfrac{Cl}{35.5}\right) + \left(\dfrac{SO_4}{48}\right)\right] / \left(\left[\dfrac{CO_3 + HCO_3}{50}\right]\right)$	<1 = non-corrosive >1=corrosive	Hwang et al. (2016), Patel et al. (2016) and Kaur et al. (2017)

(Kumar et al. 2012). Moreover, Puckorius developed an index that predicts the scaling tendency more accurately. Apart from these the corrosivity ratio, defines the susceptibility of water to corrosion only, based on the alkaline earth and saline salts (Kaur et al. 2017).

5.4.2 Health Risk Assessment

The contaminated water poses health hazard to the depended population. The expensive and tedious standard medical diagnosis does not always able to forecast health risk (Aitman 2001). Therefore, few approaches were developed to assess the health risk. The first assessment technique is two parameter-based approaches, which compares the existing water quality standards with the measured water quality. This approach assumes that the water quality complying with the standard is safe for health as the water quality standards are recommended based on the human health. However, this approach only deal with the drinking water quality. It neither evident the impact of water quality on dermal exposure, nor it defines the risk as cancerous or non-cancerous. Therefore, quantative health risk was developed to predict risk more accurately. The United States Environmental Protection Agency in its publication of the carcinogenic risk assessment guidelines in 1976 has defined the guidelines and empirical models for health risk assessment, which consists of the following steps:

Hazard identification: It identifies the hazardous components, which adversely affect human health. The hazardous chemicals are divided according to their mechanism of action and effects such as toxic chemicals, carcinogens and endocrine or endocrine disrupting compounds.

Exposure assessment: It measures the frequency, intensity and duration of human exposed to the hazard. It also identifies the exposure pathway, i.e. through oral ingestion or dermal contact.

Dose-response assessment: It quantifies the toxicity level of the hazardous agents. The toxicity data are derived from the experiments in which the laboratory animals were exposed to increasingly higher concentrations or doses and observe their corresponding effects. The result is interpreted through various mathematical models such as one-hit model, the multistage model, the multihit model, and the probit model, that establish the correlation between the exposure of the human to the concentration of hazardous element and the risk of an adverse result from that dose (Gerba 2000). However, environmental protection agency prefers linear multistage model (Fig. 5.2), a modified form of the multistage prototype, to determine the toxicity dose because it considers carcinogens, cocarcinogens, and promoters that probable knocks cancer to human. The curve generated by this mathematical model gives a dose-response Eq. (5.1) for a carcinogen as:

$$Lifetime\ Risk = Average\ Dose \times Slope\ factor \qquad (5.1)$$

The slope factor is the slope at low dose of the dose-response curve and average dose is the risk produced by a lifetime average dose of 1 mg/kg/day (Navoni et al. 2014; Giri and Singh 2015).

Moreover, dose-response curve also determine the threshold dose below which there is no observable toxic effect (NOAEL) or a lowest observed adverse effect level (LOAEL) of a substance where probable health effect can be predicted (Fig. 5.3).

These thresholds are represented by the reference dose of a substance that has adverse effect on human health by its exposure. The reference dose is expressed as intake of the hazardous substance per unit body weight per day (Eq 5.2):

$$Reference\ Dose = \frac{No\ Observable\ Toxic\ Effect}{\left(Uncertainty\ factor\right)_1 \times \left(Uncertainty\ factor\right)_2 \times \left(Uncertainty\ factor\right)_n}$$

$$(5.2)$$

A tenfold of uncertainty factor is used for sensitive individuals such as pregnant women, young children, and the elderly population. Moreover another factor of 10 is added when 'no observable toxic effect' is extrapolated for human from the data of animals.

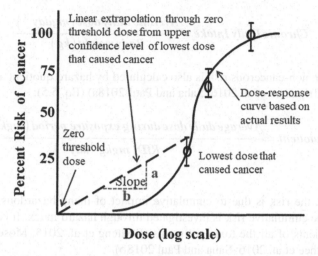

Fig. 5.2. Dose-response graph developed by the multistage model to determine the toxicity dose while considering carcinogens, co-carcinogens, and promoters that probable knocks cancer to human. (Source: Modified after Toxicologicalschools 2018)

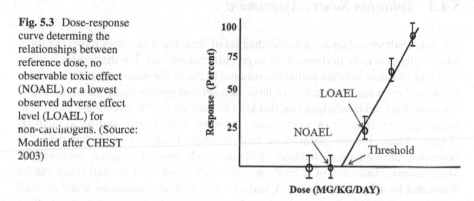

Fig. 5.3 Dose-response curve determing the relationships between reference dose, no observable toxic effect (NOAEL) or a lowest observed adverse effect level (LOAEL) for non-carcinogens. (Source: Modified after CHEST 2003)

Risk characterization: The assessment is done by integration the exposure and dose–response output. It accesses the risk to human under specific exposure conditions with appropriate media and pathways. The incremental lifetime risk of cancer is estimated as (Eq. 5.3):

$$Incremental\ lifetime\ risk\ of\ cancer = Chronic\ Daily\ Intake \times Slope\ Factor$$

$$(5.3)$$

The chronic daily intake is defined as (Muhammad et al. 2011; Bhutiani et al. 2016; Rahman et al. 2017) (Eq. 5.4),

$$Chronic\ Daily\ Intake = \frac{Averahe\ daily\ dose\left(mgday^{-1}\right)}{Body\ weight\left(kg\right)} \quad (5.4)$$

Moreover, non-cancerous risk is also calculated by hazard quotient as (Bhutiani et al. 2016; Tripathee et al. 2016; Saha and Paul 2018a) (Eq. 5.5):

$$Hazard\ Quiotient = \frac{Average\ daily\ dose\ during\ exposure\ period\left(mgkg^{-1}day^{-1}\right)}{RfD\left(mgkg^{-1}day^{-1}\right)}$$

$$(5.5)$$

However, the risk is due to cumulative impact of many hazardous chemicals; therefore, the cumulative risk is investigated through hazard index. It is the sum of hazard quotients of all the toxic pollutants (Boateng et al. 2015; Moses and Etuk 2015; Tripathee et al. 2016; Saha and Paul 2018b).

5.4.3 Pollution Source Assessment

The data analysis is easy for a diminished set of data, but it becomes rather complex when analysis is to be performed for larger information sets. For the comprehensive appraisal of water, in order to find the solution of the major issues related to environment and health requires dealing of large data set comprising many variables. Many scientists have acknowledged that this kind to data set can be effectively transfer to meaningful information by using some advanced sophisticated statistical methodologies. Many hidden phenomena and inherent hydro-chemical behaviour is expressed through these methods without much loss of original information. Simultaneous evaluation of several variables with respect to time and space can be illustrated by these procedures. A variety of statistical techniques were devised since the early 1920s for investigation of system to find origin of threat and effect on water environment.

5.4.3.1 Bivariate Statistics

The most elementary kind of statistical tool that defines the empirical relation between two variables (often denoted as X, Y) is bivariate statistics. Most common measure for bivariate analysis that is used by various researchers related to field of hydrochemistry is Pearson correlation analysis, which measures the familiarity between a pair of data (Mondal et al. 2005; Bhattacharyya et al. 2014). Sir Karl Pearson developed the concept of correlation proposed by Sir Francis Galton into statistical form and published his research on correlation and regression in the Philosophical Transactions of the Royal Society of London in 1846 (Stanton 2001).

Table 5.7 Broad interpretations of Pearson's correlation based on the value of correlation coefficient, r

Value of r	Interpretation
0.7 to 1.0	Strong linear association
0.5 to 0.7	Moderate linear association
0.3 to 0.5	Weak linear association
0 to 0.3	Little or no linear association

Source: ARMCANZ and ANZECC (2000)

The correlation coefficient varied from −1 to +1, where closeness to +1 or −1, reflects the perfect linear relationship between bivariates (ARMCANZ and ANZECC 2000). Table 5.7 demonstrate the detailed interpretation of correlation coefficient. The correlation coefficient, r_{xy} between two variables x and y with mean values $\bar{x}$ and $\bar{y}$ is determined by the following Eq. (5.6):

$$r_{xy} = \frac{\sum_{i=1}^{n}(x_i - \bar{x})(y_i - \bar{y})}{\sqrt{\sum_{i=1}^{n}\left(x_i - \bar{x}\right)^2 \left(y_i - \bar{y}\right)^2}} \tag{5.6}$$

5.4.3.2 Multivariate Statistics

A monitoring database consist of large number of sampling sites with numerous parameters for different time-period, which cannot be effectively evaluated by using mathematical models because of the interrelationship among the parameters, sampling sites and time period. Multivariate statistical method has overcome this limitation by giving a better interpretation of this type of complex database. The multivariate statistical approaches were successfully applied in a number of hydrogeo-chemical studies (Simeonov et al. 2003; Singh et al. 2005; Kowalkowski et al. 2006; Boyacioglu and Boyacioglu 2008). The basic multivariate statistical techniques, cluster analysis (CA) and factor analysis (FA) reduces a large data set into an interpretable form that reveals the relationship between the variables based on their similarities.

Cluster Analysis (CA) It is a multivariate hydrostatical technique with a primary aim to assemble the homogeneous set of correlated parameters in the configuration of clusters, which are based on the similar features of the parameter (Bhat et al. 2014). Hierarchical cluster analysis (HCA), non-hierarchical cluster analysis (non-HCA) and combination of both methods can be applied to the dataset for the analysis of results. The hierarchical cluster analysis confirms its application on any type of research question with a guarantee of optimum clustering of observations. However, separate interpretation is required for understanding the relation in each cluster in non-hierarchical cluster analysis, whereas a single interpretation will estimate the outcome of all clusters in hierarchical cluster analysis. The main objective of cluster analysis is to determine the data structure by clubbing the most similar observations into groups. Similarity measurement, cluster formation and group formation are the three basic steps for cluster analysis. Similarity measurement is executed by a set of rule, which exemplifies the level of correspondence among parameters used in the analysis. Correlational measure is relatively less used technique of similarity measurement, where correlation coefficients show the similarity between the parameters, whereas, the distance measure is the most popular method to assess similarity. Euclidean distance, Square Eucledian distance, Chebychev distance, etc. measure distance with specific characteristics. The most popular distance method used by researchers in the field of hydrology is Euclidean distance, which is expressed by the following Eq. (5.7).

$$d_{ij} = \sqrt{\left[\sum_{l=1}^{q} \left(x_{il} - x_{jl} \right)^2 \right]}$$ (5.7)

Where, d_{ij} = Euclidean distance for two individuals i and j, each measured on q variables, x_{ij}, x_{jl}, i = 1…….q.

Once the distance between observations are calculated, then the two most similar or closest observations are combined together to form cluster through an algorithm. The algorithm in hierarchical procedure is designed to move in stepwise mode. Further, the algorithm is categorized as divisive and agglomerative. Divisive algorithm begins with whole objects in one cluster, subsequently at each step divided into more clusters that comprise the most dissimilar objects and looped until it forms cluster 1 to n sub clusters, whereas, agglomerative hierarchical procedure begins where each observation is stated as a cluster and then combining the similar clusters successively until all observations are in a single cluster, i.e., from n sub clusters to 1 cluster. Average-linkage, centroid method, complete-linkage/farthest neighbor, single-linkage/nearest neighbor, wards method are various agglomerative algorithms. The most popular algorithm used in hydrology study is wards method, which sorts variables in an easily explainable clusters by using analysis of variance approach. It also evaluates the distance between clusters, which is calculated by the total sum of square deviations from the mean of a cluster (Shrestha and Kazama 2007; Abu-Khalaf et al. 2013). The graphical representation of clustering is tree-like structure dendrogram.

Factor Analysis (FA) It is another multivariate statistical technique that diminishes and classifies a large data set into a smaller data set. Confirmatory Factor Analysis (CFA) and Exploratory Factor Analysis (EFA) are the two major categories of factor analysis (DeCoster 1998; Thompson 2004). The exploratory factor analysis is most popular among researchers as it serves to explore the research questions or large information set to identify patterns (Swisher et al. 2004; Henson and Roberts 2006). The main steps involved in factor analysis are extraction of factors and selection of rotational methods. Data extraction step reduces large number of items into factors. The factors can be extracted through Generalized least square, maximum likelihood, principal axis factoring, principal component analysis and unweighted least square methods (Pett et al. 2003). The researchers recommend to use principal component method, when no prior theory or model exists to establish preliminary solution (Williams et al. 2010). The eigen values and eigen vectors known as loading or weightings in this process are computed from the covariance or correlation square matrix. The square matrix is constructed from the original data set (Raschka 2014). Hydrologist generally uses correlation matrix for displaying the relationships between individual parameters. Eigen vector with the highest eigen value is termed as principal component or factors. Statistically, principal component are termed as the unrelated or orthogonal variables obtained by multiplying original correlated variables with eigen vectors (Vega et al. 1998; Helena et al. 2000).Then feature vector is built by picking out the selected principal components from the list of eigen vectors obtained in the former steps. The final step of the method is to multiply the transpose matrix of eigen vector with the transpose of original dataset. The following Eq. (5.8) shows principal component generated through principal component analysis:

$$y_{ij} = a_{i1}x_{1j} + a_{i2}x_{2j} + a_{i3}x_{3j} + \ldots\ldots\ldots + a_{im}x_{mj} \qquad (5.8)$$

Where
   y = component score
   a = component loading
   x = measured value of the variable,
   i = component number
   j = sample number and
   m = total number of variables.

Component score sometimes called factor score is the value of transformed variable corresponding to a particular data point. The first principal component accounts for the maximum possible proportion of the total variance in the data set followed by the second component, which accounts for the maximum of the remaining variance and so on. Merely a few numbers of principal components with eigenvalue greater than one are held back in principal component analysis (Bhat et al. 2014). The selection of principal components with eigen value greater than one is known as Kaiser criteria. The number of principal components are less than or equal to the number of original variables (Shrestha and Kazama 2007; Krishna et al. 2009). To make the interpretation of the extracted factor more easier and reliable, rotational method is applied as proposed by Thurstone in 1947 and Cattell in 1978 (Osborne 2015). Orthogonal and oblique are the two common rotation techniques, where the former generate uncorrelated factors (the angle between the axes is 90°) and the latter produce correlated factors (the axes has different angle than 90°). Researchers on the ground of their interpretation skill choose orthogonal varimax or oblique albumin rotation (Abdi 2003). The most common rotational technique used in factor analysis is orthogonal varimax rotation developed by Sir Thompson (Thompson 2004). Most of the literature has cited orthogonal varimax rotation for data analysis because of the fact that uncorrelated factors are more easily interpretable (Panda et al. 2006; Yidana et al. 2008; Bhattacharyya et al. 2014).

5.4.4 Geospatial Assessment

Reporting water quality status depending only on field study is not reliable and economic at present time. Apart from this, future demand is to determine water quality status of an area that needs a high degree of flexibility of data management. One way to achieve flexibility is to use a highly automated and intelligent information system such as geographical information system (GIS). It facilitates in reducing the understanding gap between researchers as it also provides the visual summary of water quality status of an area. It is an integrated approach to evaluate the quality of water by assimilating data sets and statistical results with geographical locations. The hydrological, temporal and spectral data can be processed using geographical information system technology (Ahn and Chon 1999; Babiker et al. 2004; Usali and Ismail 2010). In term of assessing water quality, researchers has used this technique for modelling and mapping of thematic layers (Fritch et al. 2000; Gogu and Dassargues 2000; Khan and Moharana 2002; Asadi et al. 2007). Hydrologists collect discrete data from the field sampling, which need to convert into continuous surface for further interpretation. The spatial interpolation tool provided in the software interface helps in formation of continuous surface. The classification of interpolation methods are geometrical, statistical, basic functions and artificial neural network (Sárközy 1999). Various models has been developed and integrated in the software for conversion of vector data structure (data represented by point, line or polygon) to raster data structure (data represented as surface divided into regularly-spaced grid cells, known as pixels). The deterministic models use a mathematical function to predict unknown values and further use for creation of continuous layer of the studied parameter (Xie et al. 2004). It creates surface from measured points, based on either the extent of similarity or the degree of smoothing. Inverse distance weighting, natural neighbourhood interpolation, spline interpolation and rectangular interpolation are few mathematical based models. Apart from this, statistical techniques also help in interpolation with higher accuracy. The most popular geostatistical technique is krigging that finds the relation between the observations at measured points and autocorrect statistically (Childs 2004). It follows two steps: I) creation of semivariogram, which is a discrete mathematical function. This explains the degree of relationship between spatial random points (known as spatial autocorrelation); II) prediction of unknown values through statistical estimator. Ordinary kriging, simple kriging and universal kriging are the three different methods of krigging interpolation (Childs 2004). However, the major disadvantage of using krigging interpolation is that it does not pass through any of the point values and causes interpolated values to be higher or lower than real values. According to the requirement of the research, one of the mentioned techniques can be used for spatial interpolation, which simplifies the data analysis.

5.5 Key Findings

1. The pollutants enter into an aquatic system through natural (water-soil and water-rock interaction) and anthropogenic (industries, municipal, domestic, agriculture, atmospheric precipitation of the emissions) sources.
2. These pollutants result in physical, muscular, neurological degenerative processes that can cause acute and chronic diseases.
3. The researchers and various authorities have proposed protocol that details about the water quality of an area.
4. Water quality management plan starts with the sampling of the water and analytical analysis of the water quality parameters.
5. The analytical results of periodic monitoring of numerous samples generate huge data sets that is neither understandable by the general public nor regulatory authorities, nor represents the actual status of the water quality. Therefore, it draws an attention to manage the huge data set into an understandable format that can represent the water quality of an area.
6. Researchers and scientists have developed numerous interdisciplinary techniques to minimize the large data volume that also assists in evaluation of suitability of water for various purposes, health risk assessment and pollution source identification.
7. Suitability of water for drinking is checked through comparing the measured concentration of water quality parameters with the recommended standards or water quality index. Moreover, the suitability for irrigation and industrial use is evaluated through comparative analysis or various indices based on specific parameters. Even few mathematical models have devolved to assess the human health risk.
8. Many scientists have acknowledged that various bivariate and multivariate statistical approaches such as Pearson correlation, factor analysis, cluster analysis etc helps in pollution source apportionment.
9. Geographical information system integrates the data sets and assessment results with geographical locations (geospatial assessment) that provide the visual summary of water quality status in earth surface. It helps in documentation of effective water quality management plan.

5.6 Research Gap

(a) Water quality index helps in transformation of analytical data set into a single digit that indicates the overall water quality of a water body. However, the application of this index is limited for assessing drinking water quality only. Moreover, water quality for irrigation, industries and other purposes were determined by indices relying on specific parameters. These conventional approaches do not clearly specify the suitability of water, which requires attention towards formulating an effective tool for evaluation of water quality for various purposes.

(b) The indexing techniques as their primary objective predict the overall water quality status (as polluted or non-polluted) and its health impact (as unsafe or safe for human use). There is no such assessment technique available in the literature that quantifies the concentration of toxic pollutants present in the water body that are responsible for health hazard. Even, the existing assessment techniques do not estimate the required percentage of pollutants to be removed from water bodies in order to make these safe for human health. Therefore, there is a need to a technique, which aggregates the toxicity level of pollutants in water body responsible for health impact. This technique should also aims to predict the required removal percentage of individual pollutants from the water bodies.

This review illustrate that the statistical method, indexing approach and GIS added method effectively helps in analysis of both ground and surface water.

References

Abdi H (2003) Factor rotations in factor analyses. Encyclopedia for Research Methods for the Social Sciences. Sage, Thousand Oaks, pp 792–795

Abu-Khalaf N, Khayat S, Natsheh B (2013) Multivariate data analysis to identify the groundwater pollution sources in Tulkarm area/Palestine. Sci Technol 3(4):99–104. https://doi.org/10.5923/j.scit.20130304.01

Ahn HI, Chon HT (1999) Assessment of groundwater contamination using geographic information systems. Environ Geochem Health 21(3):273–289. https://doi.org/10.1023/A:1006697512090

Aitman TJ (2001) DNA microarrays in medical practice. BMJ 323(7313):611–615. https://doi.org/10.1136/bmj.323.7313.611

Akcil A, Koldas S (2006) Acid mine drainage (AMD): causes, treatment and case studies. J Clean Prod 14(12-13):1139–1145. https://doi.org/10.1016/j.jclepro.2004.09.006

Akoto O, Bruce TN, Darko D (2008) Heavy metals pollution profiles in streams serving the Owabi reservoir. Afr J Environ Sci Technol 2(11):354–359. IISN 1996-0786

Alloway BJ (1990) Soil processes and the behavior of metals. Heavy metals in soils, Blackie & Son Ltd. Press, UK, pp 7–27

Alimonti A, Petrucci F, Krachler M, Bocca B, Caroli S (2000) Reference values for chromium, nickel and vanadium in urine of youngsters from the urban area of Rome. J Environ Monit 2(4):351–354

Amadi AN, Yisa J, Ogbonnaya IC, Dan-Hassan MA, Jacob JO, Alkali YB (2012) Quality evaluation of river chanchaga using metal pollution index and principal component analysis. J Geograph Geol 4(2):13. https://doi.org/10.5539/jgg.v4n2p13

Ameh EG, Akpah FA (2011) Heavy metal pollution indexing and multivariate statistical evaluation of hydrogeochemistry of River PovPov in Itakpe Iron-Ore mining area, Kogi State, Nigeria. Adv Appl Sci Res 2(1):33–46

ARMCANZ ANZECC (2000) Australian guidelines for water quality monitoring and reporting. Natl Water Q Manag Strat Paper, 7

Asadi SS, Vuppala P, Reddy MA (2007) Remote sensing and GIS techniques for evaluation of groundwater quality in municipal corporation of Hyderabad (Zone-V), India. Intl J Environ Res Public Health 4(1):45–52. https://doi.org/10.3390/ijerph2007010008

Ascherio A, Rimm EB, Hernan MA, Giovannucci EL, Kawachi I, Stampfer MJ, Willett WC (1998) Intake of potassium, magnesium, calcium, and fiber and risk of stroke among US men. Circulation 98(12):1198–1204. https://doi.org/10.1161/01.CIR.98.12.1198

ATSDR (2000) Agency for toxic substances and disease registry. Toxicological profile for chromium. U.S Department of Health and Human Services, Washington, DC

Babiker IS, Mohamed MA, Terao H, Kato K, Ohta K (2004) Assessment of groundwater contamination by nitrate leaching from intensive vegetable cultivation using geographical information system. Environ Intl 29(8):1009–1017. https://doi.org/10.1016/S0160-4120(03)00095-3

Ball JW, Izbicki JA (2004) Occurrence of hexavalent chromium in groundwater in the western Mojave Desert, California. Appl Geochem 19(7):1123–1135. https://doi.org/10.1016/j.apgeochem.2004.01.011

Barałkiewicz D, Siepak J (1999) Chromium, nickel and cobalt in environmental samples and existing legal norms. Pol J Environ Stud 8(4):201–208

Basappa Reddy M (2003) Status of groundwater quality in Bangalore and its Environs. Report, Dept. of Mines and Geo, Bangalore

Bawaskar HS, Bawaskar PH, Bawaskar PH (2010) Chronic renal failure associated with heavy metal contamination of drinking water: a clinical report from a small village in Maharashtra. Clin Toxicol. https://doi.org/10.3109/15563650.2010.497763

Bent S, Böhm K (1995) Copper-induced liver cirrhosis in a 13-month old boy. Gesundheitswesen (Bundesverband der Arzte des Offentlichen Gesundheitsdienstes (Germany)) 57(10):667–669

Bhat SA, Pandit AK (2014) Surface water quality assessment of Wular Lake, a Ramsar site in Kashmir Himalaya, using discriminant analysis and WQI. J Ecosyst. https://doi.org/10.1155/2014/724728

Bhat SA, Meraj G, Yaseen S (2014) Pandit AK (2014) Statistical assessment of water quality parameters for pollution source identification in Sukhnag stream: an inflow stream of lake Wular (Ramsar Site), Kashmir Himalaya. J Ecosyst. https://doi.org/10.1155/2014/898054

Bhattacharya P, Mukherjee AB, Jacks G, Nordqvist S (2002) Metal contamination at a wood preservation site: characterisation and experimental studies on remediation. Sci Total Environ 290(1–3):165–180. https://doi.org/10.1016/s0048-9697(01)01073-7

Bhattacharyya R, Ojha SN, Singh U.K (2014) Risk prediction based of heavy metals distribution in groundwater of Durgapur using GIS techniques

Bhutiani R, Kulkarni DB, Khanna DR, Gautam A (2016) Water quality, pollution source apportionment and health risk assessment of heavy metals in groundwater of an industrial area in North India. Exp Health 8(1):3–18. https://doi.org/10.1007/s12403-015-0178-2

BIS (2012) Indian standard drinking water specifications IS 10500:2012. Bureau of Indian Standards, New Delhi

Block GA, Klassen PS, Lazarus JM, Ofsthun N, Lowrie EG, Chertow GM (2004) Mineral metabolism, mortality, and morbidity in maintenance hemodialysis. J Am Soc Nephrol 15(8):2208–2218. https://doi.org/10.1097/01.ASN.0000133041.27682.A2

Boateng TK, Opoku F, Acquaah SO, Akoto O (2015) Pollution evaluation, sources and risk assessment of heavy metals in hand-dug wells from Ejisu-Juaben Municipality, Ghana. Environ Syst Res 4(1):18. https://doi.org/10.1186/s40068-015-0045-y

Borrok DM, Nimick DA, Wanty RB, Ridley WI (2008) Isotopic variations of dissolved copper and zinc in stream waters affected by historical mining. Geochimica et Cosmochimica Acta 72(2):329–344. https://doi.org/10.1016/j.gca.2007.11.014

Bouchard MF, Sauvé S, Barbeau B, Legrand M, Brodeur MÈ, Bouffard T, Limoges E, Bellinger DC, Mergler D (2010) Intellectual impairment in school-age children exposed to manganese from drinking water. Environ Health Perspect 119(1):138–143. https://doi.org/10.2307/41000694

Boyacioglu H, Boyacioglu H (2008) Water pollution sources assessment by multivariate statistical methods in the Tahtali Basin, Turkey. Environ Geol 54(2):275–282. https://doi.org/10.1007/s00254-007-0815-6

Brown RM, McClelland NI, Deininger RA, O'Connor MF (1972) A water quality index—crashing the psychological barrier. In Indicators of environmental quality. Springer, Boston, pp 173–182. https://doi.org/10.1016/b978-0-08-017005-3.50067-0

Butterworth RF (2010) Metal toxicity, liver disease and neurodegeneration. Neurotox Res 18(1):100–105. https://doi.org/10.1007/s12640-010-9185-z

Carpenter SR, Caraco NF, Correll DL, Howarth RW, Sharpley AN, Smith VH (1998) Nonpoint pollution of surface waters with phosphorus and nitrogen. Ecol Appl 8(3):559–568. https://doi.org/10.1890/1051-0761(1998)008[0559:NPOSWW]2.0.CO;2

Caruso BS, Bishop M (2009) Seasonal and spatial variation of metal loads from natural flows in the upper Tenmile Creek watershed, Montana. Mine Water Environ 28(3):166–181. https://doi.org/10.1007/s10230-009-0073-9

Cempel M, Nikel G (2006) Nickel: a review of its sources and environmental toxicology. Pol J Environ Stud 15(3). https://doi.org/10.1109/TUFFC.2008.827

Chapman PM (2007) Determining when contamination is pollution—weight of evidence determinations for sediments and effluents. Environ Intl 33(4):492–501. https://doi.org/10.1016/j.envint.2006.09.001

CHEST (2003) Children's health and the environment, risk assessment, introduction to risk assessment. http://ec.europa.eu/health/ph_projects/2003/action3/docs/2003_3_09_a23_en.pdf. 18 December 2018

Chi RA, Xiao CQ, Gao H (2006) Bioleaching of phosphorus from rock phosphate containing pyrites by Acidithiobacillus ferrooxidans. Miner Eng 19(9):979–981. https://doi.org/10.1016/j.mineng.2005.10.003

Childs C (2004) Interpolating surfaces in ArcGIS spatial analyst. ArcUser 3235:569

Chouhan S, Flora SJS (2010) Arsenic and fluoride: two major ground water pollutants. ISSN: 00195189

Cotruvo J (2011) Hardness in drinking water, background Document for development of WHO guidelines for drinking water quality. World Health Organization, Geneva

CPCB (2007) Central Pollution Control Board, guidelines for water quality monitoring, MINARS/27/2007-08

Cude CG (2001) Oregon Water Quality Index a Tool for Evaluating Water Quality Management Effectiveness. JAWRA: J Am Water Resour Assoc 37(1):125–137. https://doi.org/10.1111/j.1752-1688.2001.tb05480.x

Daly RM, Ebeling PR (2010) Is excess calcium harmful to health? Nutrients 2(5):505–522. https://doi.org/10.3390/nu2050505

Darvishi G, Kootenaei FG, Ramezani M, Lotfi E, Asgharnia H (2016) Comparative investigation of river water quality by OWQI, NSFWQI and Wilcox indexes (case study: the Talar River–IRAN). Arch Environ Protect 42(1):41–48. https://doi.org/10.1515/aep-2016-0005

DeCoster J (1998) Overview of factor analysis. http://www.stat-help.com/notes.html. Accessed 15 Feb 2016

Doneen LD (1964) Notes on water quality in agriculture. Department of Water Science and Engineering. University of California, Davis

Dung TTT, Cappuyns V, Swennen R, Phung NK (2013) From geochemical background determination to pollution assessment of heavy metals in sediments and soils. Rev Environ Sci Bio/Technol 12(4):335–353. https://doi.org/10.1007/s11157-013-9315-1

Duruibe JO, Ogwuegbu MOC, Egwurugwu JN (2007) Heavy metal pollution and human biotoxic effects. Intl J Phys Sci 2(5):112–118. https://doi.org/10.1016/j.proenv.2011.09.146

Eaton FM (1950) Significance of carbonates in irrigation waters. Soil Sci 69(2):123–134. https://doi.org/10.1097/00010694-195002000-00004

Edmunds WM, Shand P, Hart P, Ward RS (2003) The natural (baseline) quality of groundwater: a UK pilot study. Sci Total Environ 310(1-3):25–35

EU (1998) Quality of water intended for human consumption. Council of the European Union

FAO (2018) Food and Agriculture Organization, Water Quality for Agriculture, http://www.fao. org/docrep/003/T0234E/T0234E01.htm

Farooqi A, Masuda H, Siddiqui R, Naseem M (2009) Sources of arsenic and fluoride in highly contaminated soils causing groundwater contamination in Punjab, Pakistan. Arch Environ Contam Toxicolgy 56(4):693–706. https://doi.org/10.1007/s00244-008-9239-x

Fawell J, Bailey K, Chilton J, Dahi E, Fewtrell L, Magara Y (2013) Fluoride in drinking-water. Water Intell Online 12. https://doi.org/10.1007/BF01783490

Fernandez-Luqueno F, López-Valdez F, Gamero-Melo P, Luna-Suárez S, Aguilera-González EN, Martínez AI, García-Guillermo MDS, Hernández-Martínez G, Herrera-Mendoza R, Álvarez-Garza MA, Pérez-Velázquez IR (2013) Heavy metal pollution in drinking water-a global risk for human health: a review. Afr J Environ Sci Technol 7(7):567–584. https://doi.org/10.5897/AJEST12.197

Fritch TG, Mcknight CL, Yelderman JC Jr, Arnold JG (2000) An aquifer vulnerability assessment of the Paluxy aquifer, central Texas, USA, using GIS and a modified DRASTIC approach. Environ Manag 25(3):337–345. https://doi.org/10.5897/AJEST12.197

Gerba CP (2000) Risk assessment. In: Gerba CP, Maier RM, Pepper IL (eds) Environmental microbiology. Academic, London, pp 557–571

Gibert J, Danielopol D, Stanford JA (1994) Groundwater ecology, vol 1. Academic, San Diego

Giri S, Singh AK (2014) Assessment of surface water quality using heavy metal pollution index in Subarnarekha River, India. Water Qual Expo Health 5(4):173–182. https://doi.org/10.1007/s12403-013-0106-2

Giri S, Singh AK (2015) Human health risk assessment via drinking water pathway due to metal contamination in the groundwater of Subarnarekha River Basin, India. Environ Monit Assess 187(3):63. https://doi.org/10.1007/s10661-015-4265-4

Giri S, Singh G, Gupta SK, Jha VN, Tripathi RM (2010) An evaluation of metal contamination in surface and groundwater around a proposed uranium mining site, Jharkhand, India. Mine Water Environ 29(3):225–234. https://doi.org/10.1007/s10230-010-0107-3

Gleick PH (1993) Water in crisis: a guide to the worlds fresh water resources. Pacific Institute for Studies in Dev., Environment & Security, Stockholm Env. Institute. Oxford Univ. Press, Oxford. 473p, 9

Gogu RC, Dassargues A (2000) Current trends and future challenges in groundwater vulnerability assessment using overlay and index methods. Environ Geol 39(6):549–559. https://doi.org/10.1007/s002540050466

Gowd SS, Govil PK (2008) Distribution of heavy metals in surface water of Ranipet industrial area in Tamil Nadu, India. Environ Monit Assess 136(1-3):197–207. https://doi.org/10.1007/s10661-007-9675-5

Helena B, Pardo R, Vega M, Barrado E, Fernandez JM, Fernandez L (2000) Temporal evolution of groundwater composition in an alluvial aquifer (Pisuerga River, Spain) by principal component analysis. Water Res 34(3):807–816. https://doi.org/10.1016/s0043-1354(99)00225-0

Henson RK, Roberts JK (2006) Use of exploratory factor analysis in published research: common errors and some comment on improved practice. Educ Psychol Measure 66(3):393–416. https://doi.org/10.1177/0013164405282485

Hofman A, Valkenburg HA, Vaandrager GJ (1980) Increased blood pressure in schoolchildren related to high sodium levels in drinking water. J Epidemiol Communy Health 34(3):179–181. https://doi.org/10.1136/jech.34.3.179

Holdgate MW (1980) A perspective of environmental pollution. Cambridge University Press Archive, Cambridge

Horton RK (1965) An index number system for rating water quality. J Water Pollut Contrl Federation 37(3):300–306

Hoseinzadeh E, Yusefzadeh A, Rahimi N, Khorsandi H (2013) Evaluation of corrosion and scaling potential of a water treatment plant. Arch Hyg Sci 2(2):41–47. http://jhygiene.muq.ac.ir/article-1-11-en.html

Howe P, Malcolm H, Dobson S (2004) Manganese and its compounds: environmental aspects (No. 63). World Health Organization

Hsu SC, Wong GT, Gong GC, Shiah FK, Huang YT, Kao SJ, Tsai F, Lung SCC, Lin FJ, Lin II, Hung CC (2010) Sources, solubility, and dry deposition of aerosol trace elements over the East China Sea. Marine Chem 120(1-4):116–127. https://doi.org/10.1016/j.marchem.2008.10.003

Huang W, Duan D, Zhang Y, Cheng H, Ran Y (2014) Heavy metals in particulate and colloidal matter from atmospheric deposition of urban Guangzhou, South China. Marine Pollut Bull 85(2):720–726. https://doi.org/10.1016/j.marpolbul.2013.12.041

Huffmeyer N, Klasmeier J, Matthies M (2009) Geo-referenced modeling of zinc concentrations in the Ruhr river basin (Germany) using the model GREAT-ER. Sci Total Environ 407(7):2296–2305. https://doi.org/10.1016/j.scitotenv.2008.11.055

Hwang JY, Park S, Kim HK, Kim MS, Jo HJ, Kim JI, Lee GM, Shin IK, Kim TS (2016) Hydrochemistry for the assessment of groundwater quality in Korea. J Agric Chem Environ 6(01):1.. ronment, 6, 1–29. https://doi.org/10.4236/jacen.2017.61001

Ipeaiyeda AR, Dawodu M (2008) Heavy metals contamination of topsoil and dispersion in the vicinities of reclaimed. Bull Chem Soc Ethiopia 22(3). https://doi.org/10.4314/bcse.v22i3.61205

Jarup L (2003) Hazards of heavy metal contamination. Br Med Bull 6:167–182. https://doi.org/10.1093/bmb/ldg032

Jarvie HP, Whitton BA, Neal C (1998) Nitrogen and phosphorus in east coast British rivers: speciation, sources and biological significance. Sci Total Environ 210:79–109. https://doi.org/10.1016/s0048-9697(98)00109-0

Johnson DB, Hallberg KB (2005) Acid mine drainage remediation options: a review. Sci Total Environ 338(1-2):3–14. https://doi.org/10.1016/j.scitotenv.2004.09.002

Jones DC, Miller GW (2008) The effects of environmental neurotoxicants on the dopaminergic system: a possible role in drug addiction. Biochem Pharmacol 76(5):569–581. https://doi.org/10.1016/j.bcp.2008.05.010

Kang J, Choi MS, Yi HI, Song YH, Lee D, Cho JH (2011) A five-year observation of atmospheric metals on Ulleung Island in the East/Japan Sea: temporal variability and source identification. Atmos Environ 45(25):4252–4262. https://doi.org/10.1016/j.atmosenv.2011.04.083

Karbassi AR, Nouri J, Ayaz GO (2007) Flocculation of trace metals during mixing of Talar river water with Caspian Seawater. Intl J Environ Res 1(1):66–73. https://doi.org/10.5923/j.scit.20130304.01

Karbassi AR, Monavari SM, Bidhendi GRN, Nouri J, Nematpour K (2008) Metal pollution assessment of sediment and water in the Shur River. Environ Monit Assess 147(1-3):107. https://doi.org/10.1007/s10661-007-0102-8

Katyal D (2011) Water quality indices used for surface water vulnerability assessment. Intl J Environ Sci 2(1):154. Katyal, Deeksha, Water Quality Indices Used for Surface Water Vulnerability Assessment (June 11, 2011). International Journal of Environmental Sciences, Volume 2, No. 1, September 2011. https://ssrn.com/abstract=2160726

Kaur T, Bhardwaj R, Arora S (2017) Assessment of groundwater quality for drinking and irrigation purposes using hydrochemical studies in Malwa region, southwestern part of Punjab, India. Appl Water Sci 7(6):3301–3316. https://doi.org/10.1007/s13201-016-0476-2

Khan MA, Moharana PC (2002) Use of remote sensing and geographical information system in the delineation and characterization of ground water prospect zones. J Indian Soc Remote Sens 30(3):131–141. https://doi.org/10.1007/bf02990645

Kotaś J, Stasicka Z (2000) Chromium occurrence in the environment and methods of its speciation. Environ Pollut 107(3):263–283. https://doi.org/10.1016/s0269-7491(99)00168-2

Kowalkowski T, Zbytniewski R, Szpejna J, Buszewski B (2006) Application of chemometrics in river water classification. Water Res 40(4):744–752

Kowalski A, Siepak M, Boszke L (2007) Mercury contamination of surface and ground waters of Poznań, Poland. Pol J Environ Stud 16(1)

Krishna AK, Satyanarayanan M, Govil PK (2009) Assessment of heavy metal pollution in water using multivariate statistical techniques in an industrial area: a case study from Patancheru, Medak District, Andhra Pradesh, India. J Hazard Mater 167(1-3):366–373

Kumar SS, Suriyanarayanan A, Panigrahi BS (2012) Studies on performance of indices in cooling water system. ISSN: 0975-0991

Lantzy RJ, Mackenzie FT (1979) Atmospheric trace metals: global cycles and assessment of man's impact. Geochimica et Cosmochimica Acta 43(4):511–525. https://doi.org/10.1016/0016-7037(79)90162-5

Li P, Feng XB, Qiu GL, Shang LH, Li ZG (2009) Mercury pollution in Asia: a review of the contaminated sites. J Hazard Mater 168(2-3):591–601. https://doi.org/10.1016/j.jhazmat.2009.03.031

Li Y, Yang R, Zhang A, Wang S (2014) The distribution of dissolved lead in the coastal waters of the East China Sea. Marine Pollut Bull 85(2):700–709. https://doi.org/10.1016/j.marpolbul.2014.02.010

Liu CW, Lin KH, Kuo YM (2003) Application of factor analysis in the assessment of groundwater quality in a blackfoot disease area in Taiwan. Sci Total Environ 313(1-3):77–89. https://doi.org/10.1016/S0048-9697(02)00683-6

Lohani MB, Singh A, Rupainwar DC, Dhar DN (2008) Seasonal variations of heavy metal contamination in river Gomti of Lucknow city region. Environ Monit Assess 147(1-3):253–263. https://doi.org/10.1007/s10661-007-0117-1

Lumb A, Halliwell D, Sharma T (2006) Application of CCME Water Quality Index to monitor water quality: a case study of the Mackenzie River basin, Canada. Environ Monit Assess 113(1-3):411–429. https://doi.org/10.1007/s10661-005-9092-6

Lumb A, Sharma TC, Bibeault JF (2011) A review of genesis and evolution of water quality index (WQI) and some future directions. Water Q Exp Health 3(1):11–24. https://doi.org/10.1007/s12403-011-0040-0

Madsen H, Poulsen L, Grandjean P (1990) Risk of high copper content in drinking water. Ugeskrift for Laeger 152(25):1806–1809. ISSN: 0041-5782

Mebrahtu G, Zerabruk S (2011) Concentration and health implication of heavy metals in drinking water from urban areas of Tigray region, Northern Ethiopia. Momona Ethiopian J Sci 3(1):105–121. https://doi.org/10.4314/mejs.v3i1.63689

Mercer JF (2001) The molecular basis of copper-transport diseases. Trends Mol Med 7(2):64–69. https://doi.org/10.1016/S1471-4914(01)01920-7

Mirenda RJ (1986) Acute toxicity and accumulation of zinc in the crayfish, Orconectes virilis (Hagen). Bull Environ Contam Toxicol 37(1):387–394. https://doi.org/10.1007/BF01607778

Mishra S, Dwivedi SP, Singh RB (2010) A review on epigenetic effect of heavy metal carcinogens on human health. Open Nutraceut J 3:188–193. https://doi.org/10.2174/1876396001003010188

Mohan SV, Nithila P, Reddy SJ (1996) Estimation of heavy metals in drinking water and development of heavy metal pollution index. J Environ Sci Health Part A 31(2):283–289. https://doi.org/10.2174/1876396001003010188

Mohod CV, Dhote J (2013) Review of heavy metals in drinking water and their effect on human health. Int J Innov Res Sci Eng Technol 2(7):2992–2996. ISSN: 2319-8753

Mondal NC, Saxena VK, Singh VS (2005) Assessment of groundwater pollution due to tannery industries in and around Dindigul, Tamilnadu, India. Environ Geol 48(2):149–157. https://doi.org/10.1007/s00254-005-1244-z

Moses EA, Etuk BA (2015) Human health risk assessment of trace metals in water from Qua iboe river estuary, Ibeno, Nigeria. J Environ Occup Sci 4(3):151. https://doi.org/10.5455/jeos.20150714122504

Muhammad S, Shah MT, Khan S (2011) Health risk assessment of heavy metals and their source apportionment in drinking water of Kohistan region, northern Pakistan. Microchem J 98(2):334–343. https://doi.org/10.1016/j.microc.2011.03.003

Mukherjee A, Sengupta MK, Hossain MA, Ahamed S, Das B, Nayak B, Lodh D, Rahman MM, Chakraborti D (2006) Arsenic contamination in groundwater: a global perspective with emphasis on the Asian scenario. J Health Popul Nutr. ISSN: 16060997

Mullaney JR, Lorenz DL, Arntson AD (2009) Chloride in groundwater and surface water in areas underlain by the glacial aquifer system, northern United States. US Geological Survey, Reston, p 41

Murray KE, Thomas SM, Bodour AA (2010) Prioritizing research for trace pollutants and emerging contaminants in the freshwater environment. Environ Pollut 158(12):3462–3471. https://doi.org/10.1016/j.envpol.2010.08.009

Nagajyoti PC, Lee KD, Sreekanth TVM (2010) Heavy metals, occurrence and toxicity for plants: a review. Environ Chem Lett 8(3):199–216. https://doi.org/10.1007/s10311-010-0297-8

Naik PK, Dehury BN, Tiwari AN (2007) Groundwater pollution around an industrial area in the coastal stretch of Maharashtra state, India. Environ Monit Assess 132(1-3):207–233. https://doi.org/10.1007/s10661-006-9529-6

Nalawade PM, Bholay AD, Mule MB (2012) Assessment of groundwater and surface water quality indices for heavy metals nearby area of Parli thermal power plant. Univ J Environ Res Technol 2(1):47–51. ISSN: 2249-0256

Navoni JA, De Pietri D, Olmos V, Gimenez C, Mitre GB, De Titto E, Lepori EV (2014) Human health risk assessment with spatial analysis: study of a population chronically exposed to arsenic through drinking water from Argentina. Sci Total Environ 499:166–174. https://doi.org/10.1016/j.scitotenv.2014.08.058

NIH (2016) U.S National Library of Medicine, Chloride in diet. Available: https://www.nlm.nih.gov/medlineplus/ency/article/002417.htm, 17 February 2016

Nordstrom DK (2002) Worldwide occurrences of arsenic in ground water. https://doi.org/10.1007/b101867

Notcutt G, Davies F (2001) Environmental accumulation of airborne fluorides in Romania. Environ Geochem Health 23(1):43–51. https://doi.org/10.1023/A:1011062115049

Nriagu JO (1989) A global assessment of natural sources of atmospheric trace metals. Nature 338(6210):47. https://doi.org/10.1038/338047a0

Olajire AA, Imeokparia FE (2001) Water quality assessment of Osun River: studies on inorganic nutrients. Environ Monit Assess 69(1):17–28. https://doi.org/10.1023/A:1010796410829

Olías M, Nieto JM, Sarmiento AM, Cerón JC, Cánovas CR (2004) Seasonal water quality variations in a river affected by acid mine drainage: the Odiel River (South West Spain). Sci Total Environ 333(1-3):267–281. https://doi.org/10.1016/j.scitotenv.2004.05.012

Osborne JW (2015) What is rotating in exploratory factor analysis. Pract Assess Res Eval 20(2):1–7. ISSN 1531-7714

Owlad M, Aroua MK, Daud WAW, Baroutian S (2009) Removal of hexavalent chromium-contaminated water and wastewater: a review. Water Air Soil Pollut 200(1-4):59–77. https://doi.org/10.1007/s11270-008-9893-7

Panda UC, Sundaray SK, Rath P, Nayak BB, Bhatta D (2006) Application of factor and cluster analysis for characterization of river and estuarine water systems–a case study: Mahanadi River (India). J Hydrol 331(3-4):434–445. https://doi.org/10.1016/j.jhydrol.2006.05.029

Parker RL, Fleischer M (1967) Data of geochemistry. US Government Printing Office, Washington, DC, p 19

Patel P, Raju NJ, Reddy BSR, Suresh U, Gossel W, Wycisk P (2016) Geochemical processes and multivariate statistical analysis for the assessment of groundwater quality in the Swarnamukhi River basin, Andhra Pradesh, India. Environ Earth Sci 75(7):611. https://doi.org/10.1007/s12665-015-5108-x

Peter S, Sreedevi C (2012) Water Quality Assessment and GIS mapping of ground water around KMML industrial area, Chavara. In: Green Technologies (ICGT), 2012 International Conference on, pp 117–123. IEEE. https://doi.org/10.1109/ICGT.2012.6477958

Pett MA, Lackey NR, Sullivan JJ (2003) Making sense of factor analysis: the use of factor analysis for instrument development in health care research. Sage, Thousand Oaks

Popovic A, Djordjevic D, Polic P (2001) Trace and major element pollution originating from coal ash suspension and transport processes. Environ Intl 26(4):251–255. https://doi.org/10.1016/S0160-4120(00)00114-8

Post JE (1999) Manganese oxide minerals: crystal structures and economic and environmental significance. Proc Natl Acad Sci 96(7):3447–3454. https://doi.org/10.1073/pnas.96.7.3447

Purandara BK, Varadarajan N (2003) Impacts on groundwater quality by urbanization. J Indian Water Resour Soc 23(4):107–115

Qadir A, Malik RN, Husain SZ (2008) Spatio-temporal variations in water quality of Nullah Aik-tributary of the river Chenab, Pakistan. Environ Monit Assess 140(1–3):43–59. https://doi.org/10.1007/s10661-007-9846-4

Rahman MAT, Saadat AHM, Islam MS, Al-Mansur MA, Ahmed S (2017) Groundwater characterization and selection of suitable water type for irrigation in the western region of Bangladesh. Appl Water Sci 7(1):233–243. https://doi.org/10.1007/s13201-014-0239-x

Rakotondrabe F, Ngoupayou JRN, Mfonka Z, Rasolomanana EH, Abolo AJN, Ako AA (2018) Water quality assessment in the Betare-Oya gold mining area (East-Cameroon): multivariate statistical analysis approach. Sci Total Environ 610:831–844. https://doi.org/10.1016/j.scitotenv.2017.08.080

Ramakrishnaiah CR, Sadashivaiah C, Ranganna G (2009) Assessment of water quality index for the groundwater in Tumkur Taluk, Karnataka State, India. J Chem 6(2):523–530. https://doi.org/10.1155/2009/757424

Raman N, Narayanan DS (2008) Impact of solid waste effect on ground water and soil quality nearer to Pallavaram solid waste landfill site in Chennai. Rasayan J Chem 1(4):828–836. ISSN: 09760083

Raschka S (2014) Implementing a Principal Component Analysis (PCA) in Python step by step

Ravichandran S (2003) Hydrological influences on the water quality trends in Tamiraparani Basin, South India. Environ Monit Assess 87(3):293–309. https://doi.org/10.1023/A:1024818204664

Ravikumar P, Mehmood MA, Somashekar RK (2013) Water quality index to determine the surface water quality of Sankey tank and Mallathahalli lake, Bangalore urban district, Karnataka, India. Appl Water Sci 3(1):247–261. https://doi.org/10.1007/s13201-013-0077-2

Razo I, Carrizales L, Castro J, Díaz-Barriga F, Monroy M (2004) Arsenic and heavy metal pollution of soil, water and sediments in a semi-arid climate mining area in Mexico. Water Air Soil Pollut 152(1–4):129–152. https://doi.org/10.1023/B:WATE.0000015350.14520.c1

Reza R, Singh G (2010a) Assessment of ground water quality status by using water quality index method in Orissa, India. World Appl Sci J 9(12):1392–1397. ISSN 1818-4952

Reza R, Singh G (2010b) Heavy metal contamination and its indexing approach for river water. Intl J Environ Sci Technol 7(4):785–792. https://doi.org/10.1007/BF03326187

Richards LA (1954) Diagnosis and improvement of saline and alkali soils. LWW 78(2):154

Rim-Rukeh A, Ikhifa OG, Okokoyo AP (2006) Effects of agricultural activities on the water quality of Orogodo River, Agbor Nigeria. J Appl Sci Res 2(5):256–259

Ritz E, Hahn K, Ketteler M, Kuhlmann MK, Mann J (2012) Phosphate additives in food—a health risk. Dtsch Ärztebl Int 109(4):49. https://doi.org/10.3238/arztebl.2012.0049

Robles-Camacho J, Armienta MA (2000) Natural chromium contamination of groundwater at Leon Valley, Mexico. J Geochem Explor 68(3):167–181. https://doi.org/10.1016/S0375-6742(99)00083-7

Roy S (1970) Manganese-bearing silicate minerals from metamorphosed manganese formations of India. I. Juddite. Mineral Mag 37(290):708–716. https://doi.org/10.1180/minmag.1970.037.290.09

Saha P, Paul B (2018a) Assessment of heavy metal toxicity related with human health risk in the surface water of an industrialized area by a novel technique. Hum Ecol Risk Assess Int J 1–22. https://doi.org/10.1080/10807039.2018.1458595

Saha P, Paul B (2018b) Suitability assessment of surface water quality with reference to drinking, irrigation and fish culture: a human health risk perspective. Bull Environ Contam Toxicol 101(2):262–271. https://doi.org/10.1007/s00128-018-2389-2

Salem HM, Eweida AE, Farag A (2000) Heavy metals in drinking water and their environmental impact on human health. ICEHM, Cairo University, Egypt, pp 542–556

Sarkar A, Paul B (2016) The global menace of arsenic and its conventional remediation – a critical review. Chemosphere 158:37–49. https://doi.org/10.1016/j.chemosphere.2016.05.043

Sárközy F (1999) GIS functions-interpolation. Periodica Polytechnica Civ Eng 43(1):63–87

Sawka MN (2005) Dietary reference intakes for water, potassium, sodium, chloride, and sulfate. Chapter 4-Water (No. MISC04-05). Army Research Institute of Environmental Medicine Natick Ma

Sekabira K, Origa HO, Basamba TA, Mutumba G, Kakudidi E (2010) Assessment of heavy metal pollution in the urban stream sediments and its tributaries. Intl J Environ Sci Technol 7(3):435–446. https://doi.org/10.1007/BF03326153

Shankar BS, Balasubramanya N, Reddy MM (2008) Impact of industrialization on groundwater quality–a case study of Peenya industrial area, Bangalore, India. Environ Monit Assess 142(1-3):263–268. https://doi.org/10.1007/s10661-007-9923-8

Shivasharanappa Srinivas P (2013) Studies on seasonal variation of groundwater quality using multivariate analysis for BIDAR urban and its industrial area (Karnataka state, India). Int Res Eng Technol, IC – RICE Conference, pp 252–260. https://www.academia.edu/7710849/STUDIES_ON_SEASONAL_VARIATION_OF_GROUND_WATER_QUALITY_USING_MULTIVARIATE_ANALYSIS_FOR_BIDAR_URBAN_and_ITS_INDUSTRIAL_AREA_KARNATAKA-STATE_INDIA

Shrestha S, Kazama F (2007) Assessment of surface water quality using multivariate statistical techniques: a case study of the Fuji river basin, Japan. Environ Modell Software 22(4):464–475. https://doi.org/10.1016/j.envsoft.2006.02.001

Simeonov V, Stratis JA, Samara C, Zachariadis G, Voutsa D, Anthemidis A, Sofoniou M, Kouimtzis T (2003) Assessment of the surface water quality in Northern Greece. Water Res 37(17):4119–4124. https://doi.org/10.1016/s0043-1354(03)00398-1

Simsek C, Gunduz O (2007) IWQ index: a GIS-integrated technique to assess irrigation water quality. Environ Monit Assess 128(1-3):277–300. https://doi.org/10.1007/s10661-006-9312-8

Singh G, Kamal RK (2015) Assessment of groundwater quality in the mining areas of Goa, India. Indian J Sci Technol 8(6):588–595. ISSN: 0974-5645

Singh KP, Malik A, Sinha S (2005) Water quality assessment and apportionment of pollution sources of Gomti river (India) using multivariate statistical techniques—a case study. Analytica Chimica Acta 538(1-2):355–374. https://doi.org/10.1016/j.aca.2005.02.006

Singh AK, Mondal GC, Kumar S, Singh TB, Tewary BK, Sinha A (2008) Major ion chemistry, weathering processes and water quality assessment in upper catchment of Damodar River basin, India. Environ Geol 54(4):745–758. https://doi.org/10.1007/s00254-007-0860-1

Sirajudeen J, Manikandan SA, Manivel V (2013) Water quality index of ground water around Ampikapuram area near Uyyakondan channel Tiruchirappalli District, Tamil Nadu, India. Arch Appl Sci Res 5(3):21–26. ISSN : 0975-508X

Slack RJ, Gronow JR, Voulvoulis N (2005) Household hazardous waste in municipal landfills: contaminants in leachate. Sci Total Environ 337(1-3):119–137. https://doi.org/10.1016/j.scitotenv.2004.07.002

Smedley PL, Kinniburgh DG (2002) A review of the source, behaviour and distribution of arsenic in natural waters. Appl Geochem 17(5):517–568. https://doi.org/10.1016/s0883-2927(02)00018-5

Smith AH, Steinmaus CM (2009) Health effects of arsenic and chromium in drinking water: recent human findings. Ann Rev Public Health 30:107–122. https://doi.org/10.1146/annurev.publhealth.031308.100143

Spandana MP, Suresh KR, Prathima B (2013) Developing an irrigation water quality index for Vrishabavathi command area. Int J Eng Res Technol 2:821–830

Srivastava SK, Ramanathan AL (2008) Geochemical assessment of groundwater quality in vicinity of Bhalswa landfill, Delhi, India, using graphical and multivariate statistical methods. Environ Geol 53(7):1509–1528. https://doi.org/10.1007/s00254-007-0762-2

Stanton JM (2001) Galton, Pearson, and the peas: a brief history of linear regression for statistics instructors. J Stat Educ 9(3). https://doi.org/10.1080/10691898.2001.11910537

Subrahmanyam K, Yadaiah P (2001) Assessment of the impact of industrial effluents on water quality in Patancheru and environs, Medak district, Andhra Pradesh, India. Hydrogeol J 9(3):297–312. https://doi.org/10.1007/s100400000120

Sujitha K, Shankaret R, Gajendran T, Vasanthi B (2014) Invitro reduction, kinetic modelling and Optimisation of parameters for biosorption of Cr (VI) using an ecological sorbent. IOSR J Environ Sci Toxicol Food Technol 8(6):1–7. https://doi.org/10.9790/2402-08610107

Swisher LL, Beckstead JW, Bebeau MJ (2004) Factor analysis as a tool for survey analysis using a professional role orientation inventory as an example. Phys Ther 84(9):784–799. https://doi.org/10.1093/ptj/84.9.784

Terry PA, Stone W (2002) Biosorption of cadmium and copper contaminated water by Scenedesmus abundans. Chemosphere 47(3):249–255. https://doi.org/10.1016/s0045-6535(01)00303-4

The New York Times (2016) Sodium in diet. Health guide. Available: http://www.nytimes.com/health/guides/nutrition/sodium-in-diet/overview.html, 16 May, 2016

Thompson B (2004) Exploratory and confirmatory factor analysis: Understanding concepts and applications. American Psychological Association. ISBN 1-59147-093-5

Tiwari AK, Singh PK, Mahato MK (2014) GIS-based evaluation of water quality index of ground water resources in West Bokaro Coalfield, India. Curr World Environ 9(3):843. https://doi.org/10.12944/CWE.9.3.35

Toxicologicalschools (2018) Dose-response assessment. http://www.toxicologyschools.com/free_toxicology_course1/a63.htm, December 2018

Tripathee L, Kang S, Sharma CM, Rupakheti D, Paudyal R, Huang J, Sillanpää M (2016) Preliminary health risk assessment of potentially toxic metals in surface water of the Himalayan Rivers, Nepal. Bull Environ Contam Ttoxicol 97(6):855–862. https://doi.org/10.1007/s00128-016-1945-x

Tyagi S, Sharma B, Singh P, Dobhal R (2013) Water quality assessment in terms of water quality index. Am J Water Resour 1(3):34–38. https://doi.org/10.12691/ajwr-1-3-3

Usali N, Ismail MH (2010) Use of remote sensing and GIS in monitoring water quality. J Sustain Dev 3(3):228. https://doi.org/10.5539/jsd.v3n3p228

USEPA (2003) US Environmental Protection Agency, Health effects support document for manganese. EPA822R03003. U.S. Environmental Protection Agency, Washington, DC

USEPA (2009) National primary drinking water regulations. United States Environmental Protection Agency

USGAO (2000) Health effect of lead in drinking water, U.S. General Accounting Office Reports

USGS (2016) How much water is on earth?. http://water.usgs.gov/edu/gallery/global-water-volume.html

Vaheedunnisha D, Shukla DSK (2013) Water quality assessment of RoopSagarPond of SatnaUsing NSF-WQI. Intl J Innovat Res Sci Eng Technol 2(5):1386–1388. ISSN: 2319-8753

Vega M, Pardo R, Barrado E, Debán L (1998) Assessment of seasonal and polluting effects on the quality of river water by exploratory data analysis. Water Res 32(12):3581–3592. https://doi.org/10.1016/s0043-1354(98)00138-9

Viers J, Oliva P, Nonell A, Gélabert A, Sonke JE, Freydier R, Gainville R, Dupré B (2007) Evidence of Zn isotopic fractionation in a soil–plant system of a pristine tropical watershed (Nsimi, Cameroon). Chem Geol 239(1-2):124–137. https://doi.org/10.1016/j.chemgeo.2007.01.005

Vine JD, Tourtelot EB (1970) Geochemistry of black shale deposits; a summary report. Econ Geol 65(3):253–272. https://doi.org/10.2113/gsecongeo.65.3.253

Visnjic-Jeftic Z, Jaric I, Jovanovic L, Skoric S, Smederevac-Lalic M, Nikcevic M, Lenhardt M (2010) Heavy metal and trace element accumulation in muscle, liver and gills of the Pontic shad (Alosa immaculata Bennet 1835) from the Danube River (Serbia). Microchem J 95(2):341–344. https://doi.org/10.1016/j.microc.2010.02.004

Von Ehrenstein OS, Mazumder DG, Hira-Smith M, Ghosh N, Yuan Y, Windham G, Das S (2006) Pregnancy outcomes, infant mortality, and arsenic in drinking water in West Bengal, India. Am J Epidemiol 163(7):662–669. https://doi.org/10.1093/aje/kwj089

Wang S, Mulligan CN (2006) Occurrence of arsenic contamination in Canada: sources, behavior and distribution. Sci Total Environ 366(2-3):701–721. https://doi.org/10.1016/j. scitotenv.2005.09.005

Wang Q, Kim D, Dionysiou DD, Sorial GA, Timberlake D (2004) Sources and remediation for mercury contamination in aquatic systems—a literature review. Environ Pollut 131(2):323–336. https://doi.org/10.1016/j.envpol.2004.01.010

Warner NR, Jackson RB, Darrah TH, Osborn SG, Down A, Zhao K, White A, Vengosh A (2012) Geochemical evidence for possible natural migration of Marcellus Formation brine to shallow aquifers in Pennsylvania. Proc Natl Acad Sci 109(30):11961–11966. https://doi.org/10.1073/pnas.1121181109

Webb M (1979) The geochemistry, biochemistry and biology of cadmium. Elsevier/Noyth Holland Biomedical Press, Amesterdam

Wedepohl KH (1995) The composition of the continental crust. Geochimica et cosmochimica Acta 59(7):1217–1232. https://doi.org/10.1007/bf01829361

Wilcox L (1955) Classification and use of irrigation waters

Williams B, Onsman A, Brown T (2010) Exploratory factor analysis: a five-step guide for novices. Australasian J Paramed 8(3). https://doi.org/10.33151/ajp.8.3.93

Wong CSC, Li XD, Zhang G, Qi SH, Peng XZ (2003) Atmospheric deposition of heavy metals in the Pearl River Delta, China. Atmos Environ 37(6):767–776. https://doi.org/10.1016/s1352-2310(02)00929-9

World Health Organisation (WHO) (1986) Diseases caused by manganese and its toxic compounds. Early detection of occupational diseases. World Health Organization, Geneva, pp 69–73

World Health Organisation (WHO) (2009) Calcium and magnesium in drinking-water: public health significance. World Health Organization, Geneva

World Health Organisation (WHO) (2011) Guidelines for drinking water quality, 4th edn. World Health Organisation, Geneva

World Health Organisation (WHO) (2012) Effect of increased potassium intake on cardiovascular disease, coronary heart disease and stroke. World Health Organization, Geneva

World Health Organisation (WHO) (2016) Water health Sanitation, http://www.who.int/water_sanitation_health/diseases/burden/en/

Xie M, Esaki T, Zhou G (2004) GIS-based probabilistic mapping of landslide hazard using a three-dimensional deterministic model. Nat Hazards 33(2):265–282. https://doi.org/10.1023/b:nhaz .0000037036.01850.0d

Yammani SR, Reddy TVK, Reddy MRK (2008) Identification of influencing factors for groundwater quality variation using multivariate analysis. Environ Geol 55(1):9–16. https://doi.org/10.1007/s00254-007-0958-5

Yang CY (1998) Calcium and magnesium in drinking water and risk of death from cerebrovascular disease. Stroke 29(2):411–414. https://doi.org/10.1161/01.STR.29.2.411

Yidana SM, Ophori D, Banoeng-Yakubo B (2008) A multivariate statistical analysis of surface water chemistry data—the Ankobra Basin, Ghana. J Environ Manag 86(1):80–87. https://doi.org/10.1016/j.jenvman.2006.11.023

Zanoni AE (1972) Ground-water pollution and sanitary landfills—a critical review. Groundwater 10(1):3–16. https://doi.org/10.1111/j.1745-6584.1972.tb02895.x

Chapter 6
Effect of Emerging Contaminants on Crops and Mechanism of Toxicity

Bansh Narayan Singh, Akash Hidangmayum, Ankita Singh, Akankhya Guru, Bhudeo Rana Yashu, and Gopal Shankar Singh

Abstract Human activities have induced the rise of emerging pollutants in the environment, and, as a consequence, have increased the pressure on food production. This pressure is acute in developing countries where food demand exceeds the available resources. Pollutants are mainly of industrial origin and medical discharges. Pollutants can be translocated from soil to plants, and in turn be transferred to consumers. Pollutants include steroidal estrogens, organochlorine, ibuprofen, bromoform, caffeine, and methyl dihydrojasmonate. The concentration of pollutants in irrigated agricultural waters ranges from 10 to 5000 ng L^{-1}. The average concentration of pollutants in crops ranges from 1 to 7500 ng kg^{-1}. Approximately 10 μg of pollutant per person were detected in humans from consumption of fruit and vegetable. Plants and microbial-based phytoremediation and bioremediation of pollutants have been. Here we review the sources of contaminants, toxicity and remediation strategies.

Keywords Crop · Emerging contaminants · Toxicity · Remediation

B. N. Singh (✉)
Institute of Environmental and Sustainable Development, Banaras Hindu University, Varanasi, Uttar Pradesh, India

Department of Plant Physiology, Institute of Agricultural Sciences, Banaras Hindu University, Varanasi, Uttar Pradesh, India

A. Hidangmayum · A. Singh · A. Guru · B. R. Yashu
Department of Plant Physiology, Institute of Agricultural Sciences, Banaras Hindu University, Varanasi, Uttar Pradesh, India

G. S. Singh
Institute of Environmental and Sustainable Development, Banaras Hindu University, Varanasi, Uttar Pradesh, India

© Springer Nature Switzerland AG 2020
E. Lichtfouse (ed.), *Sustainable Agriculture Reviews 40*, Sustainable Agriculture
Reviews 40, https://doi.org/10.1007/978-3-030-33281-5_6

Abbreviations

CVD	Chemical vapour deposition
DDT	Dichlorodiphenyltrichloroethane
OCPs	Organochlorine pesticides
PAHs	Polycyclic aromatic hydrocarbons
PAN	Peroxyacetyl nitrate
PBDEs	Polybrominateddiphenyl ethers
Perfluorinated alkyl acids	Perfluorinated alkyl acids
Perfluorosulfonamide	Perfluorosulfonamidc
PFOA	Perfluorooctanoic acid
PFOS	Perfluorosulfonate acid
Volatile organic compounds	Volatile organic compounds

6.1 Introduction

Soil is the primary sinks of all kinds of waste chemicals which might cause adverse effects for plants and human health (Teng et al. 2014). Presence of pharmaceuticals, hormones, artificial sweeteners along with pesticides is referred as emerging contaminants (Manoli et al. 2019). These waste trace elements chemicals is either personal care, industrial or agricultural goods. These emerging contaminants in municipal wastewater has attracted for research in recent year (El-Said et al. 2018). The unregulated and mismanagement disposal of these waste trace elements could raise the effective risk to soil ecosystem such as micro-and macro-organisms and plants (Bender and Heijden 2015). Plants considered as a primary source for food production for all the terrestrial consumers. Any disturbance in the lifestyle of plants could have a latent impact of consumers at primary tropic level (producers) in the food chain (Zhuang et al. 2009). Plants have unique features for the uptake of microchemicals present in the soil. This has been established through plant-based phytoremediation for the removal of chemical contaminants from soil. Phytoremediation based removal of this contaminant (heavy metals) has been established for several years. Recently, improvement in technological advancement has allowed more attention which detecting trace amount of emerging contaminants in soil, water, food or part of plant tissue at any stage (Stefanakis and Becker 2015; Cincotta et al. 2018). Due to industrialization and commercialization, several new chemicals have been arising in the soil which harms soil biota and imbalance the ecological system (Estévez et al. 2005) (Fig. 6.1). These waste trace elements chemicals released in soil concern with major contaminants are considered as emerging contaminants.

The term emerging contaminants (Table 6.2) refers as those compounds originated via anthropogenic activities which have not occurred previously in the environment at a global scale and cause severe effects on the ecosystem or human health (Stuart et al. 2012; Firouzsalari et al. 2019). Mostly these emerging contaminants are organic compound and together with it transformed in other products. It is not

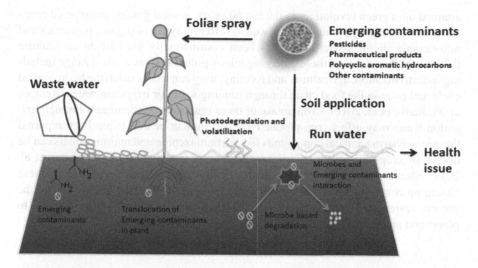

Fig. 6.1 Source and fate of contaminants in soils. Directly or indirectly, plants become exposed with contaminants through either by foliar spray or soil application. Contaminants present in soil are absorbed through the roots system

always part of regular existing contaminant but may be a runner for the future parameter. The focus also takes an account for emerging contaminants concerns, contaminants which have been present previously for a long time in the environment, but for those contaminants which have been raised recently. A neurotoxin β-N-methylamino-L-alanine compound was detected in irrigation water whose translocate in roots and shoots in *Triticum aestivum* (Contardo-Jara et al. 2014). Moreover, cyanobacteria occurred in large amount in the fresh water body. But due to anthropogenic activity, cyanobacteria population is spread in agricultural practises. Cyanotoxins a secondary metabolite secreted from cyanobacteria is account as an emerging contaminant (Bouaïcha and Corbel, 2016). Their potential toxicity is affected aquatic environment and soil body, and may also responsible to affects human health (Testai et al. 2016). Several other emerging contaminants may be existing includes personal care products, plasticizers, surfactants, pesticides, food additive, pharmaceutical products, industrial additives, herbicides and whose eco-toxicological effects are relatively unknown (Murray et al. 2010). Due to easily solubilisation and mobilization in water and mixing in soil, they adversely affect the food chain. Conventional wastewater treatment is not much effective design for complete removal of these emerging contaminants and related compounds from the environment (Kümmerer 2009; Pal et al. 2010). Therefore, these compounds persist continuously in the environment and may be bio accumulate in the existing organisms.

Reports highlighted regarding agriculture status that fulfils the demand for food as the world human population increases from 7.3 to 9.7 billion by 2050; agricultural production needs to double at present requirement. FAO (2016) stated that agricultural production increase by 70% to be meet actual requirement perhaps needs to be attained third green revolution. In the last 50 years, several genetic engineered crops variety, use of high yielding crops as well as different types of organic pesticides and non-organic chemicals fertilizers has been commercially used in the agriculture field. Since the source of these emerging micro-pollutants has a broad range including industrialization, agriculture, and forming, they can easily reach the hydrological cycle and enter in the food chain through running water or irrigation water (Kemper 2008; Radke et al. 2010). Crop uptake of these emerging contaminants through irrigation water may affect crop productivity and human health. It has been reported that some pharmaceutical compounds like carbamazepine and antimicrobials can be uptake by crops through running water, irrigation water or soil (Jones-Lepp et al. 2011; Shenker et al. 2011). Due to the flexible nature and behaviour of these Emerging contaminants in the environment, their probable risks might vary. In the present review, we explored emerging contaminants and their potential risks to plants and range of factors that affect the uptake, removal, growth.

6.2 Fate of Emerging Contaminants in Irrigated Water

Global climate change has drastically affected water availability and has been increasingly affected in almost every continent figuring about 40% of the world population. About 70% of global water withdrawals account for agricultural used and is further rises to 80% in arid and semiarid regions. Treated wastewater had evolved a reliable alternative to reused water both direct and indirect for crop irrigation and contributes trace elements strategic options in seasonal drought and weather variability that may able be cope up in peak water demand. This could be useful in farming activities which require continuous irrigation and could potentially reduce the incident of crop failure and income losses (EC 2016). Proper assessment of nutrients in treated wastewater could also decrease fertilization needs associated with reducing environmental contamination with fewer production costs (Haruvy 1997). This stands to be benefitted in peri-urban areas where farming system required high water demand and proximity to treated wastewater (Kurian et al. 2013). However, treated wastewater may contain toxic pollutants that might cause severe environmental hazards when applied as irrigation (Prosser and Sibley 2015). These pollutants include so-called contaminants of emerging concerns, synthetic chemicals or chemicals deriving from a natural source that are potentially harmful on environment and human health (Naidu et al. 2016).

Moreover, treated wastewater has also rise concern of containing heavy metals such as Cu, Zn, Fe, Pb and Ni which thereby could affect human health and agricultural productivity when such contaminants are translocated in crops (Fig.6.1) (Rattan et al. 2005). Although there are countries where such metals have been regulated due to health implications (Khan et al. 2015) however, emerging contaminants remain unregulated. Irrigation of crops with secondary treated wastewater contains a higher amount of emerging contaminants such as pharmaceuticals and personal care products (722 mg L^{-1} on average) than groundwater (31 mg L^{-1} on average) (Calderón-Preciado et al. 2013). It was reported that emerging contaminants induce phytotoxic, morphological and physiological changes in crop plants (Shahid et al. 2015; Christou et al. 2017). But effects on crop productivity have not been reported in field level. Rattan et al. (2005) reported that irrigation with treated wastewater for 20 years have significantly increased the deposition of extractable Trace elements (Trace element) such as Zn (208%), Fe (170%), Ni (63%), Pb (29%) and Cu (170%) as compared to adjacent region irrigated through tube-well water. The threshold for Trace elements in crop production depends on the crop and the element (FAO 1985). For instances, arsenic toxicity in the plant ranges from 12 mg /L for Sudan grass to 0.05 mg/L for rice. Cobalt show phytotoxicity in tomato plant at 0.1 mg/L in nutrient solution whereas, copper show phytotoxic at 0.1 to 1.0 mg/L. Zn toxicity in plant ranges widely while nickel range from 0.5 to 1.0 mg/L; in both the cases, toxicity is reduced at alkaline or neutral pH. Moreover, boron shows the positive effect at 0.2 mg/L in some crops while at 1–2 mg/L it shows phytotoxicity. Recent research has shown that irrigation with treated wastewater that contain micro and macronutrients positively affect crop production (Li et al. 2015a, b; Urbano et al. 2017). However, there is no report on the co-occurrence of emerging contaminants and Trace elements in irrigated water supply through treated wastewater and their plant responses in field conditions.

Peri-urban horticulture provides certain ecological services to nearby urban areas while performing various environmental and socioeconomic functions. They disseminate low carbon footprint such as in vegetables which favour environmental sustainability, provides recreational, landscape structure and other services (Van Veenhuizen and Danso 2007). Nevertheless, peri-urban agriculture faces certain adversities such as atmospheric and water pollution which are the result from transportation infrastructure such as airports, harbours, highways and use of reclaimed water which contain contaminants of various toxic origins like organic micro-contaminants and Trace elements leading to deposition in soil and potentially affects crop productivity (Colon and Toor 2016). In particular, the supplementation of manure and biosolids as a soil amendment in irrigated water have found to contribute to Organic micro-contaminants that negatively affects crop productivity (Eggen and Lillo 2012). On the other hand, due to climate change water scarcity has been increasing and the need for reliable option cantered towards the use of reclaimed water for crop irrigation especially in arid and semi-arid regions (WHO 1989). There are reports where reclaimed water has been contaminated by organic micro-contaminants from pharmaceuticals and personal products which range from mg/L to μg/L. Their continual release in the environment coupled with their transformation products has evolved as a "pseudo-persistent" way (Daughton and Ternes 1999). Reclaimed water use as irrigation is the sources of Trace elements along with soil amendment, bio-solid and atmospheric deposition (Liacos et al. 2012). As a result, Trace elements have contributed to negative effects on crop production in peri-urban areas (Singh and Kumar 2006). Bioaccumulation of organic micro-contaminants has continuously been observed in crops grown under field condition irrigated with treated wastewater (Wu et al. 2014). It was found that the highest concentrations were detected in the edible part of leafy vegetables (Carbamazepine, 347 ng/g dry weight in lettuce) rather than in root or fruit-bearing vegetables (Riemenschneider et al. 2016). There is limited research regarding the irrigation with treated wastewater in crop under field condition (Calderón-Preciado et al. 2011). In another research, it was found that long term irrigation with treated wastewater for three consecutive years have a significant impact on crop uptake and accumulation of organic micro-contaminants (Christou et al. 2017).

Trace elements such as Pb, Cd, and Zn have been found to accumulate higher in the road side of peri-urban than rural sides (Nabulo et al. 2010). This has created an alarming situation concerning human exposure to organic micro-contaminants and Trace elements as has been found in edible parts of plants (Pan et al. 2014). This can be incorporated with insightful research in field studies depicting the real scenario. However, there is limited information published regarding the above concern (Malchi et al. 2014). Another substantial adverse effect of the use of reclaimed water in agriculture is its high nitrate content. This nitrate is eventually taken by as a dietary need by humans and about 5% nitrate get converted to a toxic form of nitrite (Santamaria 2006). Nitrate fertilization and reuse of treated wastewater contribute to nitrate uptake by food crops (Castro et al. 2009).

The presence of organic micro-contaminants and trace elements in agricultural soil exhibits morphological and physiological changes in the exposed plants. For instance, Hurtado et al. (2017) demonstrated that the presence of organic micro-contaminants at the environmentally relevant concentration in the treated wastewater supply as irrigation leads to decrease in chlorophyll content, morphological changes, and alteration in the metabolic profile of lettuces. In zucchini plant, it has been shown to reduce biomass productivity and alteration in hormone and nutrient content due to the presence of carbamazepine and verapamil at an environmentally relevant concentration of 0.005–10 mg/L in soil (Carter et al. 2015). Increase in Pb concentration in plants can decrease chlorophyll biosynthesis and reduce carbohydrate content in vegetable (Peralta-Videa et al. 2009), while the presence of Cd cause lipid peroxidation (Monteiro et al. 2007). There are reports regarding the regulation of Trace elements and pesticides residues in EU (EFSA), USA (FDA), China, Australia and New Zealand, and FAO has published international guidelines (FAO 2016). However, there are no regulations regarding Organic micro-contaminants.

6.3 Occurrence of Pesticide Contamination

6.3.1 Toxicity

According to Pan et al. (2017), pesticides are defined as a group of chemicals which persist in soil for a longer period. Environmental contaminants such as pesticides residues pose a serious threat to humans and animals. Application of organochlorine pesticides to control weeds induces resistance to insects and pests and vector-borne diseases (Abhilash and Singh, 2009). Dichlorodiphenyltrichloroethane in soil takes a long time to degrade which ranges from 4 to 30 years, and chlorinated organochlorine pesticides may remain stable for many years after their application (Afful et al. 2010). Due to the slow degradation process, Dichlorodiphenyltrichloroethane is hampered in physical, chemical, biological and microbiological ways (Afful et al. 2010; Kumar and Singh 2017). These chemicals are fat-soluble components and can easily bioaccumulate inside the lipid domain of biota including fatty tissues, breast milk, and blood through the food chain. As a result, humans, and animals are more prone to these micro-pollutants that are originated from contaminated soil or water (Mishra and Bharagava 2016). Apart from these threats, pollutants are also highly toxic to aquatic fauna and eventually cause serious diseases in human and animals (Aiyesanmi and Idowu 2012) and soil micro-flora (Megharaj 2002).

To alleviate these environmental issues, different physicochemical methods such as chemical amendment, land-filling, incineration and composting or burning have been employed to exclude pesticide contamination from the environment over the last few decades (Wehtje et al. 2000). Using a broad spectrum of chemicals in agricultural production has dominated over the years which have provided significant protection and reduce losses of crops in both developed and developing nations. However, the continual use of these pesticides expectedly results in accumulation of residues in food and lead to potential adverse effects on the environment and human health. These damages are more prone to farm workers as they are continually engage in the field. Moreover, pesticides residue has been found in drinking water and food crops especially in fruits and vegetables which received the largest dosages of pesticides and posed a serious threat in human health (Pimentel et al. 1992). The pesticides which are currently available have a large mixture of chemical compounds which exhibits a different mode of action, absorbance by the body, metabolism, removal from the body and toxic effect on humans and other living organisms.

Some pesticides exhibit acute toxicity, but when they come in the body, they metabolized an eliminated. But some other show lower acute toxicity and have strong inclination to assemble in the body. Toxic effects may also be caused by critical ingredients and other impurities also by solvents, emulsifiers, carriers and other constituents of the formulated products. Implementation of low cost and environmentally friendly technologies such as bioremediation could be better alternatives in the present scenario (Singh and Seneviratne 2017). Presently, microbial-based bioremediation has been proposed to be safe and environmental friendly associated with sustainability in the removal of contaminated trace elements.

6.3.2 Effect of Pesticides on Plants

Application of pesticides not only affects the target plants but also non-target plants. For instance, phenoxy herbicides are toxic in trees and shrubs (Dreistadt 2016). Herbicides, sulfonylureas, sulphonamides, and imidazolinones have an intense effect on the productivity of non-targeted crops, plants and associated wildlife (Fletcher et al. 1993) (Table 6.2). Also, glyphosate herbicides can increase the susceptibility of plants to disease and infection (Brammall and Higgins 1988). Pesticides such as carbamate trace elements, pyrethroids, and organophosphate trace elements could potentially harm the population of beneficial insects such as beetles and bees. Several methods are employed such as spraying on the crop plant or the soil, mixing with soil and as granular form. These pesticides after their application dispersed in the air or degrade in the soil and leaching which eventually lead to water bodies and may be consumed by aquatic flora and fauna (Caicedo et al. 2011). Over application of fungicides could affect the functioning of soil biota and indirectly affect soil fertility. Man and Zucong (2009) reported that nitrification and de-nitrification processes are disrupted by application of fungicides chlorothalonil and dinitrophenyl.

6.4 Miscellaneous Compounds as Emerging Contaminants

6.4.1 Perfluorinated Compound

The major constituents of organofluorine compounds of a class of perfluorinated compounds contain carbon. Based on functional groups, they are classified as perfluoroalkyl or perfluorocarboxylic acids. In textile and refrigeration industries they have a wide range of applications. These compounds are presents in soil and plants can uptake it is broadly trace elements. Interesting findings had shown that the reclaimed water which is used for irrigating soil grown plants consists of perfluorinated alkyl acids. It was proven that in edible plants such as strawberry and lettuce perfluorinated alkyl acids tended to be bio-accumulated at the exposure concentration of 0.2–4.0 mg/L (Blaine et al. 2014). In tomato, cabbage and zucchini plant, it has been revealed that the potential uptake of 14 different types of perfluorinated alkyl acids compounds in addition to the hydroponics studies (Felizeter et al. 2014). The perfluorinated alkyl acids s which have more than 11 carbon atoms have higher root uptake, and in the edible parts, these long chained perfluorinated alkyl acids s were not found. It was observed that translocation part of the plant which is edible, the amount of simple structured perfluorinated alkyl acids s taken up less (Felizeter et al. 2014).

Similarly, it was observed in wheat plants for their uptake and translocation of perfluorinated alkyl acids. The uptake of perfluorohexanoic acid was highest among all these PFCAs; then with either increase or decrease in the number of carbons, it decreased significantly. Thus, in PFCS the role of several carbon atoms is significant. In other similar study, the uptake by carrot and lettuce plants was trace elements for three different PFCs: the perfluorooctanoic acid, perfluorosulfonate acid, and perfluorosulfonamide (Bizkarguenaga et al. 2016). Based on root core, root and the leaves in the case of carrots and into heart and leaves for lettuce, the plant was harvested, and the harvested plants were divided into it. The higher concentration of PFOA and perfluorosulfonamide were observed in the case of carrot leaves (0.6 and 3.4) while in the core and peel compartments its lesser value was observed. In the edible part of lettuce, the PFOA was accumulated more than PFOS (Bizkarguenaga et al. 2016). Thus, in both plants PFOA was early taken and translocate.

6.4.2 Polybrominated Diphenyl Ethers Compounds

There are other chemicals in addition to the previous chemical mentioned that are also trace elements for their uptake and translocation. For instance, it was trace elements that the uptake and translocation of polybrominateddiphenyl ethers in wheat plants. Wheat roots and straw successfully take only the decabromodiphenyl ether (BDE 209) of seven different types of compounds (Li et al. 2015a, b). Cucumber and carrot plants take dioxin-based compounds (Engwall and Hjelm 2000). Tomato plants take phthalate compounds, and it was reported that only 1% was taken by its fruits (Sablayrolles et al. 2013). The negligible amount of polychlorinated biphenyl was taken up by the plants (Mikes et al. 2009; Sablayrolles et al. 2008).

6.4.3 Steroid Compounds

For a growing global population, the major challenge for the agricultural sector is to produce more and safe food. Livestock industry products such as meat and dairy products and the growth promoters such as the use of synthetic steroid hormones (Bartelt-Hunt et al. 2012), increasing the muscle mass in the developed countries these are most adopted practices. The natural estrogen is estimated to discharge 30,000 kg/year for 7 billion of world human population and form contraceptive pills practices of synthetic estrogens (EE2) of about 700 kg/year (Adeel et al. 2017). Livestock is contributing much greater possible unit of estrogen to the environment, the U.S and European Union alone calculated that by livestock excretion annual estrogen secretion is about 83,000 kg/year, which is more than double as produced by the world human population. The relation between annual feeding operations and the detection of estrogen is an aquatic environment has been made (Shrestha et al. 2012). Hormones produced from animal and human waste trace elements are natural but use damaging effect on the environment, the widely adopted practice in modern agriculture is to apply animal manure or sludge bio-solids onto agricultural land as substituting fertilizer to organic products (Xuan et al. 2008).

The presence of many classes of organic pollutants such as steroid estrogen in reclaimed water has been well documented in the study. In all over the world in agriculture treated sewage sludge is widely used and for the next, the most adopted practice of disposal is land application (Calderón-Preciado et al. 2013; Gabet-

Giraud et al. 2014). Previous studies indicate that in beans aquatic macrophytrace elements and algae, the steroid estrogen can be taken up, accumulate in, or metabolized (Shi et al. 2010; Card et al. 2012). For instance, in terrestrial plants including leafy vegetables and fruits the animal excrement and reclaimed water were taken up by them which are derived from steroid estrogens (Karnjanapiboonwong et al. 2011; Zheng et al. 2013). The emerging pollutants are entering the food chain of terrestrial which is the result of land application of reclaimed water and animal manure. In soil, the bioavailable concentration of estrogens affect plants to be taken by up, so it tends to be estimated that the concentration is difficult to measure (Dodgen 2014). In soil, 74 ng g^{-1} 17β E2 and in ground water 69 ng L^{-1} have been found in our study (Song et al. 2018). In many physiological processes, for instance, development and reproductive process as well as management of protein, sugar, and mineral. These compounds participate in a hundred types of steroid estrogenic compounds produced by plants and animals (Janeczko et al. 2012; Sherafatmandjour et al. 2013). E1, E2 and E3 type of steroid compounds lie on interconnecting metabolic pathways. Reverse transformation of E2 to E1 occurs in microbes and afterward degraded to E3 under aerobic conditions. The inorganic contents of barley, maize, chickpea, and bean seeds can be significantly improve biochemical level by mammalian sex hormones (MSH) (Erdal et al. 2012) and chlorophyll, sugar, carotenoids, and protein in duckweed seed, lentil, soybean and fennel (Czerpak and Szamrej, 2003; Dumlupinar et al. 2011; Sherafatmandjour et al. 2013). Vegetative growth of alfalfa plants inhibited by steroid estrogen found in sewage water (Shore et al. 1993). Steroid estrogens reduced root growth at a concentration of 1 μM and also caused morphological abnormalities including epinasty in tomato plants (Guan and Roddick 1988). Hence, in various ecological environments, it is important to evaluate their disruptive potential. To date, the effects of these hormones to plants as stresses have been described in a few studies.

6.4.4 Volatile Compounds

In 'acid rain' the acid derived from inorganic gases produced by fossil fuel combustion, sulfur dioxide, and nitrogen oxides and these air pollutants mainly affect vegetation. Peroxyacetyl nitrate (PAN) and ozone are oxidants produced photo-chemically and have the deleterious effects. For over 30 years, the direct effect of these gases has been known following the classic studies in the San Bernardino mountain of California (Miller and McBride 1999). The pressure of volatile organic compounds, nitrogen oxides (NOx) and sunlight are required for the photochemical production of phytotoxic concentration of ozone. An attempt has been introduced to reduce the concentration of secondary pollutants by emission controls on both volatile organic compounds and NOx. At current ambient concentration, the volatile organic compounds have the potential to effect the plants directly largely at high concentration of volatile organic compounds based on short term exposure of vegetation. However, the relative role of volatile organic compounds becomes important as primary pollutants, as a reduction in emission on traditional inorganic pollutants. Based on laboratory and field data, this review considers the available long term evidence.

6.4.4.1 Effect of Volatile Organic Compounds in Plants

Volatile organic compounds are compounds which contain carbon atoms (such as xylene containing eight carbon atoms), nitrogen, and oxygen or halogen atoms respectively (Table 6.1). At normal temperatures, the vapour pressure gradient for organic compounds equals that of gases present in the atmosphere. Therefore, volatile organic compounds are not implicitly different from semi-volatile organic compounds. Certain compounds such as polychlorinated biphenyls, dioxins and polycyclic aromatic hydrocarbons, where a large fraction of the material in the atmosphere may exist in the particulate phase rather than as a gas, are excluded specifically from this study.

It has been observed from the previous experimental studies that there is a short term exposure of plants to large concentrations of volatile organic compounds relative to ambient air. Although there is no report on direct effects of Volatile organic compounds on plants, certain data on experiments where plants have been exposed to substances in the vapour phase has been mentioned in the United State trace elements EPA web-searchable database PHYTOTOX (http://www.epa.gov/med/databases/databases.html#-phytotox). Seven hundred ninety-three entries had been produced by this database search for the above defined volatile organic compounds. However, five published papers on examining the effects of ethylene have reported that only 23 entries involve fumigation with concentrations of 100 nl L^{-1} or less. Many experiments have been conducted on a large number of plants using high concentrations (factor of 100 or more) of Volatile organic compounds showing exposure time in hours particularly. Under these conditions, effects are likely to be

Table 6.1 Effects of different volatile organic compounds in plants

VOC	Conc (nl L^{-1})	Plant species	Effects	References
Ethylene	85	*Phaseolus vulgaris*	Epinasty, chlorosis, reduction in dry weight	van Haut and Prinz (1979)
	1–10,000	*Ricinodendro-nrautaneni*	Seed germination at 1 nl l^{-1}	Keegan et al. (1989)
	10–100	*Lilium longiflorum*	Reduced shoot, flower dry wt, earlier flowering 10 nl l^{-1}	Blankenship et al. (1993)
Tetrachloroethane	1–300	*12 species*	Reduced pod number and shoot dry wt	Franzaring et al. (2000)
Halon 1301	1	*Lepidium sativum*	Reduction in protein content	Debus and Schroder 1991
Peroxyacetyl nitrate	10–60	*Petunia hybrid*	Reduced leaf area, total dry wt. at 30 nl l^{-1}	Izuta et al. (1993)

observed on plants as well as plant parts which are more sensitive on exposure to very high concentrations as in case of chemical spillage or accident). However, it is quite uncertain whether those plant species and plant parts would resist very small concentrations for a long period or not. Abeles (1992) suggested two examples of non-linear relationships of dose-response involving Volatile organic compounds and expressing dose as a concentration-exposure time product. When Digitalis plant was exposed to 1000 ml L^{-1} over 6 h and 50 nl L^{-1} over 24 h, similar effects of ethylene on senescence of flower were observed; the same ethylene induced flowering response was observed in case of Iris bulb on exposure to different concentrations as well as time period- 80,000 nl L^{-1} for 1 h, 2000 nl L^{-1} for 6 h and 300 nl L^{-1} for 24 h respectively. Though the relationship between concentration and exposure time is non-linear, there is difficulty in the prediction of concentration which would produce a similar effect for exposure to a longer period.

Recently, *C. melo* var. *reticulatus* bearing medium sized fruit having a thin reticulated rind with sweet orange flesh, has obtained the designation of Protected Geographical Indication. As a consequence of several petrochemical activities in the south of Sicily, two areas-Augusta-Priolo and Gela with large industrial facilities have been designated as high-risk areas based on the national concern (Mudu et al. 2014). Guerriero et al. (2011) reported the existence of many chemical compounds which occur in high levels in soil, water, groundwater, air sediments, fish and shellfish from both areas. According to Mudu et al. 2014, benzene, toluene, ethylbenzene and xylene have been considered as major contaminants. However, several reports are being presented regarding the uptake of environmental contaminants by plants and the exposure of human beings through the intake of diet. Studies have shown that crops could potentially uptake environmental contaminants (Goldstein et al. 2014). The uptake of aromatic organic contaminants found in soil and their internal storage by vegetables influences the mobility of environmental contaminants (Chang et al. 2013). Studies showed on the identification of an extensive array of emerging environmental contaminants and their quantification in water, soil, air and food products (Table 6.2).

6.5   Pathways of Contamination of Emerging Contaminants

It is already known to us that poor disposal practices and over spraying of water courses are responsible for the occurrence of higher amounts of agricultural chemicals in water supplies. Generally, higher levels of agricultural chemicals attribute to percolation of contaminated run off into the pathway of nature as well as man-made pathways through soil (Fig. 6.3). Most importantly, wells- a source of water supply are being polluted due to the availability of lesser facilities and direct infiltration of contaminated runoff and this has been considered as one of the most common causes of artificial pollution. Likewise, leaching of highly cracked soil and finely cracked rocks results in groundwater contamination. While applying highly soluble and mobile herbicides during downward movement of water and when plants tran-

Table 6.2 Lists of plant parts exposed by emergent contaminants in different plant species

Class	Types of Chemical	Detected in plant part	Plant species	References
Polycyclic aromatic hydrocarbons	Pyrenne, anthracene, phenanthrene	Root and shoot	Alfalfa	Alves et al. (2017)
	Naphthalene, phenanthrene	Root, shoot and grain	Rice	Patowary et al. (2017)
	Fluoranthene, pyrene, indenopyrene, benzo-perylene	Straw and grains	Wheat	Li and Ma (2016)
	Phenanthrene, anthracene, fluoranthene, pyrene, benzo(a) anthracene, chrysene	Root and leaves	Couch grass, wood small-reed, wild buckwheat, and white goosefoot	Gworek et al. (2016)
	Fluorene, phenanthrene	Root and shoot	Wheat, alfalfa and sunflower	Salehi-Lisar et al. (2015)
Pharmaceutical and personal care products	Chlortetracycline, enrofloxacin, and sulphathiazole	Root and leaves	Radish	Chung et al. (2017)
	Tetracycline	Seedling and stem	Pea	Margas et al. (2016)
	Sulfamethoxazole, norfloxacin and doxycycline	Seed germination	Pakchoi cabbage and radish	Wang et al. (2016)
	Triclosan, triclocarban	Leaf, root skin and root core	Radish and carrot	Fu et al. (2016)
	Sulfadiazine	Root, leaf and seed	Hazel plants	Michelini et al. (2015)
	Galaxolide, Tonalide, and triclosan	Root and leaves	Carrot, barley, meadow plant	Macherius et al. (2012a)
	Triclosan	Root	Carrot	Macherius et al. (2012b)
Pesticides	Chlordecone	Leaf, root and tuber	Radish	Létondor et al. (2015)
	Aminocyclopyrachlor	Stem, leaves and root	Trumpet flower	Reis et al. (2015)
	Chlorpyrifos	Stem, leaves and root	Chinese cabbage and cabbage	Zhang et al. (2011)
Perflourinated compounds	Perfluorooctanoic acid, perfluorooctane sulfonate	Root core, root peel and leaves: Carrot heart and leaves: Lettuce	Carrot and lettuce	Bizkarguenaga et al. (2016)
	Perflourinated alkyl acids	Root, stem, leaves and edible parts	Zucchini and cabbage	Felizeter et al. (2014)

spire less, there is a possibility of leaching of these herbicides to groundwater. Generally, the applications of agricultural chemicals were done directly to plants through foliar spraying or to the soil. The concentrations of these chemicals in surface water depend on various factors such as soil characteristics, rainfall, irrigation, interception loss and rate of application (both over application and misapplication). Agricultural practices, improper disposal of excess chemicals, water resulted from the washing of equipment applied and spills heavily contaminate surface water. Chemical storage facilities nearer to water sources, chemicals treatment trace elements, trace elements for animal treatment by dipping and spraying where a high concentration of spills occur, etc. are some other examples of important sources of contamination (Fig. 6.2).

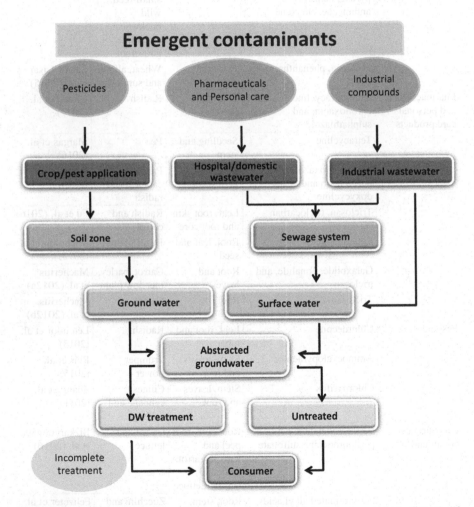

Fig. 6.2 Types of contaminants, and how these contaminants enter in consumer body through different pathways

6.6 Strategy for Removal of Emerging Contaminants

6.6.1 Photodegradation Method

Organic micro-pollutant can be phyto-degraded in the presence of sunlight either by direct or indirect ways (Rivera-Utrilla et al. 2013; Vione et al. 2014). In case of direct mechanism for the degradation of micro-pollutant; its chemical property decides the threshold level of photo energy which leads the breakdown of bonds to present between the micro-pollutants leads to degradation (Rivera-Utrilla et al. 2013). Also, photo-degradation of the micro-pollutant by indirect mechanism promote trace elements by ROS which is produced by the action of the photosensitizers such as nitrate, nitrite or chromophoric chemical substances which degrade the organic matter present in water (Niu et al. 2016). A lot of studies conducted on photo-degradation of organic micro-pollutant in laboratory condition with simulated sunlight and in natural waters which acknowledged us that photo-degradation of organic micro-pollutant is the key removal mechanism (Packer et al. 2003; Rivera-Utrilla et al. 2013).

6.6.2 Adsorbent Method

Numerous reviews showing the removal of waste water organic pollutant by using different adsorbents have been published over the last few decades (Hamdaoui and Naffrechoux, 2009). Investigations have been done on the treatment of various Emerging contaminants from water and wastewater sludge sediments and soil, pesticides, human pharmaceuticals from the environment (Kuster et al. 2005; Hernando et al. 2005; Grassi et al. 2012; Brooks and Huggett 2012). Several nano-materials have been developed in recent years which can be used as effective adsorbents for the sorption of Emerging contaminants from effluents (Kaur et al. 2017). Graphene layers are produced through the exfoliation of graphite by using oxidizing agents; methane and other hydrocarbons deposited by the chemical vapour deposition, ultrasound can also be used for graphene synthesis (Chowdhury and Balasubramanian 2014; Muthoosamy and Manickam 2017). The methods involved in the production of carbon nanotubes are mainly chemical vapour deposition, as well as combustion chemical vapour deposition, arch discharge using carbon electrodes and laser ablation using graphite (Dahman et al. 2017). Interestingly, the effective treatment of Emerging contaminants involves both graphene and carbon nanotubes (Jauris et al. 2016; Zhao et al. 2016). Furthermore, activated carbons, clay nanotubes, mesoporous silica, and tailored composite trace elements, nano composite trace elements of carbon nanotubes/nanofibers and layered silicate trace elements, biochars, etc. are also used as adsorbents.

6.7 Abiotic Factors Influencing the Uptake of Emerging Contaminants by Plants

One of the major components of soil, the humic acids, plays a significant role in the uptake of emerging contaminants. For instance, the humic content is determinant for the retention of emerging contaminants, such as PAHs, in soil (Khan et al. 2013). Whereas, in the case of the ENMs, the overall bioavailability towards plants is negatively influenced by the humic content, since ENMs tend to form aggregate trace elements in soil. Humic acids can also act as adsorbents for certain chemicals such as PCBs and PPCPs (Yu and Bi 2015).

Soil enriched with higher humus layer signifies a higher amount of PCBs (Wenzel et al. 2002). Some soil components like colloidal clays act as transport carriers for PPCPs in the soil (Xing et al. 2016). The bioavailability of emerging contaminants is influenced by the pH. If the pH is alkaline, it increases the bioavailability of emerging contaminants because of the increase in microbial growth. The adsorption of emerging contaminants on the soil was facilitated by acidic pH. This sorption is generally due to interactions between the donor and acceptor and the outer- and -inner-sphere complex in the soil (Guo et al. 2016). When emerging contaminants leaches into the lower layers of the soil, makes them less bioavailable to the roots of plants (Xu et al. 2009). Some factors including mobility and aging in soil hugely influence the emerging contaminants uptake by plants. Also, hydraulic conductivity, wettability of soil and swelling capacity affects the emerging contaminants transport in plants (Ávila et al. 2014). The divalent ions and chelating agents can significantly influence the uptake of emerging contaminants, because of the cation exchange capacity of the soil. In some cases, soil residues are formed by the potential exchange of chemicals with soil (Masih et al. 2012) (Fig. 6.3).

6.8 Conclusion

Increasing anthropogenic application of huge organic chemicals and their mismanage disposal of these chemicals in the environment rising huge questions in concerns with food safety. Biodegradation, photo-degradation and certain environmental factors eliminate the emerging contaminants at a certain extent. Also, plant species have also an important role in the removal of emerging contaminants from the environment. Although emerging contaminants are very less quantity and still persist environment that promotes trace elements plant growth and development. This emerging contaminant present in the root part, with the lowest in fruit and grain. Another aspect of the accumulation of emerging contaminant arises where chemicals are bio-accumulated in plants with a high content of lipid. This could be possible due to the solubility of emerging compounds in lipid molecules. Overall, the uptake of emerging contaminant with plants could arise a major problem for food safety and food security. Since, there is no strong legislation related to the disposal

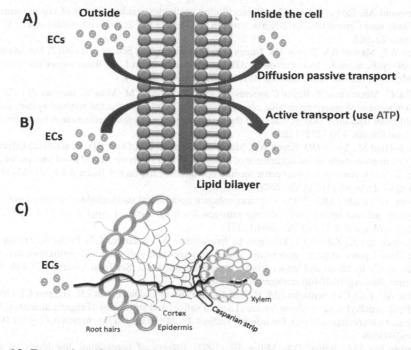

Fig. 6.3 Expected mechanism of uptake of emerging contaminants from soil to plants. Uptake of emerging contaminants from outside to inside takes place either by through passive (**a**) or active transport mechanism. Lipid bilayers of plant cell facilitate the transport of some low molecular weight and volatile polycyclic aromatic hydrocarbons through the diffusion process. (**c**) shows the route of transport of emerging contaminants in through plant cell wall

of an emerging contaminant in the soil, have been followed. As a result, it is assumed that these emerging contaminants will continuously increase. It will always be challenging in future to understand the actual concentration of emerging contaminant in plant and human and their toxic effects to achieve the best management to risks to plants, human health, and the environment.

References

Abeles FW (1992) Roles and physiological effects of ethylene in plant physiology: dormancy, growth, and development. Ethylene in plant biology

Abhilash PC, Singh N (2009) Pesticide use and application: an Indian scenario. J Hazard Mater 165(1–3):1–12. https://doi.org/10.1016/j.jhazmat.2008.10.061

Adeel M, Song X, Wang Y, Francis D, Yang Y (2017) Environmental impact of estrogens on human, animal and plant life: a critical review. Environ Int 99:107–119. https://doi.org/10.1016/j.envint.2016.12.010

Afful S, Anjm AK, Serfor-Armah Y (2010) Spectrum of organochlorine pesticide residues in fish samples from the Densu Basin. Res J Environ Earth Sci 2(3):133–138

Aiyesanmi AF, Idowu GA (2012) Organochlorine pesticides residues in soil of cocoa farms in Ondo state Central District, Nigeria. Environ Nat Resour Res 2(2):65. https://doi.org/10.5539/enrr.v2n2p65

Alves WS, Manoel EA, Santos NS, Nunes RO, Domiciano GC, Soares MR (2017) Detection of polycyclic aromatic hydrocarbons (PAHs) in Medicago sativa L. by fluorescence microscopy. Micron 95:23–30. https://doi.org/10.1016/j.micron.2017.01.004

Ávila C, Matamoros V, Reyes-Contreras C, Piña B, Casado M, Mita L, Bayona JM (2014) Attenuation of emerging organic contaminants in a hybrid constructed wetland system under different hydraulic loading rates and their associated toxicological effects in wastewater. Sci Total Environ 470:1272–1280

Bartelt-Hunt SL, Snow DD, Kranz WL, Mader TL, Shapiro CA, Donk SJV et al (2012) Effect of growth promotants on the occurrence of endogenous and synthetic steroid hormones on feedlot soils and in runoff from beef cattle feeding operations. Environ Sci Technol 46(3):1352–1360. https://doi.org/10.1021/es202680q

Bender SF, Heijden MG (2015) Soil biota enhance agricultural sustainability by improving crop yield, nutrient uptake and reducing nitrogen leaching losses. J Appl Ecol 52(1):228–239. https://doi.org/10.1111/1365-2664.12351

Bizkarguenaga E, Zabaleta I, Mijangos L, Iparraguirre A, Fernandez LA, Prieto A, Zuloaga O (2016) Uptake of perfluorooctanoic acid, perfluorooctane sulfonate and perfluorooctane sulfonamide by carrot and lettuce from compost amended soil. Sci Total Environ 571:444–451. https://doi.org/10.1016/j.scitotenv.2016.07.010

Blaine AC, Rich CD, Sedlacko EM, Hyland KC, Stushnoff C, Dickenson ER, Higgins CP (2014) Perfluoroalkyl acid uptake in lettuce (Lactuca sativa) and strawberry (Fragaria ananassa) irrigated with reclaimed water. Environ Sci Technol 48(24):14361–14368. https://doi.org/10.1021/es504150h

Blankenship SM, Bailey DA, Miller JE (1993) Effects of continuous, low-levels of ethylene on growth and flowering of Easter lily. Sci Hortic 53:311–317. https://doi.org/10.1016/0304-4238(93)90051-Q

Bouaïcha N, Corbel S (2016) Cyanobacterial toxins emerging contaminants in soils: a review of sources, fate and impacts on ecosystems, plants and animal and human health. In: Soil contamination-current consequences and further solutions. IntechOpen

Brammall RA, Higgins VJ (1988) The effect of glyphosate on resistance of tomato to Fusarium crown and root rot disease and on the formation of host structural defensive barriers. Can J Bot 66(8):1547–1555. https://doi.org/10.1139/b88-213

Brooks BW, Huggett DB (2012) Human pharmaceuticals in the environment, current and future perspectives. In: Shugart LR, Ridge O (eds) Emerging topics in ecotoxicology, principles, approaches and perspectives. Springer, TN, p 4

Caicedo P, Schröder A, Ulrich N, Schröter U, Paschke A, Schüürmann G, Richter P (2011) Determination of lindane leachability in soil–biosolid systems and its bioavailability in wheat plants. Chemosphere 84(4):397–402. https://doi.org/10.1016/j.chemosphere.2011.03.070

Calderón-Preciado D, Jiménez-Cartagena C, Matamoros V, Bayona JM (2011) Screening of 47 organic microcontaminants in agricultural irrigation waters and their soil loading. Water Res 45(1):221–231. https://doi.org/10.1016/j.watres.2010.07.050

Calderón-Preciado D, Matamoros V, Savé R, Muñoz P, Biel C, Bayona JM (2013) Uptake of microcontaminants by crops irrigated with reclaimed water and groundwater under real field greenhouse conditions. Environ Sci Pollut Res 20(6):3629–3638. https://doi.org/10.1007/s11356-013-1509-0

Card ML, Schnoor JL, Chin YP (2012) Uptake of natural and synthetic estrogens by maize seedlings. J Agric Food Chem 60(34):8264–8271. https://doi.org/10.1021/jf3014074

Carter LJ, Williams M, Böttcher C, Kookana RS (2015) Uptake of pharmaceuticals influences plant development and affects nutrient and hormone homeostases. Environ Sci Technol 49(20):12509–12518. https://doi.org/10.1021/acs.est.5b03468

Castro E, Manas MP, De Las Heras J (2009) Nitrate content of lettuce (Lactuca sativa L.) after fertilization with sewage sludge and irrigation with treated wastewater. Food Addit Contam 26(2):172–179. https://doi.org/10.1080/02652030802425334

Chang ML, Wang MJ, Kuo DTF, Shih YH (2013) Sorption of selected aromatic compounds by vegetables. Ecol Eng 61:74–81. https://doi.org/10.1016/j.ecoleng.2013.09.059

Chowdhury S, Balasubramanian R (2014) Recent advances in the use of graphene-family nano-adsorbents for removal of toxic pollutants from wastewater. Adv Colloid Interf Sci 204:35–56. https://doi.org/10.1016/j.cis.2013.12.005

Christou A, Karaolia P, Hapeshi E, Michael C, Fatta-Kassinos D (2017) Long-term wastewater irrigation of vegetables in real agricultural systems: concentration of pharmaceuticals in soil, uptake and bioaccumulation in tomato fruits and human health risk assessment. Water Res 109:24–34. https://doi.org/10.1016/j.watres.2017.04.065

Chung HS, Lee YJ, Rahman MM, El-Aty AA, Lee HS, Kabir MH et al (2017) Uptake of the veterinary antibiotics chlortetracycline, enrofloxacin, and sulphathiazole from soil by radish. Sci Total Environ 605:322–331. https://doi.org/10.1016/j.scitotenv.2017.06.231

Cincotta F, Verzera A, Tripodi G, Condurso C (2018) Volatile emerging contaminants in melon fruits, analysed by HS-SPME-GC-MS. Food Add Contamin Part A 35(3):512–518. https://doi.org/10.1080/19440049.2017.1401738

Colon B, Toor GS (2016) A review of uptake and translocation of pharmaceuticals and personal care products by food crops irrigated with treated wastewater. Advances in Agronomy. Elsevier Inc. https://doi.org/10.1016/bs.agron.2016.07.001

Contardo-Jara V, Schwanemann T, Pflugmacher S (2014) Uptake of a cyanotoxin, β-Nmethylamino-l-alanine, by wheat (Triticum aestivum). Ecotoxicol Environ Saf 104:127–131. https://doi.org/10.1016/j.ecoenv.2014.01.039

Czerpak R, Szamrej IK (2003) The effect of β-estradiol and corticosteroids on chlorophylls and carotenoids content in Wolffia arrhiza (L.) Wimm.(Lemnaceae growing in municipal Bialystok tap tap T water). Pol J Environ Stud 12(6):677–684

Dahman Y, Bayan A, Volynets B, Ghaffari N (2017) Chapter 7 – Nanotubes. In: Dahman Y (ed) Nanotechnology and functional materials for engineers. Elsevier, Amsterdam, pp 145–173

Daughton CG, Ternes TA (1999) Pharmaceuticals and personal care products in the environment: agents of subtle change? Environ Health Perspect 107(suppl 6):907–938. https://doi.org/10.1021/bk-2001-0791

Debus R, Schröder P (1991) Responses ofPetunia hybrid and Phaseolus vulgaristo fumigation with difluorochlorobromomethane. Chemosphere 21:1499–1505

Dodgen L (2014) Behavior and fate of PPCP/EDCs in soil-plant systems. Doctoral dissertation, UC Riverside

Dreistadt SH (2016) Pests of landscape trees and shrubs: an integrated pest management guide, vol 3359. UCANR Publications

Dumlupinar R, Genisel M, Erdal S, Korkut T, Taspinar MS, Taskin M (2011) Effects of progesterone, β-estradiol, and androsterone on the changes of inorganic element content in barley leaves. Biol Trace Elem Res 143(3):1740–1745. https://doi.org/10.1007/s12011-011-8980-6

EC (2016).Water reuse – background and policy context

Eggen T, Lillo C (2012) Antidiabetic II drug metformin in plants: uptake and translocation to edible parts of cereals, oily seeds, beans, tomato, squash, carrots, and potatoes. J Agric Food Chem 60(28):6929–6935. https://doi.org/10.1021/jf301267c

El-Said WA, El-Khouly ME, Ali MH, Rashad RT, Elshehy EA, Al-Bogami AS (2018) Synthesis of mesoporous silica-polymer composite for the chloridazon pesticide removal from aqueous media. J Environ Chem Eng 6:2214–2221. https://doi.org/10.1016/j.jece.2018.03.027

Engwall M, Hjelm K (2000) Uptake of dioxin-like compounds from sewage sludge into various plant species–assessment of levels using a sensitive bioassay. Chemosphere 40(9–11):1189–1195. https://doi.org/10.1016/S0045-6535(99)00368-9

Erdal S, Genisel M, Turk H, Gorcek Z (2012) Effects of progesterone application on antioxidant enzyme activities and K+/Na+ ratio in bean seeds exposed to salt stress. Toxicol Ind Health 28(10):942–946. https://doi.org/10.1177/0748233711430975

Estévez MC, Font H, Nichkova M, Salvador JP, Varela B, Sánchez-Baeza F, Marco MP (2005) Immunochemical determination of industrial emerging pollutants. In: Barceló D (ed) Water pollution. Springer, Berlin, pp 119–180. https://doi.org/10.1007/b98609

FAO (1985) Water quality for agriculture. FAO Irrigation and Drainage Paper No. 29, Rome

FAO (2016) Codex Alimentarius. http://www.fao.org/fao-whocodexalimentarius/codextexts/guidelines/en/. Accessed date: 16 Apr 2018

Felizeter S, McLachlan MS, De Voogt P (2014) Root uptake and translocation of perfluorinated alkyl acids by three hydroponically grown crops. J Agric Food Chem 62(15):3334–3342. https://doi.org/10.1021/jf500674j

Firouzsalari NZ, Shakerkhatibi M, Pourakbar M, Yadeghari A, Safari GH, Sarbakhsh P (2019) Pyrethroid pesticide residues in a municipal wastewater treatment plant: occurrence, removal efficiency, and risk assessment using a modified index. J Wat Proc Eng 29:100793. https://doi.org/10.1016/j.jwpe.2019.100793

Fletcher JS, Pfleeger TG, Ratsch HC (1993) Potential environmental risks associated with the new sulfonylurea herbicides. Environ Sci Technol 27(10):2250–2252

Franzaring J, van der Eerden LJM, van Dijk CJ, Helsper JPFG, Pikaar PJJ, Nijhuis EH, Tonneijck AEG (2000) Assessing the chronic effects of atmospheric tetrachloroethylene (PER) on plants. (report no). Plant Res Intl, Wageningen

Fu Q, Wu X, Ye Q, Ernst F, Gan J (2016) Biosolids inhibit bioavailability and plant uptake of triclosan and triclocarban. Water Res 102:117–124. https://doi.org/10.1016/j.watres.2016.06.026

Gabet-Giraud V, Miège C, Jacquet R, Coquery M (2014) Impact of wastewater treatment plants on receiving surface waters and a tentative risk evaluation: the case of estrogens and beta blockers. Environ Sci Pollut Res 21(3):1708–1722. https://doi.org/10.1007/s11356-013-2037-7

Goldstein M, Shenker M, Chefetz B (2014) Insights into the uptake processes of wastewater-borne pharmaceuticals by vegetables. Environ Sci Technol 48(10):5593–5600

Grassi M, Kaykioglu G, Belgiorno V, Lofrano G (2012) Removal of emerging contaminants from water and wastewater by adsorption process. In: Lofrano G (ed) Emerging compounds removal from wastewater. Springer, Dordrecht, pp 15–37. https://doi.org/10.1007/978-94-007-3916-1_2

Guan M, Roddick JG (1988) Epibrassinolide-inhibition of development of excised, adventitious and intact roots of tomato (Lycopersicon esculentum): comparison with the effects of steroidal estrogens. Physiol Plant 74(4):720–726. https://doi.org/10.1111/j.1399-3054.1988.tb02043.x

Guerriero C, Bianchi F, Cairns J, Cori L (2011) Policies to clean up toxic industrial contaminated sitrace elements of Gela and Priolo: a cost-benefit analysis. Environ Health 10(1):68. https://doi.org/10.1186/1476-069X-10-68

Guo X, Yin Y, Yang C, Zhang Q (2016) Remove mechanisms of sulfamethazine by goethite: the contributions of pH and ionic strength. Res Chem Intermed Elem 42(7):6423–6435. https://doi.org/10.1007/s11164-016-2472-4

Gworek B, Klimczak K, Kijeńska M, Gozdowski D (2016) Comparison of PAHs uptake by selected Monocotyledones and Dicotyledones from municipal and industrial sewage sludge. Environ Sci Pollut Res 23(19):19461–19470. https://doi.org/10.1007/s11356-016-7130-2

Hamdaoui O, Naffrechoux E (2009) Adsorption kinetics of 4-chlorophenol onto granular activated carbon in the presence of high frequency ultrasound. Ultrason Sonochem 16(1):15–22. https://doi.org/10.1016/j.ultsonch.2008.05.008

Haruvy N (1997) Agricultural reuse of wastewater: nation-wide cost-benefit analysis. Agric Ecosyst Environ 66(2):113–119. https://doi.org/10.1016/S0167-8809(97)00046-7

van Haut H, Prinz B (1979) Beurteilung der relativen Pflanzenscha̎-dlichkeit organischer Luftverunreinigungen im LIS-Kurzzeittest. Staub—Reinhaltung der Luft 39:408–414

Hernando MD, Ferrer I, Agüera A, Fernandez-Alba AR (2005) Evaluation of pesticides in wastewaters. A combined (chemical and biological) analytical approach. In Water pollution, pp 53–77

Hurtado C, Parastar H, Matamoros V, Piña B, Tauler R, Bayona JM (2017) Linking the morphological and metabolomic response of Lactuca sativa L exposed to emerging contaminants using GC× GC-MS and chemometric tools. Sci Rep 7(1):6546. https://doi.org/10.1038/s41598-017-06773-0

Izuta T, Matsumura H, Ohashi T, Miyake H, Totsuka T (1993) Effects of peroxyacetyl nitrate on the growth of petunia, kidney bean and radish plants. Environ Sci 2:63–75

Janeczko A, Kocurek M, Marcińska I (2012) Mammalian androgen stimulatrace elements photosynthesis in drought-stressed soybean. Open Life Sci 7(5):902–909. https://doi.org/10.2478/s11535-012-0068-6

Jauris IM, Matos CF, Saucier C, Lima EC, Zarbin AJG, Fagan SB et al (2016) Adsorption of sodium diclofenac on graphene: a combined experimental and theoretical study. Phys Chem Chem Phys 18(3):1526–1536

Jones-Lepp TA, Alvarez D, Loganathan B (2011) On the Frontier: Analytical chemistry and the occurrence of illicit drugs in surface waters in the United States. Illicit Drugs in the Environment, 171. https://doi.org/10.1002/9781118000816.ch9

Karnjanapiboonwong A, Chase DA, Canas JE, Jackson WA, Maul JD, Morse AN, Anderson TA (2011) Uptake of 17α-ethynylestradiol and triclosan in pinto bean, Phaseolus vulgaris. Ecotoxicol Environ Saf 74(5):1336–1342. https://doi.org/10.1016/j.ecoenv.2011.03.013

Kaur Y, Bhatia Y, Chaudhary S, Chaudhary GR (2017) Comparative performance of bare and functionalize ZnO nanoadsorbents for pesticide removal from aqueous solution. J Mol Liq 234:94–103. https://doi.org/10.1016/j.molliq.2017.03.069

Keegan AB, Kelly KM, van Staden J (1989) Ethylene involvement in dormancy release of Ricinodendron rautanenii seeds. Ann Bot 63:229–234. https://doi.org/10.1093/oxfordjournals.aob.a087737

Kemper N (2008) Veterinary antibiotics in the aquatic and terrestrial environment. Ecol Indic 8(1):1–13. https://doi.org/10.1016/j.ecolind.2007.06.002

Khan NA, Hasan Z, Jhung SH (2013) Adsorptive removal of hazardous materials using metal-organic frameworks (MOFs): a review. J Hazard Mater 244:444–456. https://doi.org/10.1016/j.jhazmat.2012.11.011

Khan A, Khan S, Khan MA, Qamar Z, Waqas M (2015) The uptake and bioaccumulation of heavy metals by food plants, their effects on plants nutrients, and associated health risk: a review. Environ Sci Pollut Res 22:13772–13799. https://doi.org/10.1007/s11356-015-4881-0

Kumar A, Singh JS (2017) Cyanoremediation: a green-clean tool for decontamination of synthetic pesticides from agro-and aquatic ecosystems. In: Sing JS, Seneviratne G (eds) Agro-environmental sustainability. Springer, Cham, pp 59–83

Kümmerer K (2009) The presence of pharmaceuticals in the environment due to human use—present knowledge and future challenges. J Environ Manag 90:2354–2366. https://doi.org/10.1016/j.jenvman.2009.01.023

Kurian M, Reddy VR, Dietz T, Brdjanovic D (2013) Wastewater re-use for peri-urban agriculture: a viable option for adaptive water management? Sustain Sci 8(1):47–59. https://doi.org/10.1007/s11625-012-0178-0

Kuster M, de Alda MJL, Barceló D (2005) Estrogens and progestogens in wastewater, sludge, sediments, and soil. In: Water pollution. Springer, Berlin, pp 1–24. https://doi.org/10.1007/b98605

Létondor C, Pascal-Lorber S, Laurent F (2015) Uptake and distribution of chlordecone in radish: different contamination routrace elements in edible roots. Chemosphere 118:20–28. https://doi.org/10.1016/j.chemosphere.2014.03.102

Li H, Ma Y (2016) Field study on the uptake, accumulation, translocation and risk assessment of PAHs in a soil-wheat system with amendments of sewage sludge. Sci Total Environ 560:55–61. https://doi.org/10.1016/j.scitotenv.2016.04.017

Li H, Qu R, Yan L, Guo W, Ma Y (2015a) Field study on the uptake and translocation of PBDEs by wheat (Triticum aestivum L.) in soils amended with sewage sludge. Chemosphere 123:87–92. https://doi.org/10.1016/j.chemosphere.2014.12.045

Li P, Hu C, Qi X, Zhou Y, Jianfeng Z, Zia Z (2015b) Effect of reclaimed municipal wastewater irrigation and nitrogen fertilization on yield of tomato and nitrogen economy. Bangladesh J Bot 44(5):699–708

Liacos JW, Kam W, Delfino RJ, Schauer JJ, Sioutas C (2012) Characterization of organic, metal and trace element PM2. 5 species and derivation of freeway-based emission rates in Los Angeles, CA. Sci Total Environ 435:156–166. https://doi.org/10.1016/j.scitotenv.2012.06.106

Macherius A, Eggen T, Lorenz W, Moeder M, Ondruschka J, Reemtsma T (2012a) Metabolization of the bacteriostatic agent triclosan in edible plants and its consequences for plant uptake assessment. Environ Sci Technol 46(19):10797–10804. https://doi.org/10.1021/es3028378

Macherius A, Eggen T, Lorenz WG, Reemtsma T, Winkler U, Moeder M (2012b) Uptake of galaxolide, tonalide, and triclosan by carrot, barley, and meadow fescue plants. J Agric Food Chem 60(32):7785–7791. https://doi.org/10.1021/jf301917q

Malchi T, Maor Y, Tadmor G, Shenker M, Chefetz B (2014) Irrigation of root vegetables with treated wastewater: evaluating uptake of pharmaceuticals and the associated human health risks. Environ Sci Technol 48(16):9325–9333. https://doi.org/10.1021/es5017894

Man LANG, Zucong CAI (2009) Effects of chlorothalonil and carbendazim on nitrification and denitrification in soils. J Environ Sci 21(4):458–467

Manoli K, Morrison LM, Sumarah MW, Nakhla G, Ray AK, Sharma VK (2019) Pharmaceuticals and pesticides in secondary effluent wastewater: identification and enhanced removal by acid-activated ferrate(VI). Water Res 148:272–280. https://doi.org/10.1016/j.watres.2018.10.056

Margas M, Piotrowicz-Cieślak AI, Ziółkowska A, Adomas B (2016) Tetracycline accumulation in pea seedlings and its effects on proteome and enzyme activities. Int J Agric Biol 18(4). https://doi.org/10.17957/IJAB/15.0166

Masih A, Masih J, Taneja A (2012) Study of air–soil exchange of polycyclic aromatic hydrocarbons (PAHs) in the north-central part of India–a semi arid region. J Environ Monit 14(1):172–180. https://doi.org/10.1039/C1EM10567A

Megharaj M (2002) Heavy pesticide use lowers soil health. Farming Ahead 121:37–38

Michelini L, Meggio F, Reichel R, Thiele-Bruhn S, Pitacco A, Scattolin L et al (2015) Sulfadiazine uptake and effects in common hazel (Corylus avellana L.). Environ Sci Pollut Res 22(17):13362–13371. https://doi.org/10.1007/s11356-015-4560-1

Mikes O, Cupr P, Trapp S, Klanova J (2009) Uptake of polychlorinated biphenyls and organochlorine pesticides from soil and air into radishes (Raphanus sativus). Environ Pollut 157(2):488–496. https://doi.org/10.1016/j.envpol.2008.09.007

Miller PR, McBride J (1999) Oxidant air pollution impacts in the montane forests of southern California: a case study of the San Bernardino mountains, vol 134. Springer

Mishra S, Bharagava RN (2016) Toxic and genotoxic effects of hexavalent chromium in environment and its bioremediation strategies. J Environ Sci Health C 34(1):1–32. https://doi.org/10.1080/10590501.2015.1096883

Monteiro M, Santos C, Mann RM, Soares AM, Lopes T (2007) Evaluation of cadmium genotoxicity in Lactuca sativa L. using nuclear microsatellitrace elements. Environ Exp Bot 60(3):421–427. https://doi.org/10.1016/j.envexpbot.2006.12.018

Mudu P, Terracini B, Martuzzi M (2014) Human health in areas with industrial contamination. WHO Regional Office for Europe, Europe

Murray KE, Thomas SM, Bodour AA (2010) Prioritizing research for trace pollutants and emerging contaminants in the freshwater environment. Environ Pollut 158(12):3462–3471. https://doi.org/10.1016/j.envpol.2010.08.009

Muthoosamy K, Manickam S (2017) State of the art and recent advances in the ultrasound-assisted synthesis, exfoliation and functionalization of graphene derivatives. Ultrason Sonochem 39:478–493

Nabulo G, Young SD, Black CR (2010) Assessing risk to human health from tropical leafy vegetables grown on contaminated urban soils. Sci Total Environ 408(22):5338–5351. https://doi.org/10.1016/j.scitotenv.2010.06.034

Naidu R, Espana VAA, Liu Y, Jit J (2016) Emerging contaminants in the environment: risk-based analysis for better management. Chemosphere 154:350–357. https://doi.org/10.1016/j.chemosphere.2016.03.068

Niu XZ, Busetti F, Langsa M, Croué JP (2016) Roles of singlet oxygen and dissolved organic matter in self-sensitized photo-oxidation of antibiotic norfloxacin under sunlight irradiation. Water Res 106:214–222. https://doi.org/10.1016/j.watres.2016.10.002

Packer JL, Werner JJ, Latch DE, McNeill K, Arnold WA (2003) Photochemical fate of pharmaceuticals in the environment: naproxen, diclofenac, clofibric acid, and ibuprofen. Aquat Sci 65(4):342–351. https://doi.org/10.1007/s00027-003-0671-8

Pal A, Gin KYH, Lin AYC, Reinhard M (2010) Impacts of emerging organic contaminants on freshwater resources: review of recent occurrences, sources, fate and effects. Sci Total Environ 408(24):6062–6069. https://doi.org/10.1016/j.scitotenv.2010.09.026

Pan M, Wong CK, Chu LM (2014) Distribution of antibiotics in wastewater-irrigated soils and their accumulation in vegetable crops in the Pearl River Delta, Southern China. J Agric Food Chem 62(46):11062–11069. https://doi.org/10.1021/jf503850v

Pan H, Lei H, He X, Xi B, Xu Q (2017) Spatial distribution of organochlorine and organophosphorus pesticides in soil-groundwater systems and their associated risks in the middle reaches of the Yangtze River Basin. Environ Geochem Health 41:1–13. https://doi.org/10.1007/s10653-017-9970-1

Patowary R, Patowary K, Devi A, Kalita MC, Deka S (2017) Uptake of total petroleum hydrocarbon (TPH) and polycyclic aromatic hydrocarbons (PAHs) by Oryza sativa L. grown in soil contaminated with crude oil. Bull Environ Contam Toxicol 98(1):120–126. https://doi.org/10.1007/s00128-016-1990-5

Peralta-Videa JR, Lopez ML, Narayan M, Saupe G, Gardea-Torresdey J (2009) The biochemistry of environmental heavy metal uptake by plants: implications for the food chain. Int J Biochem Cell Biol 41(8–9):1665–1677. https://doi.org/10.1016/j.biocel.2009.03.005

Pimentel D, Acquay H, Biltonen M, Rice P, Silva M, Nelson J et al (1992) Environmental and economic costs of pesticide use. Bioscience 42(10):750–760. https://www.jstor.org/stable/1311994

Prosser RS, Sibley PK (2015) Human health risk assessment of pharmaceuticals and personal care products in plant tissue due to biosolids and manure amendments, and waste water irrigation. Environ Int 75:223–233. https://doi.org/10.1016/j.envint.2014.11.020

Radke M, Ulrich H, Wurm C, Kunkel U (2010) Dynamics and attenuation of acidic pharmaceuticals along a river stretch. Environ Sci Technol 44(8):2968–2974. https://doi.org/10.1021/es903091z

Rattan RK, Datta SP, Chhonkar PK, Suribabu K, Singh AK (2005) Long-term impact of irrigation with sewage effluents on heavy metal content in soils, crops and groundwater—a case study. Agric Ecosyst Environ 109(3–4):310–322. https://doi.org/10.1016/j.agee.2005.02.025

Reis FCD, Tornisielo VL, Cason JB, Dias ACR, Freitas M, Sotomayor JF, Barroso AA, Filho RV (2015) Uptake, translocation, and control of trumpetflower (Tecoma stans) with aminocyclo pyrachlor. J Environ Sci Health B 50(10):727–733. https://doi.org/10.1080/03601234.2015.1048106

Riemenschneider C, Al-Raggad M, Moeder M, Seiwert B, Salameh E, Reemtsma T (2016) Pharmaceuticals, their metabolitrace elements, and other polar pollutants in field-grown vegetables irrigated with treated municipal wastewater. J Agric Food Chem 64(29):5784–5792. https://doi.org/10.1021/acs.jafc.6b01696

Rivera-Utrilla J, Sánchez-Polo M, Ferro-García MÁ, Prados-Joya G, Ocampo-Pérez R (2013) Pharmaceuticals as emerging contaminants and their removal from water. A review. Chemosphere 93(7):1268–1287. https://doi.org/10.1016/j.chemosphere.2013.07.059

Sablayrolles C, Montréjaud-Vignoles M, Vialle C, Silvestre J, Lhoutellier C, Poitrenaud M (2008) PCB uptake by carrots after sludge and compost application. Compost and digestate: sustainability, benefits, impacts for the environment and for plant production, 27, 29th

Sablayrolles C, Silvestre J, Lhoutellier C, Montrejaud-Vignoles M (2013) Phthalates uptake by tomatoes after biosolids application: worst case and operational practice in greenhouse conditions. Fresenius Environ Bull 22(4A):1061–1069

Salehi-Lisar SY, Deljoo S, Harzandi AM (2015) Fluorene and phenanthrene uptake and accumulation by wheat, alfalfa and sunflower from the contaminated soil. Int J Phytoremediation 17(12):1145–1152. https://doi.org/10.1080/15226514.2015.1045123

Santamaria P (2006) Nitrate in vegetables: toxicity, content, intake and EC regulation. J Sci Food Agric 86(1):10–17. https://doi.org/10.1002/jsfa.2351

Shahid M, Khalid S, Abbas G, Shahid N, Nadeem M, Sabir M et al (2015) Heavy metal stress and crop productivity. In: Hakeem KR (ed) Crop production and global environmental issues. Springer, Cham, pp 1–25. https://doi.org/10.1007/978-3-319-23162-4_1

Shenker M, Harush D, Ben-Ari J, Chefetz B (2011) Uptake of carbamazepine by cucumber plants–a case study related to irrigation with reclaimed wastewater. Chemosphere 82(6):905–910. https://doi.org/10.1016/j.chemosphere.2010.10.052

Sherafatmandjour A, Khorshidi M, Abavisani A (2013) Effect of estradiol on photosynthetic pigments, proline and sugars in fennel. Int J Farm Allied Sci 2(17):567–571

Shi W, Wang L, Rousseau DP, Lens PN (2010) Removal of estrone, 17α-ethinylestradiol, and 17ß-estradiol in algae and duckweed-based wastewater treatment systems. Environ Sci Pollut Res 17(4):824–833. https://doi.org/10.1007/s11356-010-0301-7

Shore LS, Gurevitz M, Shemesh M (1993) Estrogen as an environmental pollutant. Bull Environ Contam Toxicol 51(3):361–366

Shrestha SL, Casey FX, Hakk H, Smith DJ, Padmanabhan G (2012) Fate and transformation of an estrogen conjugate and its metabolitrace elements in agricultural soils. Environ Sci Technol 46(20):11047–11053. https://doi.org/10.1021/es3021765

Singh S, Kumar M (2006) Heavy metal load of soil, water and vegetables in peri-urban Delhi. Environ Monit Assess 120(1–3):79–91. https://doi.org/10.1007/s10661-005-9050-3

Singh JS, Seneviratne G (2017) Agro-environmental sustainability: volume 2: managing environmental pollution. Springer, Cham, pp 1–251

Song J, Nagae M, Soyano K (2018) Standardization of plasma vitellogenin level for surveying environmental estrogen pollution using the Japanese common goby Acanthogobius flavimanus. J Appl Toxicol 38(7):1018–1025

Stefanakis AI, Becker JA (2015) A review of emerging contaminants in water: classification, sources, and potential risk. In: AE MK, Bugyi G (eds) Impact of water pollution on human health and environmental sustainability. IGI Global, Hershey, pp 55–80

Stuart M, Lapworth D, Crane E, Hart A (2012) Review of risk from potential emerging contaminants in UK groundwater. Sci Total Environ 416:1–21. https://doi.org/10.1016/j.scitotenv.2011.11.072

Teng Y, Wu J, Lu S, Wang Y, Jiao X, Song L (2014) Soil and soil environmental quality monitoring in China: a review. Environ Int 69:177–199. https://doi.org/10.1016/j.envint.2014.04.014

Testai E, Buratti FM, Funari E, Manganelli M, Vichi S, Arnich N et al (2016) Review and analysis of occurrence, exposure and toxicity of cyanobacteria toxins in food Autorité européenne de sécurité des aliments. EFSA Support Publ.. 2016:EN(998):309 p. https://doi.org/10.2903/sp.efsa.2016.EN-998

Urbano VR, Mendonça TG, Bastos RG, Souza CF (2017) Effects of treated wastewater irrigation on soil properties and lettuce yield. Agric Water Manag 181:108–115. https://doi.org/10.1016/j.agwat.2016.12.001

Van Veenhuizen R, Danso G (2007) Profitability and sustainability of urban and periurban agriculture, vol 19. Food & Agriculture Org, Rome

Vione D, Minella M, Maurino V, Minero C (2014) Indirect photochemistry in sunlit surface waters: photoinduced production of reactive transient species. Chem Eur J 20(34):10590–10606. https://doi.org/10.1002/chem.201400413

Wang J, Lin H, Sun W, Xia Y, Ma J, Fu J et al (2016) Variations in the fate and biological effects of sulfamethoxazole, norfloxacin and doxycycline in different vegetable–soil systems following manure application. J Hazard Mater 304:49–57. https://doi.org/10.1016/j.jhazmat.2015.10.038

Wehtje G, Walker RH, Shaw JN (2000) Pesticide retention by inorganic soil amendments. Weed Sci 48(2):248–254. https://doi.org/10.1614/0043-1745(2000)048[0248:PRBISA]2.0.CO;2

Wenzel KD, Manz M, Hubert A, Schüürmann G (2002) Fate of POPs (DDX, HCHs, PCBs) in upper soil layers of pine forests. Sci Total Environ 286(1–3):143–154. https://doi.org/10.1016/S0048-9697(01)00972-X

WHO (1989) Health Guidelines for the Use of Wastewater in Agriculture and Aquaculture: Report of a WHO Scientific Group [Meeting Held in Geneva from 18 to 23 November 1987]

Wu X, Conkle JL, Ernst F, Gan J (2014) Treated wastewater irrigation: uptake of pharmaceutical and personal care products by common vegetables under field conditions. Environ Sci Technol 48(19):11286–11293. https://doi.org/10.1021/es502868k

Xing Y, Chen X, Chen X, Zhuang J (2016) Colloid-mediated transport of pharmaceutical and personal care products through porous media. Sci Rep 6:35407

Xu J, Chen W, Wu L, Green R, Chang AC (2009) Leachability of some emerging contaminants in reclaimed municipal wastewater-irrigated turf grass fields. Environ Toxicol Chem 28(9):1842–1850. https://doi.org/10.1897/08-471.1

Xuan R, Blassengale AA, Wang Q (2008) Degradation of estrogenic hormones in a silt loam soil. J Agric Food Chem 56(19):9152–9158. https://doi.org/10.1021/jf8016942

Yu C, Bi E (2015) Roles of functional groups of naproxen in its sorption to kaolinite. Chemosphere 138:335–339

Zhang ZY, Shan WL, Song WC, Gong Y, Liu XJ (2011) Phytotoxicity and uptake of chlorpyrifos in cabbage. Environ Chem Lett 9(4):547–552. https://doi.org/10.1007/s10311-011-0320-8

Zhao H, Liu X, Cao Z, Zhan Y, Shi X, Yang Y et al (2016) Adsorption behavior and mechanism of chloramphenicols, sulfonamides, and non-antibiotic pharmaceuticals on multi-walled carbon nanotubes. J Hazard Mater 310:235–245. https://doi.org/10.1016/j.jhazmat.2016.02.045

Zheng W, Wiles K, Holm N (2013, September) Uptake, translocation, and accumulation of pharmaceutical and hormone contaminants in vegetables. In: Abstracts of papers of the American Chemical Society, vol 246. American Chemical Society, Washington, DC

Zhuang P, Huiling ZOU, Wensheng SHU (2009) Biotransfer of heavy metals along a soil-plant-insect-chicken food chain: field study. J Environ Sci 21(6):849–853. https://doi.org/10.1016/S1001-0742(08)62351-7

Printed in the United States
By Bookmasters